THE PINCH TECHNIQUE AND ITS APPLICATIONS TO NON-ABELIAN GAUGE THEORIES

Non-Abelian gauge theories, such as quantum chromodynamics (QCD) or electroweak theory, are best studied with the aid of Green's functions that are gauge invariant off-shell, but unlike for the photon in quantum electrodynamics, conventional graphical constructions fail. The pinch technique provides a systematic framework for constructing such Green's functions and has many useful applications.

Beginning with elementary one-loop examples, this book goes on to extend the method to all orders, showing that the pinch technique is equivalent to calculations in the background-field Feynman gauge. The pinch technique Schwinger-Dyson equations are derived and used to show how a dynamical gluon mass arises in QCD. Applications are given to the center vortex picture of confinement, the gauge-invariant treatment of resonant amplitudes, the definition of non-Abelian effective charges, high-temperature effects, and even supersymmetry. This book is ideal for elementary particle theorists and graduate students. This title, first published in 2011, has been reissued as an Open Access publication on Cambridge Core.

JOHN M. CORNWALL is Distinguished Professor of Physics Emeritus in the Department of Physics and Astronomy, University of California, Los Angeles. Inventor of the pinch technique, he has made many other contributions to the formalism and applications of quantum field theory, as well as to space plasma physics. He has contributed to the technical analysis of many public policy issues, ranging from ballistic missile defense to the human genome.

JOANNIS PAPAVASSILIOU is a researcher in the Department of Theoretical Physics and IFIC, the University of Valencia–CSIC. A large part of his work has been devoted to the development of the pinch technique, both its formal foundation and its many applications, and he has published articles on quantum field theory and particle phenomenology.

DANIELE BINOSI is a researcher at the European Centre for Theoretical Studies in Nuclear Physics and Related Areas (ECT*) and Fondazione Bruno Kessler. In addition to his work on extending the pinch technique and its applications, he leads several policy-related European projects on the development of the vision and sustainability of quantum information foundations and technologies.

CAMBRIDGE MONOGRAPHS ON PARTICLE PHYSICS, NUCLEAR PHYSICS AND COSMOLOGY

General Editors: T. Ericson, P. V. Landshoff

1. K. Winter (ed.): *Neutrino Physics*
2. J. F. Donoghue, E. Golowich and B. R. Holstein: *Dynamics of the Standard Model*
3. E. Leader and E. Predazzi: *An Introduction to Gauge Theories and Modern Particle Physics, Volume 1: Electroweak Interactions, the 'New Particles' and the Parton Model*
4. E. Leader and E. Predazzi: *An Introduction to Gauge Theories and Modern Particle Physics, Volume 2: CP-Violation, QCD and Hard Processes*
5. C. Grupen: *Particle Detectors*
6. H. Grosse and A. Martin: *Particle Physics and the Schrödinger Equation*
7. B. Anderson: *The Lund Model*
8. R. K. Ellis, W. J. Stirling and B. R. Webber: *QCD and Collider Physics*
9. I. I. Bigi and A. I. Sanda: *CP Violation*
10. A. V. Manohar and M. B. Wise: *Heavy Quark Physics*
11. R. K. Bock, H. Grote, R. Frühwirth and M. Regler: *Data Analysis Techniques for High-Energy Physics, Second edition*
12. D. Green: *The Physics of Particle Detectors*
13. V. N. Gribov and J. Nyiri: *Quantum Electrodynamics*
14. K. Winter (ed.): *Neutrino Physics, Second edition*
15. E. Leader: *Spin in Particle Physics*
16. J. D. Walecka: *Electron Scattering for Nuclear and Nucleon Scattering*
17. S. Narison: *QCD as a Theory of Hadrons*
18. J. F. Letessier and J. Rafelski: *Hadrons and Quark-Gluon Plasma*
19. A. Donnachie, H. G. Dosch, P. V. Landshoff and O. Nachtmann: *Pomeron Physics and QCD*
20. A. Hoffmann: *The Physics of Synchroton Radiation*
21. J. B. Kogut and M. A. Stephanov: *The Phases of Quantum Chromodynamics*
22. D. Green: *High P_T Physics at Hadron Colliders*
23. K. Yagi, T. Hatsuda and Y. Miake: *Quark-Gluon Plasma*
24. D. M. Brink and R. A. Broglia: *Nuclear Superfluidity*
25. F. E. Close, A. Donnachie and G. Shaw: *Electromagnetic Interactions and Hadronic Structure*
26. C. Grupen and B. A. Shwartz: *Particle Detectors, Second edition*
27. V. Gribov: *Strong Interactions of Hadrons at High Energies*
28. I. I. Bigi and A. I. Sanda: *CP Violation, Second edition*
29. P. Jaranowski and A. Królak: *Analysis of Gravitational-Wave Data*
30. B. L. Ioffe, V. S. Fadin and L. N. Lipatov: *Quantum Chromodynamics: Perturbative and Nonperturbative Aspects*
31. J. M. Cornwall, J. Papavassiliou, and D. Binosi: *The Pinch Technique and Its Applications to Non-Abelian Gauge Theories*

THE PINCH TECHNIQUE AND ITS APPLICATIONS TO NON-ABELIAN GAUGE THEORIES

JOHN M. CORNWALL

University of California at Los Angeles, USA

JOANNIS PAPAVASSILIOU

University of Valencia–CSIC, Spain

DANIELE BINOSI

European Centre for Theoretical Studies in Nuclear Physics and Related Areas, Italy

CAMBRIDGE
UNIVERSITY PRESS

CAMBRIDGE
UNIVERSITY PRESS

Shaftesbury Road, Cambridge CB2 8EA, United Kingdom

One Liberty Plaza, 20th Floor, New York, NY 10006, USA

477 Williamstown Road, Port Melbourne, VIC 3207, Australia

314–321, 3rd Floor, Plot 3, Splendor Forum, Jasola District Centre, New Delhi – 110025, India

103 Penang Road, #05-06/07, Visioncrest Commercial, Singapore 238467

Cambridge University Press is part of Cambridge University Press & Assessment,
a department of the University of Cambridge.

We share the University's mission to contribute to society through the pursuit of
education, learning and research at the highest international levels of excellence.

www.cambridge.org
Information on this title: www.cambridge.org/9781009402446

DOI: 10.1017/9781009402415

First published 2011
Reissued as OA 2023

A catalogue record for this publication is available from the British Library.

ISBN 978-1-009-40244-6 Hardback
ISBN 978-1-009-40243-9 Paperback

To our families and friends

Contents

Introduction: Why the pinch technique?   *page* xi

1   **The pinch technique at one loop**   **1**
1.1   A brief history   1
1.2   Notation and conventions   3
1.3   The basic one-loop pinch technique   6
1.4   Another way to the pinch technique   17
1.5   Pinch technique vertices   20
1.6   The pinch technique in the light-cone gauge   31
1.7   The absorptive pinch technique construction   34
1.8   Positivity and the pinch technique gluon propagator   42
         References   43

2   **Advanced pinch technique: Still one loop**   **45**
2.1   The pinch technique and the operator product expansion: Running
         mass and condensates   45
2.2   The pinch technique and gauge-boson mass generation   47
2.3   The pinch technique today: Background-field Feynman gauge 62
2.4   What to expect beyond one loop   72
         References   73

3   **Pinch technique to all orders**   **75**
3.1   The *s*-*t* cancellation to all orders   75
3.2   Quark-gluon vertex and gluon propagator to all orders 79
         References   85

vii

4   The pinch technique in the Batalin–Vilkovisky framework   86
4.1   An overview of the Batalin–Vilkovisky formalism   88
4.2   Examples   93
4.3   Pinching in the Batalin–Vilkovisky framework   100
         References   102

5   The gauge technique   104
5.1   The original gauge technique for QED   105
5.2   Massless longitudinal poles   108
5.3   The gauge technique for NAGTs   109
         References   113

**6   Schwinger–Dyson equations in the pinch technique
         framework   114**
6.1   Lattice studies of gluon mass generation   115
6.2   The need for a gauge-invariant truncation scheme for the
         Schwinger–Dyson equations of NAGTs   117
6.3   The pinch technique algorithm for Schwinger–Dyson equations   119
6.4   Pinch technique Green's functions from Schwinger–Dyson equations   120
6.5   Solutions of the pinch technique Schwinger–Dyson equations and
         comparison with lattice data   131
6.6   The QCD effective charge   134
         References   141

7   Nonperturbative gluon mass and quantum solitons   144
7.1   Notation   144
7.2   Introduction   144
7.3   The quantum solitons   150
7.4   The center vortex soliton   152
         References   165

8   Nexuses, sphalerons, and fractional topological charge   167
8.1   Introduction to nexuses and junctions   167
8.2   Nexuses in $SU(N)$ 170
8.3   The QCD sphaleron   181
8.4   Chiral symmetry breakdown, nexuses, and fractional topological charge 186
         References   188

9   A brief summary of $d = 3$ NAGTs   190
9.1   Introduction   190

9.2	Perturbative infrared instability	193
9.3	The exact form of the zero-momentum effective action	193
9.4	The dynamical gauge-boson mass	197
9.5	The functional Schrödinger equation	201
9.6	Dynamical gluon mass versus the Chern–Simons mass: Two phases	209
9.7	Compactness and the Chern–Simons number of YMCS solitons	216
	References	223

10 | **The pinch technique for electroweak theory** | **226**
10.1	General considerations	227
10.2	The case of massless fermions	229
10.3	Nonconserved currents and Ward identities	242
10.4	The all-order construction	246
	References	248

11 | **Other applications of the pinch technique** | **250**
11.1	Introduction	250
11.2	Non-Abelian effective charges	250
11.3	Physical renormalization schemes versus $\overline{\text{MS}}$	255
11.4	Gauge-independent off-shell form factors	256
11.5	Resummation formalism for resonant transition amplitudes	263
11.6	The pinch technique at finite temperature	270
11.7	Basic principles of thermal field theory	271
11.8	Hints of supersymmetry in the pinch technique Green's functions	276
	References	278

	Appendix: Feynman rules	281
	Index	285

Introduction: Why the pinch technique?

Non-Abelian gauge theories (NAGTs) have dominated the world of experimentally accessible particle physics for more than three decades in the form of the standard model with its $SU(2) \times U(1)$ (electroweak theory) and $SU(3)$ (quantum chromodynamics: QCD) components. NAGTs are also the ingredients of grand unified theories and technicolor theories and play critical roles in supersymmetry and string theory. It is no wonder that thousands of papers have been written on them. But many of these papers violate the principle of gauge invariance, resulting in calculations of propagators, vertices, and other off-shell form factors that are valid only in the particular gauge chosen. Until these are combined into a gauge-invariant expression, they have very limited, if any, physical meaning. The reason for such violation of gauge invariance is that standard and widely used Feynman graph techniques generate gauge-dependent Green's functions (proper self-energies, three-point vertices, etc.) for the gauge bosons.

Of course, there is one combination of off-shell Green's functions – the on-shell S-matrix – that is gauge invariant no matter which gauge is used for the propagators and vertices that go into it. Thus, authors who (correctly) insist on calculating only gauge-invariant quantities often restrict themselves to dealing only with the S-matrix. This is fine as long as the question at hand can be answered with perturbation theory, but to calculate gauge invariantly a nonperturbative feature using only S-matrix elements is not easy. Some of the processes involving off-shell Green's functions include confinement and chiral symmetry breakdown in QCD; higher-order radiative corrections to gauge-boson decay widths in electroweak theory; nonperturbative effects in the magnetic sector of high-temperature QCD and electroweak theory; and physical quantities embedded in the electroweak S-matrix such as the magnetic dipole and electric quadrupole moments of the W-boson, the top-quark magnetic moment, and the neutrino electric charge radius. These are physical, measurable phenomena, and to approximate them with gauge-dependent calculations really makes no sense.

The Abelian predecessor of NAGTs, quantum electrodynamics (QED), is also a theory with local gauge invariance. But in QED, it is much easier than in an NAGT to enforce this crucial property. For example, the gauge-boson (photon) propagator in QED is gauge invariant for any momentum and any choice of gauge for internal lines, whereas the gauge-boson propagator in an NAGT, if calculated with standard Feynman graph techniques, is not. Enforcing gauge invariance on off-shell Green's functions for an NAGT is difficult even in perturbation theory, let alone in attempts to understand nonperturbative effects such as confinement in QCD. For NAGTs in covariant gauges, one also has to deal with Faddeev–Popov ghosts, which are absent from QED.

Long ago, a new method for enforcing gauge invariance in off-shell Green's functions, and the Schwinger–Dyson equations that couple them, emerged; this was later termed the *pinch technique*. This is a technique to disassemble and reassemble the Feynman graphs of a gauge-invariant object, such as the S-matrix, into new off-shell Green's functions such as gauge-boson proper self-energies that are strictly gauge invariant, just as in QED. It works by systematically identifying parts of n-point Feynman graphs in that S-matrix element that actually belong to Green's functions with fewer legs through the use of elementary Ward identities. Moreover, requiring that the new pinch technique proper self-energy have the right analytic properties in momentum as well as other physically reasonable properties uniquely specified this new self-energy. Although the actual calculation was not carried out until considerably later, it was clear that any kind of proper gauge-boson vertex, such as the three- or four-point vertex, could also be made completely gauge invariant through application of the same ideas. Uniqueness of the new self-energy and vertices was assured by requiring them to obey Ward identities having an essentially Abelian structure, with no explicit ghost contributions, rather than the much more complex Slavnov–Taylor identities of conventional NAGTs. Another technique, called the *gauge technique* (described later), allows for the construction of nonperturbative proper three-point vertices for the gauge bosons that are approximately valid for small gauge-boson momentum but *exactly* gauge invariant and that satisfy the correct pinch technique Ward identities.

This is simple enough to do at the one-loop level, but at higher orders, a more systematic approach is needed. The surprising and far from obvious outcome is that the simple prescription for calculating gauge-invariant pinch technique Green's functions to any order is to use the conventional Feynman graphs in the background-field Feynman gauge.

The pinch technique – or its algorithmic equivalent, graphs in the background-field Feynman gauge – even shows up in string theory. Usually we think of string theory as yielding only on-shell amplitudes, but a consistent extrapolation of string amplitudes off shell in the field-theory (or zero Regge slope) limit shows that the resulting off-shell gauge-boson amplitudes are automatically presented in the background-field Feynman gauge.

The purpose of this book is to describe the pinch technique and its evolution from simple one-loop beginnings to a systematic method at all orders of perturbation theory and then to fully gauge-invariant Schwinger–Dyson equations, leading to the many applications of the pinch technique that have been developed over the years. The pinch technique has led to the clarification of a number of problems in NAGTs that were essentially unsolvable by techniques depending on conventional (Feynman–graph–derived) off-shell Green's functions because these are gauge dependent.

A quick outline of the book

This book begins with two introductory chapters (Chapters 1 and 2) that derive two-and three-gluon pinch technique Green's functions at the one-loop level, showing how a dynamical mass is demanded by infrared slavery for QCD-like gauge theories and outlining the full scope of the pinch technique program. Then Chapters 3 through 6 go into considerable detail to derive the pinch technique Green's functions to all orders; show that they are the same as the background-field Feynman-gauge Green's functions; derive the Schwinger–Dyson equations for the pinch technique in QCD-like theories; apply a special gauge-invariant method called the *gauge technique* to truncate these equations; and show how the pinch technique Schwinger–Dyson equations give rise to a dynamical gluon mass in these QCD-like theories. Chapters 7, 8, and 9 then cover applications of these results to QCD-like theories. Chapter 10 develops the pinch technique for electroweak theory to all orders, and Chapter 11 describes several applications of the pinch technique to electroweak physics, thermal gauge theories, and supersymmetric gauge theories.

Once we get past one loop, it is hard slogging through all the details, so the reader who wants a tour d'horizon of the pinch technique can find it in Chapters 1 and 2. There we hint at the nonperturbative ideas used in later chapters, making some remarks on the Schwinger–Dyson equation for the pinch technique propagator at the one-dressed-loop level. The reader interested in the inner workings of the pinch technique should read the next four chapters. The remaining chapters are on applications, so readers who are more interested in applications of the pinch technique than in its derivation can spend their time on Chapter 7 and onward after reading introductory Chapters 1 and 2. Although the book is not intended to be a textbook, it should be accessible, with perhaps a little extra work on such techniques as the background-field method and the Batalin–Vilkovisky formalism, to advanced graduate students as well as to researchers in gauge theories.

Some uses of the pinch technique

In QCD-like theories, the most important uses of the pinch technique come from the not-unexpected discovery in solving the Schwinger–Dyson equations that *asymptotic freedom*, or more accurately,[1] *infrared slavery*, requires a dynamically generated gluon mass to resolve the otherwise intractable infrared singularities of infrared slavery. From this simple result – repeatedly confirmed by lattice simulations – follows a cascade of topological quantum solitons that plausibly explain confinement and other nonperturbative phenomena of QCD-like theories and that are also seen in lattice simulations, as we briefly discuss at the end of this section and in detail in Chapters 7 and 8.

The pinch technique for NAGTs has considerable physical interest in three dimensions as well as four. One reason is that $d = 3$ NAGTs carry all of the critical nonperturbative phenomena, stemming from infrared slavery, associated with the sector of finite-temperature NAGTs having zero Matsubara frequency and therefore (perturbatively) massless magnetic gluons. There is such a sector for QCD-like theories and also for electroweak theory above the crossover temperature at which the standard-model Higgs field has zero vacuum expectation value. Another reason is the instructive differences in the conditions and techniques used in $d = 3$; for example, there is no asymptotic freedom, but there is infrared slavery that gives worse infrared divergences in $d = 3$ than in $d = 4$. We cover $d = 3$ NAGTs in Chapter 9 and a few other applications of the pinch technique to finite-temperature gauge theories in Chapter 11.

The pinch technique has also been developed to all orders in electroweak theory as the prime example of symmetry breaking in NAGTs, as we recount in Chapter 10. Here the pinch technique makes possible many applications given in Chapter 11, including gauge-invariant, non-Abelian, off-shell charges and form factors; the neutrino charge radius; and gauge-invariant constructions of resonance widths that are gauge dependent in the usual Feynman-graph formalism.

Finally, applications to NAGTs are even embedded in supersymmetric theories in which the contributions of scalars and fermions to the off-shell gluonic Green's

---

[1] An NAGT in $d = 3$ cannot be asymptotically free, but it can show infrared slavery, that is, severe infrared singularities coming from wrong-sign phenomena; in general, this book applies both to $d = 3$ and $d = 4$ NAGTs.

functions of the pinch technique confirm, by explicit calculation, well-known super-symmetry relations that would not hold for conventional Feynman-graph definitions of the Green's functions. These, too, are discussed in Chapter 11.

Constructing pinch technique Green's functions and equations

Because it is far from obvious, we will spend considerable time on the demon-stration that the systematic all-orders construction of pinch technique Green's functions gives the same Green's functions as would be calculated in the back-ground field method exhibited in the Feynman gauge. The first suggestions that the pinch technique and the background-field Feynman gauge are related came from a one-loop calculation showing that the pinch technique gives the same results as the background-field method in the Feynman gauge; we give our version of these one-loop results in Chapters 1 and 2.

This gauge, then, transcends its usual meaning as just another gauge because it incorporates – for kinematic reasons that we will explain – quite a different algo-rithm: that of the gauge-invariant pinch technique. In the pinch technique (as well as in the background-field method), the Ward identities relating a Green's function of n gauge potentials and no ghosts to those of fewer than n legs are not the standard Slavnov–Taylor identities, which involve ghosts, but rather elementary Ward iden-tities of QED type, making no reference to ghosts. There are, as is surely necessary in any covariant gauge-invariant description of an NAGT, ghost contributions and Green's functions with ghost legs, but in a pinch technique Green's function involv-ing only gauge potentials, the internal ghost loops are not explicitly distinguishable from gluons in the final product. We describe these developments in detail in Chapters 3 and 4.

The next step is to consider the infinite tower of equations, the Schwinger–Dyson equations, that couple all the pinch technique Green's functions. For example, the pinch technique propagator depends on the three- and four-point gluon vertex func-tions, which in turn depend on higher-point functions – and so on. We introduce closure to this infinite tower of equations by using the *gauge technique*, in which we construct *approximate* n-point pinch technique Green's functions that depend only on fewer-point pinch technique Green's functions and that *exactly* satisfy the correct Ward identities. Gauge technique Green's functions, which differ from their exact counterparts by completely conserved quantities, more and more accurately represent the exact Green's functions as momenta on the legs become smaller (because in the presence of a mass gap, completely conserved quantities have kinematic zeroes at vanishing momenta of higher order than any in the gauge tech-nique Green's functions). In consequence, the gauge technique Green's functions describe rather well the nonperturbative properties of gauge theories with infrared slavery but require correction in the ultraviolet; fortunately, such corrections are straightforward because of asymptotic freedom. We cover the gauge technique in Chapter 5, from its simple beginnings long ago in QED to considerably more complex forms for NAGTs.

The final step in moving toward applications is to derive the Schwinger–Dyson equations by considering the all-order perturbative results; this is done in Chapter 6. These Schwinger–Dyson equations include those for ghosts, although strictly speaking, these are not necessary (e.g., in the pinch technique Schwinger–Dyson equations as derived from light-cone gauge graphology). We should and do derive the Schwinger–Dyson equations without reference to any particular closure scheme, although the only practical way to use these Schwinger–Dyson equations is with the gauge technique, with ultraviolet corrections added perturbatively. Of course, the solutions to the Schwinger–Dyson equations closed with the gauge technique are only approximate, as will always be the case however NAGTs are approached, but they have the great virtue of local gauge invariance, quite unlike the usual Schwinger–Dyson equations based on conventional Feynman graphology.

By far the most interesting feature of the solutions of these equations is that the infrared slavery singularities are tamed by gluons' becoming massive, yet without any violation of local gauge invariance (cf. the $d = 2$ Schwinger model). It is critical that this be a *running* mass that vanishes (up to logarithms) like $\langle G^2 \rangle q^{-2}$ at large momenta, where $\langle G^2 \rangle$ is the usual vacuum expectation value of the squared field strengths.[2] This is precisely analogous to the dynamical (constituent) quark mass, vanishing like $\langle \bar{\psi}\psi \rangle q^{-2}$. If the mass did not run, it would not be a dynamical mass but a bare mass (like the current mass of quarks), and the renormalizability of an NAGT with a bare gluon mass is problematical. This vanishing of the running mass is very roughly analogous to the vanishing of a Higgs field along the axis of a soliton in symmetry-breaking theories.

The mere technical solution of the infrared-singularity problem by generation of a dynamical gluon mass is perhaps interesting but, taken alone, seems not to bear on the great issues of QCD such as confinement and chiral symmetry breaking. This is far from true because gauge-invariant mass generation in an NAGT has two important implications. First, there must be massless pure-gauge, longitudinally coupled excitations, much like Goldstone fields, and explicit in the gauge technique Green's functions. Second, these pure-gauge parts, which are indispensable parts of solitons of the massive gluon effective action, carry the long-range

---

[2] In conventional Feynman graphs for the propagator, there are condensate contributions to something that might have been identified with a running mass, except that there are contributions from non-gauge-invariant condensates. Only in the pinch technique propagator is there only a gauge-invariant condensate contribution.

topological properties essential for confinement and chiral symmetry breakdown. In themselves, the massless parts would have Dirac singularities (lying on closed sheets in $d = 4$ and on closed strings in $d = 3$), but every topological soliton has another contribution to the gauge potential that is massive and not pure gauge, giving rise only to short-range gauge field strengths. This massive part also has a Dirac singularity that exactly cancels that of the massless pure-gauge part, yielding a soliton that is essentially nonsingular and of finite action. There are many solitons, but the three most interesting classes are the center vortex, essentially a closed two-surface (responsible for confinement via a Bohm–Aharonov effect coming from linkage of a Wilson loop to the closed surface of the massless long-range, pure-gauge part of the center vortex); its closely related descendant, the nexus (a monopole whose world line must lie on the center-vortex surface, thereby dividing it into regions of different orientation and providing for the generation of nonintegral topological charge); and the sphaleron (rather like the electroweak sphaleron, interpretable not only as a cross section of a topology-changing configuration but also as an unstable glueball). These solitons and their consequences will be discussed in much more detail in Chapters 7 and 8, with Chapter 9 devoted to the special case of three dimensions, which has a number of interesting features not found in $d = 4$.

1

The pinch technique at one loop

In this chapter, we present in detail the pinch technique (PT) construction at one loop for a QCD-like theory, where there is no tree-level symmetry breaking (no Higgs mechanism). The analysis applies to any gauge group ($SU(N)$, exceptional groups, etc.); however, for concreteness, we will adopt the QCD terminology of quarks and gluons.

This introductory chapter and Chapter 2 go into both conventional technology and the pinch technique only at the one-loop level. Here, the reader will find an almost self-contained guide to the one-loop pinch technique with many calculational details plus some hints at the nonperturbative ideas used in later chapters (where nonperturbative effects will be studied by dressing the loops, i.e., using a skeleton expansion).

1.1 A brief history

Non-Abelian gauge theories (NAGTs) had been around for a long time when the pinch technique came into play [1, 2, 3, 4]. Their first use was in defining the one-loop PT gauge-boson propagator as a construct taken from some gauge-invariant object by combining parts of conventional Feynman graphs while preserving gauge invariance and other physical properties. The term *pinch technique* was introduced later [4], in a paper that extended the one-loop pinch technique to the three-gluon vertex. The name comes from a characteristic feature of the pinch technique, in which the needed parts of some Feynman graphs look as though a particular propagator line had been pinched out of existence. In all these early papers, only one-loop phenomena were studied, including a one-dressed-loop Schwinger–Dyson equation for the PT propagator. This equation showed how the infrared singularities arising because of asymptotic freedom (= infrared slavery) require dynamical gluon mass generation. Of course, the pinch technique should lead to unique results. These

1

considerations followed from five requirements for all PT Green's functions not involving ghosts:

1. All Green's functions are independent of any gauge-fixing parameters.
2. All Green's functions are independent of the particular *S*-matrix process used to define them.
3. All Green's functions obey Ward identities of QED type, not involving ghosts.
4. All Green's functions obey dispersion relations in which there are no identifiable ghost contributions or threshholds.
5. The discontinuities (imaginary parts) of Green's functions can be calculated with the usual Cutkosky rules, consistent with unitarity for the *S*-matrix.

All these are properties of Green's functions in the background-field Feynman gauge, later shown to be equivalent to the pinch technique.

One remark concerning the imaginary parts and unitarity is in order. The photon propagator of QED satisfies a Källen–Lehmann representation with a positive spectral function, a property intimately related to the positivity of the beta function of QED. Because this beta function is negative for an asymptotically free theory, it is impossible to find a NAGT gauge-boson propagator with a positive spectral function, so unitarity holds in a generalized form, with some negative contributions to spectral functions. However, as pointed out in Section 1.7, special properties of the PT propagator allow its factorization into two terms, each obeying the Källen–Lehmann representation.[1] This factorization allows the rearrangement of PT Schwinger–Dyson equations into a form in which all necessary positivity constraints are realized.

At the beginning, how to extend the pinch technique to higher orders of perturbation theory was far from clear; the pioneering technology defined in the first papers would have been forbiddingly difficult for graphs with two or more loops. Fortunately, the problem of the all-order pinch technique has a solution that can be stated with remarkable simplicity: all that has to be done, as was shown [5, 6], is to calculate conventional Feynman graphs using the background-field methodology [7] in the Feynman gauge. The original proof was for NAGTs such as QCD, but it was extended [8] to all orders of electroweak theory. This work was inspired by remarks [9, 10, 11] to the effect that the original pinch technique and the background-field Feynman gauge gave exactly the same results at one loop in perturbation theory. This, of course, could have been a coincidence without much

---

[1] The product of two functions obeying the Källen–Lehmann representation need not obey it.

meaning, but the all-order proof showed constructively how the PT requirements were satisfied at all orders in the background-field Feynman gauge.[2]

In roughly the same time period, string-theory workers [12] studied the off-shell extrapolation of string-theory amplitudes in the field theory, or zero Regge slope, limit. By imposing a consistent implementation of modular invariance, these workers showed that the off-shell gauge-theory amplitudes derived from string theory were automatically given in the background-field Feynman gauge–equivalent to the pinch technique.

The results showing the equivalence of the pinch technique and the background-field Feynman gauge set the stage for nonperturbative applications of the pinch technique, including the Schwinger–Dyson equations of the pinch technique and their consequences. The output of any PT calculation is not only independence of any gauge-fixing parameter but also freedom from contamination by unphysical objects. For example, if one tries to find the contributions of gauge-invariant condensates such as $\langle \mathrm{Tr}\, G_{\mu\nu} G^{\mu\nu} \rangle$ to the usual gauge-boson propagator, one discovers that they are inextricably bound with nonphysical and gauge-dependent condensates involving the ghost fields. But for the PT propagator, only the gauge-invariant condensate, field-strength condensate emerges; there are no ghost contributions [13].

1.2 Notation and conventions

Unless explicitly stated otherwise, we adopt the conventions of Peskin and Schröeder [14]. Sometimes, such as in Chapters 7–9 and parts of Chapter 11, it is convenient to work in Euclidean space. The canonical gauge potential $A_\mu^a(x)$ is often combined in the Hermitian matrix form

$$A_\mu(x) = A_\mu^a(x) t^a, \tag{1.1}$$

where t^a are the $SU(N)$ generators satisfying the commutation relations

$$[t^a, t^b] = \mathrm{i} f^{abc} t^c, \tag{1.2}$$

with f^{abc} being the group's totally antisymmetric structure constants. The generators are normalized according to

$$\mathrm{Tr}(t^a t^b) = \frac{1}{2} \delta^{ab}. \tag{1.3}$$

In the case of QCD, the fundamental representation is given by $t^a = \lambda^a/2$, where λ^a are the Gell–Mann matrices.

---

[2] And in no other background-field gauge; for other than the Feynman gauge, the original PT pinching rules would have to be applied to the background-field Green's functions to get those of the PT.

In Chapters 7, 8, and 9, dealing with nonperturbative phenomena, we combine the gauge potentials in the anti-Hermitean matrix form

$$A_\mu(x) = -ig A_\mu^a(x) t^a,$$

in which case the matrix potential has a unit mass dimension in all space-time dimensions. The changes in all other definitions are trivial. This definition has many advantages when we go beyond perturbation theory.

The Lagrangian density for a general $SU(N)$ non-Abelian gauge theory is given by

$$\mathcal{L} = \mathcal{L}_I + \mathcal{L}_{GF} + \mathcal{L}_{FPG}. \tag{1.4}$$

$\mathcal{L}_I$ represents the gauge invariant Lagrangian, namely,

$$\mathcal{L}_I = -\frac{1}{4} G_a^{\mu\nu} G_{\mu\nu}^a + \bar{\psi}_f^i \left(i\gamma^\mu \mathcal{D}_\mu - m \right)_{ij} \psi_f^j, \tag{1.5}$$

where $a = 1, \ldots, N^2 - 1$ (respectively, $i, j = 1, \ldots, N$) is the color index for the adjoint (respectively, fundamental) representation, and f is the flavor index. The matrix-covariant derivative and field strength are defined according to

$$\mathcal{D}_\mu = \partial_\mu - ig A_\mu \tag{1.6}$$

$$\left[\mathcal{D}_\mu, \mathcal{D}_\nu \right] = -ig G_{\mu\nu}^a t^a, \tag{1.7}$$

or, more explicitly,

$$(\mathcal{D}_\mu)_{ij} = \partial_\mu(I)_{ij} - ig A_\mu^a(t^a)_{ij} \tag{1.8}$$

$$G_{\mu\nu}^a = \partial_\mu A_\nu^a - \partial_\nu A_\mu^a + g f^{abc} A_\mu^b A_\nu^c, \tag{1.9}$$

with g being the (strong) coupling constant. Under a local (finite) gauge transformation $V = \exp[-i\theta]$,

$$A_\mu \to V \frac{i}{g} \partial_\mu V^\dagger + V A_\mu V^\dagger; \qquad G_{\mu\nu} \to V G_{\mu\nu} V^\dagger; \qquad \psi \to V\psi, \tag{1.10}$$

from which the invariance of $\mathcal{L}_I$ follows. In terms of infinitesimal local gauge transformations,

$$\delta A_\mu^a = -\frac{1}{g} \partial_\mu \theta^a + f^{abc} \theta^b A_\mu^c; \qquad \delta_\theta \psi_f^i = -i\theta^a (t^a)_{ij} \psi_f^j$$

$$\delta_\theta \bar{\psi}_f^i = i\theta^a \bar{\psi}_f^j (t^a)_{ji}, \tag{1.11}$$

where $\theta^a(x)$ are the local infinitesimal parameters corresponding to the $SU(N)$ generators t^a.

To quantize the theory, the gauge invariance needs to be broken; this breakup is achieved through a (covariant) gauge-fixing function $\mathcal{F}^a$, giving rise to the (covariant) gauge-fixing Lagrangian $\mathcal{L}_{\text{GF}}$ and its associated Faddeev–Popov ghost term $\mathcal{L}_{\text{FPG}}$. The most general way of writing these terms is through the Becchi–Rouet–Stora–Tyutin (BRST) operator s [15, 16] and the Nakanishi–Lautrup multipliers B^a [17, 18], which represent auxiliary, nondynamical fields that can be eliminated through their (trivial) equations of motion. Then, one gets

$$\mathcal{L}_{\text{GF}} = -\frac{\xi}{2}(B^a)^2 + B^a \mathcal{F}^a \tag{1.12}$$

$$\mathcal{L}_{\text{FPG}} = -\bar{c}^a s \mathcal{F}^a, \tag{1.13}$$

where

$$\delta_{\text{BRST}} \Phi = \epsilon s \Phi, \tag{1.14}$$

with ϵ being a Grassmann constant parameter and s being the BRST operator acting on the QCD fields according to

$$
\begin{aligned}
s A_\mu^a &= \partial_\mu c^a + g f^{abc} A_\mu^b c^c; & s c^a &= -\tfrac{1}{2} g f^{abc} c^b c^c \\
s \psi_{\text{f}}^i &= i g c^a (t^a)_{ij} \psi_{\text{f}}^j; & s \bar{c}^a &= B^a \\
s \bar{\psi}_{\text{f}}^i &= -i g c^a \bar{\psi}_{\text{f}}^j (t^a)_{ji}; & s B^a &= 0.
\end{aligned}
\tag{1.15}
$$

From the preceding transformations, it is easy to show that the BRST operator is nilpotent: $s^2 = 0$. In addition, as a result, the sum of the gauge-fixing and Faddev–Popov terms can be written as a total BRST variation:

$$\mathcal{L}_{\text{GF}} + \mathcal{L}_{\text{FPG}} = s \left(\bar{c}^a \mathcal{F}^a - \frac{\xi}{2} \bar{c}^a B^a \right). \tag{1.16}$$

This, of course, is expected because of the well-known property that total BRST variations cannot appear in the physical spectrum of the theory, which in turn implies the ξ independence of the S-matrix elements and physical observables.

As far as the gauge-fixing function is concerned, there are several possible choices. The ubiquitous R_ξ gauges correspond to the covariant choice

$$\mathcal{F}_{R_\xi}^a = \partial^\mu A_\mu^a. \tag{1.17}$$

In this case, one has

$$\mathcal{L}_{\text{GF}} = \frac{1}{2\xi} (\partial^\mu A_\mu^a)^2 \tag{1.18}$$

$$\mathcal{L}_{\text{FPG}} = \partial^\mu \bar{c}^a \partial_\mu c^a + g f^{abc} (\partial^\mu \bar{c}^a) A_\mu^b c^c; \tag{1.19}$$

the Feynman rules corresponding to such a gauge are reported in the appendix. One can also consider noncovariant gauge-fixing functions such as

$$\mathcal{F}_n^a = \frac{n^\mu n^\nu}{n^2} \partial_\mu A_\nu^a, \tag{1.20}$$

where n^μ is an arbitrary but constant four vector. In general, we can classify these gauges by the different values of n^2, i.e., $n^2 < 0$ (axial gauges), $n^2 = 0$ (light-cone gauge), and finally, $n^2 > 0$ (Hamilton or time-like gauge). Clearly, the gauge-fixing form of Eq. (1.20) does not work for the light-cone gauge, which needs a separate treatment, given in Section 1.6. In the other cases,

$$\mathcal{L}_{\mathrm{GF}} = \frac{1}{2\xi(n^2)^2}(n^\mu n^\nu \partial_\mu A_\nu^a)^2 \tag{1.21}$$

$$\mathcal{L}_{\mathrm{FPG}} = \frac{n^\mu n^\nu}{n^2}\left[\partial_\mu \bar{c}^a \partial_\nu c^a + g f^{abc}(\partial^\mu \bar{c}^a)A_\nu^b c^c\right]. \tag{1.22}$$

Notice that these noncovariant gauges, as well as the light-cone gauge, are ghost free because the ghosts decouple completely from the S-matrix in dimensional regularization.

Finally, because of the correspondence [9, 10, 11] between the PT and the particular class of gauges known as background field gauges [7], the latter will be described in depth in Chapter 2.

We end this section observing that when dealing with loop integrals, we will use dimensional regularization and employ the shorthand notation

$$\int_k \equiv \mu^\epsilon (2\pi)^{-d} \int d^d k, \tag{1.23}$$

where $d = 4 - \epsilon$ is the dimension of space-time and μ is the 't Hooft mass scale, introduced to guarantee that the coupling constant is dimensionless in d dimensions. In addition, the standard result,

$$\int_k \frac{1}{k^2} = 0, \tag{1.24}$$

will be used often to set various terms appearing in the PT procedure to zero.

1.3 The basic one-loop pinch technique

We begin with some notation for propagators and a special decomposition for the free three-gluon vertex, a decomposition that also occurs in the background-field method.

1.3.1 Origin of the longitudinal momenta

Consider the S-matrix element for the quark-quark elastic scattering process $q(p_1)q(r_1) \rightarrow q(p_2)q(r_2)$ in QCD. We have that $p_1 + r_1 = p_2 + r_2$ and set $q = r_2 - r_1 = p_1 - p_2$, with $s = q^2$ being the square of the momentum transfer. The longitudinal momenta responsible for triggering the kinematical re-arrangements characteristic of the pinch technique stem either from the bare gluon propagator $\Delta_{\alpha\beta}^{(0)}(k)$ or from the *external* bare (tree-level) three-gluon vertices, i.e., the vertices where the physical momentum transfer q is entering.

To study the origin of the longitudinal momenta in detail, first consider the gluon propagator $\Delta_{\alpha\beta}(k)$; after factoring out the trivial color factor δ^{ab}, in the R_ξ gauges, it takes the form

$$i\Delta_{\alpha\beta}(q, \xi) = P_{\alpha\beta}(q)\Delta(q^2, \xi) + \xi \frac{q_\alpha q_\beta}{q^4}, \tag{1.25}$$

with $P_{\alpha\beta}(q)$ being the dimensionless transverse projector, defined as

$$P_{\alpha\beta}(q) = g_{\alpha\beta} - \frac{q_\alpha q_\beta}{q^2}. \tag{1.26}$$

The scalar function $\Delta(q^2, \xi)$ is related to the all-order gluon, self-energy

$$\Pi_{\alpha\beta}(q, \xi) = P_{\alpha\beta}(q)\Pi(q^2, \xi), \tag{1.27}$$

through

$$\Delta(q^2, \xi) = \frac{1}{q^2 + i\Pi(q^2, \xi)}. \tag{1.28}$$

Because $\Pi_{\alpha\beta}$ has been defined in Eq. (1.28) with the imaginary factor i factored out in front, it is simply given by the corresponding Feynman diagrams in Minkowski space. The inverse of $\Delta_{\alpha\beta}$ can be found by requiring that

$$\Delta_{\alpha\mu}^{am}(q, \xi)(\Delta^{-1})_{mb}^{\mu\beta}(q, \xi) = \delta^{ab} g_\alpha^\beta, \tag{1.29}$$

and it is given by

$$-i\Delta_{\alpha\beta}^{-1}(q, \xi) = P_{\alpha\beta}(q)\Delta^{-1}(q^2, \xi) + \frac{1}{\xi} q_\alpha q_\beta. \tag{1.30}$$

At tree level,

$$i\Delta_{\alpha\beta}^{(0)}(q, \xi) = d(q^2)\left[g_{\alpha\beta} - (1 - \xi)\frac{q_\alpha q_\beta}{q^2} \right] \tag{1.31}$$

$$d(q^2) = \frac{1}{q^2}. \tag{1.32}$$

Evidently, the longitudinal (pinching) momenta are proportional to the combination $\lambda = 1 - \xi$ and vanish for the particular choice $\xi = 1$ (Feynman gauge) so that the free propagator is simply proportional to $g_{\alpha\beta} d(q^2)$. This is a particularly important feature of the Feynman gauge, which, as we will see, makes PT computations much easier. In this gauge, only longitudinal momenta from vertices can contribute to pinching at the one-loop level. The popular case $\xi = 0$ (Landau gauge) gives rise to a transverse $\Delta_{\alpha\beta}^{(0)}(k)$, which may have its advantages but really complicates the PT procedure at this level.

Next, we consider the conventional three-gluon vertex, to be denoted by $\Gamma_{\alpha\mu\nu}^{amn}(q, k_1, k_2)$, given by the following manifestly Bose-symmetric expression (all momenta are incoming, i.e., $q + k_1 + k_2 = 0$):

$$i\Gamma_{\alpha\mu\nu}^{amn}(q, k_1, k_2) = g f^{amn} \Gamma_{\alpha\mu\nu}(q, k_1, k_2) \tag{1.33}$$

$$\Gamma_{\alpha\mu\nu}(q, k_1, k_2) = g_{\mu\nu}(k_1 - k_2)_\alpha + g_{\alpha\nu}(k_2 - q)_\mu + g_{\alpha\mu}(q - k_1)_\nu.$$

This vertex satisfies the standard Ward identities:

$$q^\alpha \Gamma_{\alpha\mu\nu}(q, k_1, k_2) = k_2^2 P_{\mu\nu}(k_2) - k_1^2 P_{\mu\nu}(k_1) \tag{1.34}$$

$$k_1^\mu \Gamma_{\alpha\mu\nu}(q, k_1, k_2) = q^2 P_{\alpha\nu}(q) - k_2^2 P_{\alpha\nu}(k_2) \tag{1.35}$$

$$k_2^\nu \Gamma_{\alpha\mu\nu}(q, k_1, k_2) = k_1^2 P_{\alpha\mu}(k_1) - q^2 P_{\alpha\mu}(q). \tag{1.36}$$

Unfortunately, the right-hand side is not the difference of inverse propagators, a defect that shows up in higher orders as the appearance of ghost terms in the identities, now called the Slavnov–Taylor identities.

But it is possible to decompose the vertex in a special way into two pieces, one of which satisfies a Ward identity of an elementary (ghost-free) type and the other contains the only longitudinal momenta capable of generating pinches [1, 19]. In the general ξ gauge, this decomposition, as applied to the vertex of Figure 1.1(*b*), is

$$\Gamma_{\mu\nu\alpha}(q, k_1, k_2) = \Gamma_{\mu\nu\alpha}^{\xi} + \Gamma_{\mu\nu\alpha}^{P\xi}, \tag{1.37}$$

where

$$\Gamma_{\mu\nu\alpha}^{\xi}(q, k_1, k_2) = (k_1 - k_2)_\alpha g_{\mu\nu} - 2q_\mu g_{\nu\alpha} + 2q_\nu g_{\mu\alpha}$$

$$+ \left(1 - \frac{1}{\xi}\right) [k_{2\nu} g_{\alpha\mu} - k_{1\mu} g_{\alpha\nu}], \tag{1.38}$$

and

$$\Gamma_{\mu\nu\alpha}^{P\xi}(q, k_1, k_2) = \frac{1}{\xi} [k_{2\nu} g_{\alpha\mu} - k_{1\mu} g_{\alpha\nu}]. \tag{1.39}$$

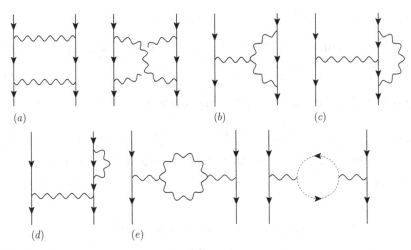

Figure 1.1. The diagrams contributing to the one-loop quark elastic scattering S-matrix element. (a) box contributions, (b) non-Abelian and (c) Abelian vertex contributions (two similar diagrams omitted), (d) quark self-energy corrections (three similar diagrams omitted), and (e) gluon self-energy contributions.

It is easy to check that Γ^ξ obeys the elementary Ward identity:

$$q^\alpha \Gamma^\xi_{\mu\nu\alpha}(q, k_1, k_2) = \Delta^{-1}_{\mu\nu}(k_2, \xi) - \Delta^{-1}_{\mu\nu}(k_1, \xi), \qquad (1.40)$$

and that $\Gamma^{P\xi}$ is the only part of the vertex that triggers pinches. In the pinch technique, (a trivial modification of) this ghost-free Ward identity holds to all orders and has, as a consequence, as in QED, the equality of the gluon wave function and vertex renormalization constants – a relation of great importance for further developments. Note that the vertex $\Gamma^\xi_{\alpha\mu\nu}(q, k_1, k_2)$ is Bose symmetric only with respect to the μ and ν legs. Evidently, the preceding decomposition assigns a special role to the q-leg, which is attached to two on-shell lines. In fact, this vertex Γ^ξ also occurs in the background-field method (see the appendix).[3]

It would be possible to carry out the (one-loop) PT manipulations with this vertex decomposition for any ξ, but, just as for the propagator, things simplify in the Feynman gauge, where a substantial part of Γ^ξ vanishes. Because we will use this gauge extensively, we record its vertex decomposition using the notation $\Gamma^F = \Gamma^{\xi=1}$, $\Gamma^{P\xi=1} = \Gamma^P$. Then,

$$\Gamma_{\alpha\mu\nu}(q, k_1, k_2) = \Gamma^F_{\alpha\mu\nu}(q, k_1, k_2) + \Gamma^P_{\alpha\mu\nu}(q, k_1, k_2), \qquad (1.41)$$

---

[3] Actually, in both the pinch technique and the background-field method, there are two kinds of vertices; at the one-loop level, only the one used here matters.

with

$$\Gamma^{\text{F}}_{\alpha\mu\nu}(q, k_1, k_2) = (k_1 - k_2)_\alpha g_{\mu\nu} + 2q_\nu g_{\alpha\mu} - 2q_\mu g_{\alpha\nu}, \tag{1.42}$$

$$\Gamma^{\text{P}}_{\alpha\mu\nu}(q, k_1, k_2) = k_{2\nu} g_{\alpha\mu} - k_{1\mu} g_{\alpha\nu}, \tag{1.43}$$

and this allows $\Gamma^{\text{F}}_{\alpha\mu\nu}(q, k_1, k_2)$ to satisfy the Ward identity

$$q^\alpha \Gamma^{\text{F}}_{\alpha\mu\nu}(q, k_1, k_2) = (k_2^2 - k_1^2)g_{\mu\nu}, \tag{1.44}$$

where the right-hand side is the difference of two inverse propagators in the Feynman gauge.

1.3.2 The basic pinch operation

The term *pinch* arises from the operation of longitudinal momenta, such as in Γ^{P}, on vertices, which triggers Ward identities that lead to the cancellation of a preexisting propagator by an inverse propagator coming from the Ward identity. The resulting graph looks like a Feynman graph from which one line has been removed, as if it had been pinched out.

Whether acting on a vertex or a box diagram, as in Figure 1.1, the effect of the pinching momenta, regardless of their origin (gluon propagator or three-gluon vertex), is to trigger the elementary Ward identity

$$k_\nu \gamma^\nu = (\slashed{k} + \slashed{p} - m) - (\slashed{p} - m), \tag{1.45}$$

where the right-hand side (rhs) is the difference of two inverse tree-level quark propagators. The first of these terms cancels (pinches out) the internal tree-level fermion propagator $S^{(0)}(k + p)$, and the second term on the rhs vanishes when hitting the on-shell external leg. Diagrammatically speaking, an unphysical effective vertex appears in the place where $S^{(0)}(k + p)$ was, i.e., a vertex that does not appear in the original Lagrangian; as we will see, all such vertices cancel in the full, gauge-invariant amplitude.

First of all, it is immediate to verify the cancellation of the ξ-dependent terms at tree level. After extracting a kinematic factor of the form

$$i\mathcal{V}^{a\alpha}(p_1, p_2) = \bar{u}(p_1)ig t^a \gamma^\alpha u(p_2), \tag{1.46}$$

the tree-level amplitude reads

$$T^{(0)} = i\mathcal{V}^{a\alpha}(r_1, r_2)i\Delta^{(0)}_{\alpha\beta}(q)i\mathcal{V}^{a\beta}(p_1, p_2). \tag{1.47}$$

Then, because the on-shell spinors satisfy the equations of motion

$$\bar{u}(p)(\slashed{p} - m) = 0 = (\slashed{p} - m)u(p), \tag{1.48}$$

the longitudinal part coming from $\Delta^{(0)}_{\alpha\beta}$ vanishes, and we obtain

$$T^{(0)} = i\mathcal{V}^{a\alpha}(r_1, r_2)d(q^2)\mathcal{V}^a_\alpha(p_1, p_2). \tag{1.49}$$

Next, let us concentrate on the two box diagrams, direct and crossed, shown in Figure 1.1(a). The sum of the two graphs gives

$$(a) = g^2 \int_k \bar{u}(r_1)\gamma^\alpha t^a S^{(0)}(r_2 - k)\gamma^\rho t^r u(r_2)\Delta^{(0)}_{\alpha\beta}(k - q)\Delta^{(0)}_{\rho\sigma}(k)$$

$$\times g^2 \bar{u}(p_1)\left\{\gamma^\beta t^a S^{(0)}(p_2 + k)\gamma^\sigma t^r + \gamma^\sigma t^r S^{(0)}(p_1 - k)\gamma^\beta t^a\right\} u(p_2). \tag{1.50}$$

To see how the pinch technique works, we now study the action of the longitudinal momenta appearing in the product $\Delta^{(0)}_{\alpha\beta}(k - q)\Delta^{(0)}_{\rho\sigma}(k)$. Look, for example, at the term $k_\rho k_\sigma$ coming from $\Delta^{(0)}_{\rho\sigma}(k)$. Using Eqs. (1.45) and (1.48), we find that the contraction of k_σ with the term contained in the brackets in the second line on the rhs of Eq. (1.50) gives rise to the expression

$$g^2\bar{u}(p_1)k_\sigma\{\cdots\}^{\beta\sigma} u(p_2) = g^2\bar{u}(p_1)\gamma^\beta \left\{t^a t^r - t^r t^a\right\} u(p_2)$$

$$= ig^2 f^{arn}\bar{u}(p_1)\gamma^\beta t^n u(p_2)$$

$$= g f^{arn} P^\beta_\nu(q)\bar{u}(p_1)ig\gamma^\nu t^n u(p_1)$$

$$= \left[g f^{arn} P^\beta_\nu(q)\right]i\mathcal{V}^{n\nu}(p_1, p_2). \tag{1.51}$$

Notice that in the second step, we have used the commutation relation of Eq. (1.2), and in the third step, we have used the fact that for the on-shell process, longitudinal pieces proportional to $q_\beta q_\nu$ may be added without consequence since they vanish because of current conservation, thus converting g^β_ν to $P^\beta_\nu(q)$. The term in the last line of Eq. (1.51) couples to the external on-shell quarks as a propagator; evidently all reference to the internal (off-shell) quarks inside the brackets has disappeared. To continue the calculation, (1) multiply the result by k_ρ, (2) contract k_ρ with γ^ρ in the first line of Eq. (1.50), (3) employ again the Ward identity of Eq. (1.45), and (4) use the relation $i f^{abc} t^a t^b = -1/2C_A t^c$, where C_A is the Casimir eigenvalues of the adjoint fundamental representation.[4] The final result is a purely propagator-like term, i.e., a term that only depends on q (even though it originates from a box diagram) and couples to the external on-shell quarks as a propagator

---

[4] We denote with C_A (C_f) the Casimir eigenvalue of the adjoint (fundamental) representations. For $SU(N)$, $C_A = N$ and $C_f = (N^2 - 1)/2N$.

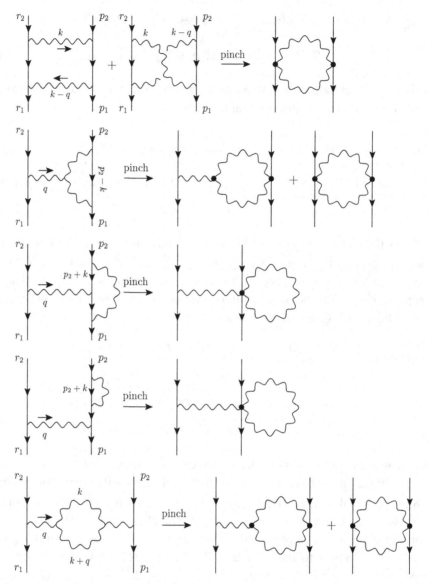

Figure 1.2. The pinching contributions coming from the different one-loop *S*-matrix diagrams.

(see Figure 1.2). Armed with these observations, it is relatively easy to track down the action of all terms proportional to $(1 - \xi)$; in fact, we can write the two boxes as follows:

$$(a) = (a)_{\xi=1} + i\mathcal{V}_\alpha^a(r_1, r_2)id(q^2)\Pi_{\text{box}}^{\alpha\beta}(q, \lambda)id(q^2)i\mathcal{V}_\beta^a(p_1, p_2), \qquad (1.52)$$

Table 1.1. *Contributions of the box, vertex, and self-energy diagrams to the different ξ-dependent structures appearing in the various $\Pi^{\alpha\beta}_{(i)}(q, \lambda)$ terms generated during the PT process*

	$\lambda^2 \int_k \frac{k_\alpha k_\beta}{k^4(k+q)^4}$	$\lambda \int_k \frac{k_\alpha k_\beta}{k^4(k+q)^2}$	$\lambda \int_k \frac{1}{k^2(k+q)^4}$	$\lambda \int_k \frac{1}{k^4}$
$\Pi_{(a)}$	$\frac{1}{2}C_A$	0	$-C_A$	0
$2\Pi_{(b)}$	0	0	0	$C_A - 2C_f$
$2\Pi_{(c)}$	$-C_A$	$2C_A$	$2C_A$	$-2C_A$
$4\Pi_{(d)}$	0	0	0	$2C_f$
$\Pi_{(e)}$	$\frac{1}{2}C_A$	$-2C_A$	$-C_A$	C_A
Total	0	0	0	0

where the ξ-dependent propagator-like term $\Pi^{\alpha\beta}_{\text{box}}$ is given by

$$\Pi^{\alpha\beta}_{\text{box}}(q, \lambda) = \lambda g^2 C_A q^4 \left[\frac{\lambda}{2} P^{\alpha\mu}(q) P^{\beta\nu}(q) \int_k \frac{k_\mu k_\nu}{k^4(k+q)^4} \right.$$
$$\left. - P^{\alpha\beta}(q) \int_k \frac{1}{k^4(k+q)^2} \right]. \tag{1.53}$$

It turns out that all the ξ-dependent parts isolated using the PT procedure are effectively propagator-like, whether they come from box-, vertex-, or propagator-like diagrams. So the general result is

$$(i) = (i)_{\xi=1} + i\mathcal{V}^a_\alpha(r_1, r_2) id(q^2) \Pi^{\alpha\beta}_{(i)}(q, \lambda) id(q^2) i\mathcal{V}^a_\beta(p_1, p_2), \tag{1.54}$$

where i runs over all possible different topologies appearing in Figure 1.1 ($i = a, b, c, d, e$). The value of the corresponding self-energy-like piece is shown in Table 1.1. The sum of each of its columns is zero, which explicitly shows at one loop the ξ-independence property of S-matrix elements.

In the PT framework, the ξ-dependent terms are eliminated in a very particular way. All ξ-dependent pieces turn out to be propagator-like so that all ξ-dependence has canceled, giving rise to subamplitudes that maintain their original kinematic identity (boxes, vertices, and self-energies) and are, in addition, individually ξ-independent. It is important to appreciate that the explicit cancellation carried out amounts effectively to choosing the Feynman gauge, $\xi = 1$, from the beginning. Of course, there is no doubt that this can be done for the entire physical amplitude considered; the point is that, thanks to the pinch technique, one may move from general ξ to $\xi = 1$ without compromising the notion of individual topologies. Such a notion would have been lost if, for instance, the demonstration of the ξ-independence involved integration over virtual momenta; had the integrations

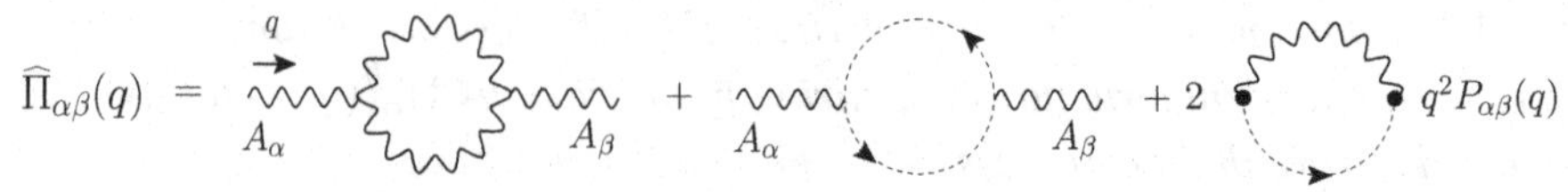

$$\widehat{\Pi}_{\alpha\beta}(q) \quad = \quad \underset{A_\alpha}{\overset{q}{\leftarrow}} \cdots \underset{A_\beta}{\cdots} \quad + \quad \underset{A_\alpha}{\cdots} \bigcirc \underset{A_\beta}{\cdots} \quad + 2 \quad \cdots \bigcirc \cdots \quad q^2 P_{\alpha\beta}(q)$$

Figure 1.3. Diagrammatic representation of the one-loop pinch technique gluon self-energy $\widehat{\Pi}_{\alpha\beta}$ as the sum of the conventional gluon self-energy terms and the pinch contributions coming from the vertex.

been done first, one would have eventually succeeded in demonstrating the ξ-independence of the entire S-matrix element but would have missed out on the ability to identify ξ-independent subamplitudes, as we did. In addition, this result suggests that there is no loss of generality in choosing $\xi = 1$ from the beginning, thereby eliminating a major source of longitudinal pieces that are bound to cancel anyway through the special pinching procedure outlined earlier.

1.3.3 The pinch technique gluon self-energy at one loop

Next, we construct the PT gluon self-energy, to be denoted by $\widehat{\Pi}_{\alpha\beta}(q)$. It is given by the sum of the conventional and self-energy-like parts extracted from the two vertices, as shown in Figure 1.3, i.e.,

$$\widehat{\Pi}_{\alpha\beta}(q) = \Pi_{\alpha\beta}(q) + 2\Pi^{\mathrm{P}}_{\alpha\beta}(q). \tag{1.55}$$

Specifically, in a closed form,

$$\widehat{\Pi}_{\alpha\beta}(q) = \frac{1}{2}g^2 C_A \left[\int_k \frac{\Gamma_{\alpha\mu\nu}\Gamma^{\mu\nu}_\beta}{k^2(k+q)^2} - \int_k \frac{k_\alpha(k+q)_\beta + k_\beta(k+q)_\alpha}{k^2(k+q)^2} \right]$$
$$+ 2g^2 C_A \int_k \frac{q^2 P_{\alpha\beta}(q)}{k^2(k+q)^2}, \tag{1.56}$$

where we have symmetrized the ghost contribution for later convenience and neglected the fermion contribution.

It would be elementary to compute $\widehat{\Pi}_{\alpha\beta}$ directly from the rhs of Eq. (1.56). It is very instructive, however, to identify exactly the parts of the conventional $\Pi_{\alpha\beta}$ that combine with (and eventually cancel) the term $\Pi^{\mathrm{P}}_{\alpha\beta}$. To make this cancellation manifest, one may carry out the following rearrangement of the two elementary three-gluon vertices appearing in Eq. (1.56):

$$\Gamma\Gamma = \Gamma^{\mathrm{F}}\Gamma^{\mathrm{F}} + \Gamma^{\mathrm{P}}\Gamma + \Gamma\Gamma^{\mathrm{P}} - \Gamma^{\mathrm{P}}\Gamma^{\mathrm{P}}, \tag{1.57}$$

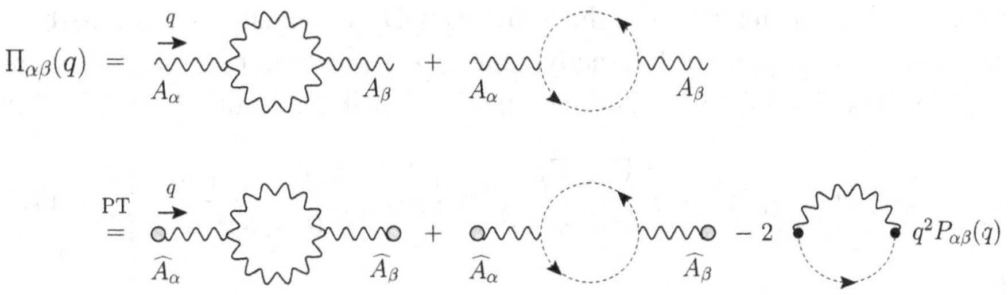

Figure 1.4. The conventional one-loop gluon self-energy before (first line) and after (second line) the pinch technique rearrangement. A shaded circle at the end of an external gluon line denotes that the corresponding gluon behaves as if it were a background gluon.

where, in this symbolic equation, all Lorentz indices have been suppressed, and the product of any two Γs means

$$\Gamma\Gamma \rightarrow \Gamma_{\alpha\mu\nu}\Gamma_{\beta}^{\mu\nu}. \tag{1.58}$$

Dropping terms leading to tadpolelike diagrams (which vanish by dimensional regularization), one then finds

$$\Gamma^P\Gamma + \Gamma\Gamma^P = -4P_{\alpha\beta}(q)q^2 - 2k_{\alpha}k_{\beta} - 2(k+q)_{\alpha}(k+q)_{\beta} \tag{1.59}$$

$$\Gamma^P\Gamma^P = 2k_{\alpha}k_{\beta} + (k_{\alpha}q_{\beta} + q_{\alpha}k_{\beta}). \tag{1.60}$$

We see that the conventional gluon self-energy can be cast in the following form (see also Figure 1.4):

$$\Pi_{\alpha\beta}^{(1)}(q) = \frac{1}{2}g^2 C_A \left[\int_k \frac{\Gamma_{\alpha\mu\nu}^F \Gamma_{\beta}^{F\,\mu\nu}}{k^2(k+q)^2} - 2 \int_k \frac{(2k+q)_{\alpha}(2k+q)_{\beta}}{k^2(k+q)^2} \right]$$

$$- 2g^2 C_A \int_k \frac{q^2 P_{\alpha\beta}(q)}{k^2(k+q)^2}. \tag{1.61}$$

It is easy to prove, using the vanishing of one-loop tadpoles, that each term appearing in the preceding equation is individually conserved so that we have the first ghost-free Ward identity:

$$q^{\alpha} \widehat{\Pi}_{\alpha\beta} = 0. \tag{1.62}$$

The PT re-arrangement has created three manifestly transverse structures, all admitting a unique diagrammatic representation and field theoretical interpretation: the first two terms have in fact precisely the structure of the background-field Feynman gauge at one loop (studied in Chapter 2), whereas the last term represents the one-loop version of a very special auxiliary function that will be thoroughly studied when extending the algorithm to the Schwinger-Dyson equations of non-Abelian gauge theories (Chapter 6). It exactly cancels the pinching contribution coming from the vertex (see Eq. (1.56)) so that one is left with the result

$$\widehat{\Pi}_{\alpha\beta}(q) = \frac{1}{2}g^2 C_A \left[\int_k \frac{\Gamma^F_{\alpha\mu\nu}\Gamma^{F\,\mu\nu}_\beta}{k^2(k+q)^2} - \int_k \frac{2(2k+q)_\alpha(2k+q)_\beta}{k^2(k+q)^2} \right]. \tag{1.63}$$

Using

$$\Gamma^F_{\alpha\mu\nu}\Gamma^{F\,\mu\nu}_\beta = d(2k+q)_\alpha(2k+q)_\beta + 8q^2 P_{\alpha\beta}(q), \tag{1.64}$$

and

$$\int_k \frac{(2k+q)_\alpha(2k+q)_\beta}{k^2(k+q)^2} = -\left(\frac{1}{d-1}\right) \int_k \frac{q^2 P_{\alpha\beta}(q)}{k^2(k+q)^2}, \tag{1.65}$$

the one-loop PT propagator can be written in the simple form

$$\widehat{\Pi}_{\alpha\beta}(q) = \frac{1}{2}g^2 C_A \left(\frac{7d-6}{d-1}\right) \int_k \frac{q^2 P_{\alpha\beta}(q)}{k^2(k+q)^2}, \tag{1.66}$$

valid at $d = 3, 4$. Writing

$$\widehat{\Pi}_{\alpha\beta}(q) = P_{\alpha\beta}(q)\widehat{\Pi}(q^2), \tag{1.67}$$

and following the standard integration rules for the Feynman integral, we obtain the following for the unrenormalized $\widehat{\Pi}$ in $d = 4$:

$$\widehat{\Pi}(q^2) = ibg^2 q^2 \left[\frac{2}{\epsilon} + \ln 4\pi - \gamma_E - \ln \frac{q^2}{\mu^2} + \frac{67}{33}\right], \tag{1.68}$$

where γ_E is the Euler–Mascheroni constant ($\gamma_E \approx 0.57721$) and

$$b = \frac{11 C_A}{48\pi^2} \tag{1.69}$$

is the gauge-invariant one-loop coefficient of the β function of QCD ($\beta = -bg^3$) in the absence of quark loops[5] (for the $d = 3$ case, see Chapter 9).

The gauge-invariant constant b in front of the logarithm corresponds to an analogous gauge-invariant number in the vacuum polarization of QED, where the corresponding coefficient is $-\alpha/3\pi$; of course, the difference in the sign occurs because QCD is asymptotically free whereas QED is not. That the PT gluon propagator captures the leading renormalization group (RG) logarithms is a direct consequence of the Ward identity Eq. (1.92) and the consequent relation $\widehat{Z}_1 = \widehat{Z}_2$. (This means [3] that

---

[5] For comparison, the standard one-loop Feynman self-energy replaces b by $(C_A/32\pi^2)(13/3 - \xi)$, which is obviously gauge dependent and not yielding the correct coefficient b even in the Feynman gauge.

the product of the unrenormalized charge g_0^2 and the unrenormalized PT propagator is the same as this product of renormalized quantities – again, just as in QED – and so this product defines a coupling constant-propagator combination that is not only gauge invariant but renormalization group invariant.) Indeed, if $\widehat{Z}_1 = \widehat{Z}_2$, then the charge renormalization constant, Z_g, and the wave function renormalization of the PT gluon self-energy, $\widehat{Z}_A$, are related by $Z_g = \widehat{Z}_A^{-1/2}$, exactly as in QED.

Finally, notice that since both the $\Gamma^F \Gamma^F$ term and the ghostlike term are separately conserved, the ghostlike term is no longer needed to satisfy the Ward identity for the proper self-energy (Eq. (1.62)). One might ask what quantitative difference it makes to drop the ghostlike term, which, in a Schwinger–Dyson context, would amount to a *truncation* of the series. The answer it that without the ghostlike term, the proper self-energy has exactly the same transverse form but is 10/11 times the self-energy with the ghostlike term. Interestingly, the pinch technique already offers at one loop the ability to truncate gauge invariantly, i.e., preserving the transversality of the truncated answer. This will have profound consequences when addressing the issue of devising a gauge-invariant truncation scheme for the Schwinger–Dyson equations of QCD (Chapter 6).

1.4 Another way to the pinch technique

So far, to develop the pinch technique, we have used a specific S-matrix element for quark-quark scattering. The question naturally arises: is the PT result found this way independent of the process used to define it? This is surely what we must expect from any sensible definition of Green's functions. The answer is yes, as we indicate in the next subsection. Another natural question, given the answer to the first question, is whether one can define the pinch technique in an intrinsically process-independent way. Again, the answer is yes. We do not make any reference to background-field techniques, where the answer to both questions is clearly yes.

1.4.1 Process independence

It is important to stress at this point that the only completely off-shell Green's function involved in the previous construction was the gluon self-energy; instead, the quark-gluon vertex has the incoming gluon off shell and the two quarks on shell, whereas the box has all four incoming quarks on shell. These latter quantities were also made ξ independent in the process of constructing the fully off-shell, ξ-independent gluonic two-point function. Similarly, the construction of a fully off-shell PT quark self-energy requires its embedding in a process such as quark-gluon elastic scattering. The generalization of the methodology is now clear; for

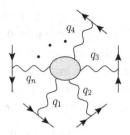

Figure 1.5. *S*-matrix embedding necessary for constructing a ξ-independent, fully off-shell gluonic *n*-point function.

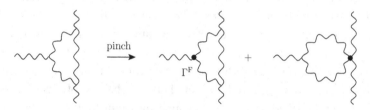

Figure 1.6. The pinching procedure when the embedding particles are *on-shell* gluons. Despite appearances, the vertex by which the pinching contribution is connected to the external gluons is a three-gluon vertex.

example, for constructing a ξ-independent, fully off-shell gluonic *n*-point function (i.e., with *n* off-shell gluons), one must consider the entire ξ-independent process consisting of *n* pairs of quarks, $q(p_1)q(k_1)$, $q(p_2)q(k_2)$, ..., $q(p_n)q(k_n)$ and hook each gluon A_i to one pair of test quarks; the off-shell momentum transfer q_i of the *i*th gluonic leg will be $q_i = p_i - k_i$ (see Figure 1.5). Note, however, that one may equally well use gluons as external test particles or as even fictitious scalars carrying color. Provided that the embedding process is ξ-independent, the answer that the pinch technique furnishes for a given fully off-shell *n*-point function is unique; that is, it is independent of the embedding process. The PT Green's functions are process independent or universal – a property of Green's functions that can hardly be violated. The universality of the one-loop gluon self-energy has been demonstrated through explicit computations using a variety of external test particles. For example, when gluons are used as external test particles, the pinching isolates propagator-like pieces that are attached to the external gluons through a tree-level three-gluon vertex (see Figure 1.6). In this case, the analog of the quark-gluon vertex $\widehat{\Gamma}^a_\alpha$ is the one-loop vertex with one off-shell and two on-shell gluons, which, as we will see in Section 1.5.2, is the one-loop generalization of Γ^F. This latter vertex should not be confused with the PT three-gluon vertex with all three gluons off shell, which can be constructed by embedding it into a six-quark process (one pair for each leg), to be discussed in Section 1.4.2. The distinction between these two three-gluon vertices is crucial and will be made more explicit later on; in addition, a more precise field-theoretic notation will be adopted that will allow us to distinguish the three-gluon vertices unambiguously.

We emphasize that the PT construction is not restricted to the use of on-shell S-matrix amplitudes and works equally well inside, for example, a gauge-invariant current correlation function or a Wilson loop. This fact is particularly relevant for the correct interpretation of the correspondence between the pinch technique and the background-field method to be discussed in Chapter 2. Actually, in the first PT calculation ever [3], the set of one-loop Feynman diagrams studied were the ones contributing to the *gauge-invariant* Green's function:

$$G(x, y) = \left\langle 0 \left| T \left\{ \text{Tr} \left[\Phi(x) \Phi^\dagger(x) \right] \text{Tr} \left[\Phi(y) \Phi^\dagger(y) \right] \right\} \right| 0 \right\rangle, \qquad (1.70)$$

where $\Phi(x)$ is a matrix describing a set of scalar test particles in an appropriate representation of the gauge group. In this case, the special momentum with respect to which the vertex decomposition of Eq. (1.41) should be carried out (i.e., the equivalent of q in that same equation) is the momentum transfer between the two sides of the scalar loop (i.e., one should count loops as if the Φ loop had been opened at x and y). The advantage of using an S-matrix amplitude is purely operational: the PT construction becomes more expeditious because several terms can be set to zero directly owing to the equation of motion of the on-shell test particles. Instead, in the case of a Wilson loop, one would have to carry out the additional step of demonstrating explicitly their cancellation against other similar terms.

1.4.2 Intrinsic pinch technique

The central achievement of the previous sections has been the construction of the ξ-independent, off-shell gluon self-energy, $\widehat{\Pi}_{\alpha\beta}$, through its embedment into a physical S-matrix element, corresponding to quark-quark elastic scattering. This was accomplished by identifying propagator-like pieces from the vertices and boxes contributing to the embedding process and reassigning them to the conventional gluon self-energy, $\Pi_{\alpha\beta}$. This procedure has been carried out for a general value of the gauge-fixing parameter ξ, leading to a unique answer that is most economically reached by choosing the Feynman gauge from the beginning. Thus $\widehat{\Pi}_{\alpha\beta}$ is obtained by adding to $\Pi_{\alpha\beta}$ the propagator-like pieces $2\Pi^P_{\alpha\beta}$ extracted from the vertices, as shown in Eq. (1.56). In the analysis following Eq. (1.56), it became clear that these latter terms cancel very precise terms of the conventional self-energy $\Pi_{\alpha\beta}$, furnishing, finally, $\widehat{\Pi}_{\alpha\beta}$. Specifically, after the vertex decomposition of Eq. (1.57), the terms Γ^P acted on the corresponding Γ, triggering the Ward identities (1.34), (1.35), and (1.36): the term $2\Pi^P_{\alpha\beta}$ cancels against the terms of the Ward identities that are proportional to $q^2 P_{\alpha\beta}$. This observation motivates the following, more expeditious, course of action: instead of identifying the propagator-like pieces from the various graphs, focus on $\Pi_{\alpha\beta}$, carry out the decomposition of Eq. (1.57), and discard the terms coming from the Ward identities that are proportional to $q^2 P_{\alpha\beta}$; what is left is then the PT answer.

This alternative, and completely equivalent, approach to pinching was first introduced in [4] and is known as the *intrinsic pinch technique*. Its main virtue is that it avoids as much as possible the embedment of the Green's function under construction into a physical amplitude. As we shall see later, the intrinsic approach is particularly suited for extending the PT construction in the context of the Schwinger–Dyson equation.

1.5 Pinch technique vertices

After the propagator, the next step is, of course, three-leg vertices. We can extract one of the vertices, the quark-gluon vertex, from the graphs of Figure 1.1 in much the same way that we found the pinch technique gluon propagator. The other vertex under consideration here, the three-gluon vertex, is much more complicated and needs a nontrivial extension of the work we have already done. We begin with the quark-gluon vertex.

1.5.1 The one-loop pinch technique quark-gluon vertex and its Ward identity

Let us now turn to the longitudinal terms contained in the pinching part $\Gamma^P_{\alpha\mu\nu}$ of the three-gluon vertex (see Eq. (1.43)) appearing in the non-Abelian vertex graph and two such vertices inside the gluon self-energy graph. One may ask at this point, what is the purpose of carrying the PT decomposition of the vertex given that one has already achieved ξ-independent structures? The answer is that the effect of the pinching momenta of $\Gamma^P_{\alpha\mu\nu}$ is to make the effective ξ-independent Green's functions satisfy ghost-free Ward identities instead of the usual Slavnov–Taylor identities.

This is best seen in the case of the one-loop quark-gluon vertex $\Gamma^a_\alpha(p_1, p_2)$, composed by the graphs of Figures 1.1(*b*) and 1.1(*c*), now written (after the gauge-fixing parameter cancellations described earlier) in the Feynman gauge. It is well known that the QED counterpart of $\Gamma^a_\alpha(p_1, p_2)$, namely, the photon-electron vertex $\Gamma_\alpha(p_1, p_2)$, satisfies to all orders (and for every ξ) the Ward identity

$$q^\alpha \Gamma_\alpha(p_1, p_2) = ie \left[S_e^{-1}(p_1) - S_e^{-1}(p_2) \right], \qquad (1.71)$$

$$iH^a(p,q) = -gt^a \quad +$$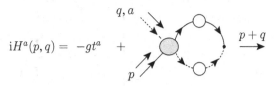

Figure 1.7. The auxiliary function H appearing in the quark-gluon vertex Slavnov–Taylor identity. The shaded blob represents the (connected) ghost-fermion kernel appearing in the usual QCD skeleton expansion.

where S_e is the (all-order) electron propagator; Eq. (1.71) is the naive, all-order generalization of the tree-level identity (1.45).

The quark-gluon vertex $\Gamma_\alpha^a(p_1, p_2)$ also obeys the Ward identity of Eq. (1.45) at tree level (multiplied by t^a):

$$q^\alpha \Gamma_\alpha^a(p_1, p_2) = igt^a \left[S^{-1}(p_1) - S^{-1}(p_2) \right]. \qquad (1.72)$$

However, at higher orders, it obeys a Slavnov–Taylor identity that is not the naive generalization of this tree-level Ward identity. Instead, $\Gamma^\alpha_a(p_1, p_2)$ satisfies the Slavnov–Taylor identity [20]

$$q^\alpha \Gamma_\alpha^a(p_1, p_2) = \left[q^2 D^{aa'}(q) \right] \left[S^{-1}(p_2) H^{a'}(q, p_1) + \bar{H}^{a'}(p_1, q) S^{-1}(p_2) \right], \qquad (1.73)$$

where $D^{aa'}(q)$ and $S(p)$ represent the full ghost and quark propagator, respectively, and H^a is a composite operator defined as (see also Figure 1.7)

$$iS(p)iD^{aa'}(q)iH^a(p,q) = -gt^d \int d^4x \int d^4y \, e^{ip\cdot x} \, e^{iq\cdot y}$$

$$\times \left\langle 0 \left| T \left\{ \bar{q}(x) \bar{c}^{a'}(y) \left[c^d(0) q(0) \right] \right\} \right| 0 \right\rangle, \qquad (1.74)$$

where T denotes the time-ordered product of fields and $\bar{H}$ is the Hermitean conjugate of H. At tree level, H_{ij}^a reduces to $H_{ij}^{(0)a} = t_{ij}^a$.

After these general considerations, let us carry out the decomposition of Eq. (1.41) to the non-Abelian vertex of Figure 1.1(b). Then let us write, suppressing again the color indices,

$$(b)_{\xi=1} = iV_a^\alpha id(q^2) \bar{u}(p_1) i\widetilde{\Gamma}_\alpha^a(p_1, p_2) u(p_2), \qquad (1.75)$$

and concentrate on the (one-loop) non-Abelian contribution to the quark-gluon vertex $\widetilde{\Gamma}_\alpha^a$. We have

$$i\widetilde{\Gamma}_\alpha^a(p_1, p_2) = \frac{1}{2} g^3 C_A t^a \int_k \frac{\left[\Gamma_{\alpha\mu\nu}^{F} + \Gamma_{\alpha\mu\nu}^{P} \right] \gamma^\nu S^{(0)}(p_2 - k) \gamma^\mu}{k^2 (k+q)^2}, \qquad (1.76)$$

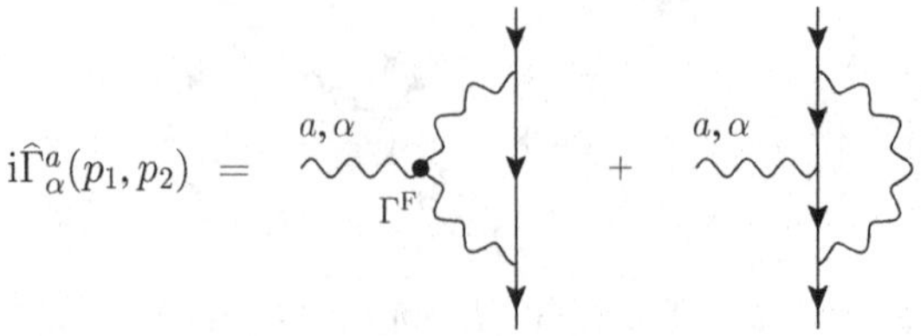

Figure 1.8. Diagrammatic representation of the pinch technique quark-gluon vertex at one loop.

where, in this case,

$$\Gamma^{\mathrm{F}}_{\alpha\mu\nu} = g_{\mu\nu}(2k + q)_\alpha + 2q_\nu g_{\alpha\mu} - 2q_\mu g_{\alpha\nu} \tag{1.77}$$

$$\Gamma^{\mathrm{P}}_{\alpha\mu\nu} = -(k + q)_\nu g_{\alpha\mu} - k_\mu g_{\alpha\nu}. \tag{1.78}$$

Despite appearances, if we use the Dirac equations of motion, the part of the vertex graph containing Γ^{P} is in fact purely propagator-like:

$$\int_k \frac{\Gamma^{\mathrm{P}}_{\alpha\mu\nu}\gamma^\nu S^{(0)}(p_2 - k)\gamma^\mu}{k^2(k + q)^2} \xrightarrow[\text{Dirac Eq.}]{\text{PT}} 2\gamma_\alpha \int_k \frac{1}{k^2(k + q)^2}. \tag{1.79}$$

So also, in this case, one obtains from the one-loop, quark-gluon vertex a propagator-like contribution given by

$$\Pi^{\mathrm{P}}_{\alpha\beta}(q) = g^2 C_A \int_k \frac{q^2 P_{\alpha\beta}(q)}{k^2(k + q)^2}. \tag{1.80}$$

As before, this term plus the equal one coming from the mirror vertex ought to be re-assigned to the PT self-energy. Let us then concentrate on the remaining terms in the vertex. In fact, the part of the vertex graph containing Γ^{F} remains unchanged because it has no longitudinal momenta. Adding it to the usual Abelian-like graph, we obtain the one-loop PT quark-gluon vertex, to be denoted by $\widehat{\Gamma}^a_\alpha$, given by (see Figure 1.8)

$$i\widehat{\Gamma}^a_\alpha(p_1, p_2) = g^3 t^a \left[\frac{1}{2} C_A \int_k \frac{\Gamma^{\mathrm{F}}_{\alpha\mu\nu}\gamma^\nu S^{(0)}(p_2 - k)\gamma^\mu}{k^2(k + q)^2} \right.$$

$$\left. + \left(C_f - \frac{C_A}{2} \right) \int_k \frac{\gamma^\mu S^{(0)}(p_1 + k)\gamma_\alpha S^{(0)}(p_2 + k)\gamma_\mu}{k^2} \right]. \tag{1.81}$$

Now it is easy to derive the Ward identity that $\widehat{\Gamma}_\alpha^a(p_1, p_2)$ satisfies. Using Eq. (1.44), we get

$$q^\alpha \widehat{\Gamma}_\alpha^a(p_1, p_2) = -ig^3 C_f t^a \left[\int_k \frac{\gamma^\mu S^{(0)}(p_2 + k)\gamma_\mu}{k^2} - \int_k \frac{\gamma^\mu S^{(0)}(p_1 + k)\gamma_\mu}{k^2} \right]$$
$$= igt^a \left\{ \widehat{\Sigma}(p_1) - \widehat{\Sigma}(p_2) \right\}. \tag{1.82}$$

Clearly, Eq. (1.82) is the naive generalization of Eq. (1.72) at one loop, i.e., the Ward identity satisfied by Γ_α^a at tree level; this makes the analogy with Eq. (1.71) fully explicit. An immediate consequence of Eq. (1.82) is that the renormalization constants of $\widehat{\Gamma}_\alpha^a$ and $\widehat{\Sigma}$, to be denoted by $\widehat{Z}_1$ and $\widehat{Z}_2$, respectively, are related by the relation $\widehat{Z}_1 = \widehat{Z}_2$, which is none other than the textbook relation $Z_1 = Z_2$ of QED realized in a non-Abelian context.

A direct comparison of the Slavnov–Taylor identity (1.73), satisfied by the conventional vertex Γ_α^a, with the Ward identity (1.82), satisfied by the PT vertex $\widehat{\Gamma}_\alpha^a$, suggests a connection between the terms removed from Γ_α^a during the process of pinching and the ghost-related quantities D^{ab} and H_{ij}^a. As we will see in detail in Chapter 4, such a connection indeed exists and is, in fact, of central importance for the generalization of the pinch technique to all orders.

1.5.2 The one-loop, three-gluon vertex and its Ward identity

The S-matrix construction The same general principles used for the propagator also apply to the proper three-gluon (or four-gluon) vertex: choose a convenient S-matrix element in which the vertex is embedded, in our case, at one loop. This S-matrix element has not only the one-loop vertex that we want but many other graphs. Extract the pinch graphs from these and add them to the conventional vertex. The resulting proper vertex $\widehat{\Gamma}_{\alpha\mu\nu}$ is completely gauge invariant (independent of ξ in an R_ξ gauge) and satisfies ghost-free Ward identities involving the gluon PT inverse propagator. It also satisfies all other properties that we could demand of a three-point vertex: complete Bose symmetry, conventional analytic properties with physical threshholds only, and independence of the S-matrix process used to create the vertex.

Figure 1.9 shows a three-quark S-matrix element, with all quark momenta p_i, $(p - q)_i$ on shell, from which we will find the PT proper three-gluon vertex. We denote the one-loop corrections by $\widehat{\Lambda}_{\mu\nu\alpha}$ and find the full one-loop PT vertex by adding the bare vertex. There are two ways to do this: one is to add the conventional fully symmetric bare vertex $\Gamma_{\alpha\mu\nu}^{(0)}$ and the other is to add the free vertex with one line singled out, as in Eq. (1.42). This is immaterial because the only difference is

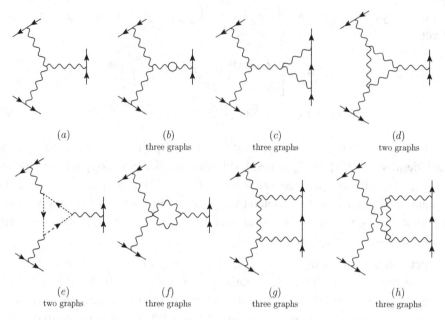

<table>
<tr><td>(a)</td><td>(b)
three graphs</td><td>(c)
three graphs</td><td>(d)
two graphs</td></tr>
<tr><td>(e)
two graphs</td><td>(f)
three graphs</td><td>(g)
three graphs</td><td>(h)
three graphs</td></tr>
</table>

Figure 1.9. *S*-matrix graphs from which the one-loop pinch technique vertex is derived.

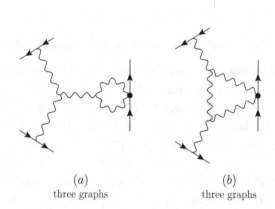

(a)	(b)
three graphs	three graphs

Figure 1.10. Pinched graphs for the vertex.

in gauge-dependent terms, receiving no radiative corrections. The unique gauge-invariant, ghost-free Ward identities relate the radiative correction term $\widehat{\Lambda}_{\alpha\mu\nu}$ to the PT proper self-energy; we give these subsequently.

The pinch parts from these graphs are shown in Figure 1.10. Actually, what we construct by pinching is an *improper vertex* that has PT propagators hooked on[6]:

$$\widehat{F}_{\alpha\mu\nu}(q_1, q_2, q_3) = \widehat{\Delta}_\alpha^\lambda(q_1)\widehat{\Delta}_\mu^\rho(q_2)\widehat{\Delta}_\nu^\sigma(q_3)\widehat{\Gamma}_{\lambda\rho\sigma}(q_1, q_2, q_3). \qquad (1.83)$$

As with the PT propagator, we will work in the R_ξ Feynman gauge ($\xi = 1$). For any ξ, the longitudinal terms in the propagators of Eq. (1.83) strike the external quark lines and give no contribution, so we recover $\widehat{\Gamma}$ from $\widehat{F}$ by truncating, for example, all propagators to the form $g_\mu^\lambda d^{-1}(q_1)$. It is not necessary to calculate $\widehat{F}_{\mu\nu\alpha}$ and then truncate it; instead, the truncation is done by omitting the normal propagator graphs of Figure 1.4(b) and subtracting the pinch parts shown in Figure 1.10(a) rather than adding them. Subtraction rather than omission follows from a straightforward evaluation of combinatoric factors.

After a very lengthy computation, using the decomposition of the bare vertex into Γ^F and pinch parts and a recombination of three vertex terms analogous to the term used for the proper self-energy in Eq. (1.57) (see Eq. (1.95)), we find (with the momenta assignment shown in Figure 1.12)

$$\widehat{\Gamma}_{\alpha\mu\nu}(q_1, q_2, q_3) = -\frac{i}{2}g^2 C_A \int_{k_1} \frac{1}{k_1^2 k_2^2 k_3^2}\widehat{N}_{\alpha\mu\nu} + 8\widehat{B}_{\alpha\mu\nu}, \qquad (1.84)$$

where

$$\widehat{N}_{\alpha\mu\nu} = \Gamma^F_{\alpha\lambda\rho}(q_1, k_3, -k_1)\Gamma^F_{\mu\sigma\lambda}(q_2, k_2, -k_3)\Gamma^F_{\nu\rho\sigma}(q_3, k_1, -k_2)$$
$$\qquad - 2(k_1 + k_3)_\alpha (k_2 + k_3)_\mu (k_1 + k_2)_\nu \qquad (1.85)$$
$$\widehat{B}_{\alpha\mu\nu} = \left(g_{\alpha\nu}q_{1\mu} - g_{\alpha\mu}q_{1\nu}\right) I(q_1) + \left(g_{\mu\nu}q_{2\alpha} - g_{\alpha\mu}q_{2\nu}\right) I(q_2)$$
$$\qquad + \left(g_{\alpha\nu}q_{3\mu} - g_{\mu\nu}q_{3\alpha}\right) I(q_3), \qquad (1.86)$$

and the integrals $I(q_i)$ are given by

$$I(q_i) = \frac{i}{2}g^2 C_A \int_k \frac{1}{k^2(k + q_i)^2}. \qquad (1.87)$$

The first term in Eq. (1.85) is the vertex analog of the $\Gamma^F\Gamma^F$ term in the numerator of the proper self-energy (Eq. (1.63)); the other term comes from ghosts and pinches.

It is now interesting to compare the Slavnov–Taylor identity satisfied by the conventional and PT vertex when contracted with one of its momenta (say, q_1; the other identities are obtained by cyclic permutation of the indices and momenta). At tree level, the identity satisfied by the conventional R_ξ vertex has been derived in

---

[6] Of greater physical significance is a half-proper vertex function $\widehat{G}(q_1, q_2, q_3)$, defined [4] by $\widehat{G}_{\lambda\rho\sigma}(q_1, q_2, q_3) = g^{-2}\bar{g}(q_1)\bar{g}(q_2)\bar{g}(q_2)\widehat{\Gamma}_{\lambda\rho\sigma}(q_1, q_2, q_3)$, where $\bar{g}(q)$ is the pinch technique running charge. This half-proper vertex is not only gauge invariant but also renormalization group invariant.

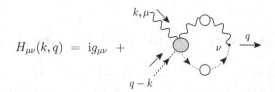

$$H_{\mu\nu}(k, q) = ig_{\mu\nu} +$$

Figure 1.11. The auxiliary function H appearing in the three-gluon vertex Slavnov–Taylor identity. Shaded blobs represent the (connected) Schwinger–Dyson kernel corresponding to the ghost-gluon kernel appearing in the usual QCD skeleton expansion.

Eq. (1.34). At higher orders, the derivation of this identity is a textbook exercise: one starts from the trivial identity

$$\langle T\left[\bar{c}^a(x)A_{\mu}^m(y)A_{\nu}^n(z)\right]\rangle = 0 \tag{1.88}$$

and re-expresses it in terms of the BRST-transformed fields, making use of the equal-time commutation relations. The result is [21]

$$q_1^{\alpha}\Gamma_{\alpha\mu\nu}^{amn}(q_1, q_2, q_3) = igf^{amn}\left[q_1^2 D(q_1)\right]\left\{\Delta^{-1}(q_3^2)P_{\nu}^{\gamma}(q_3)H_{\mu\gamma}(q_2, q_3)\right.$$
$$\left. - \Delta^{-1}(q_2^2)P_{\mu}^{\gamma}(q_2)H_{\nu\gamma}(q_3, q_2)\right\}. \tag{1.89}$$

The function H, shown in Figure 1.11, is the (amputated) one-particle irreducible (1PI) part[7] of the function ($q_1 + q_2 + q_3 = 0$):

$$L_{\mu\nu}^{amn}(q_2, q_3) = gf^{nrs}\int d^4y\int d^4z\, e^{-iq_2\cdot y}e^{-iq_3\cdot z}\left\langle T\left[\bar{c}^a(0)A_{\mu}^m(y)A_{\nu}^r(z)c^s(z)\right]\right\rangle, \tag{1.90}$$

which naturally appears when following the described procedure. The kernel appearing in this function is the conventional connected ghost-gluon kernel appearing in the usual QCD skeleton expansion; in addition, the function $H_{\mu\nu}(k, q)$ is related to the full gluon-ghost vertex $\Gamma_{\mu}(k, q)$ (with k being the gluon and q being the antighost momentum) by the identity

$$q^{\nu}H_{\mu\nu}(k, q) = -i\Gamma_{\mu}(k, q), \tag{1.91}$$

where, at tree level, $H_{\mu\nu}^{(0)} = ig_{\mu\nu}$ and $\Gamma_{\mu}^{(0)}(k, q) = \Gamma_{\mu}(k, q) = -q_{\mu}$.

For the PT vertex, by contracting Eq. (1.84) with q_1, we instead immediately get the result

$$q_1^{\alpha}\widehat{\Gamma}_{\alpha\mu\nu}^{amn}(q_1, q_2, q_3) = gf^{amn}\left\{\widehat{\Delta}^{-1}(q_2)P_{\mu\nu}(q_2) - \widehat{\Delta}^{-1}(q_3)P_{\mu\nu}(q_3)\right\}, \tag{1.92}$$

---

[7] Let us recall that a diagram is called 1PI if it cannot be split into two disjoined pieces by cutting a single internal line; when this is not the case, it is called one-particle reducible (1PR).

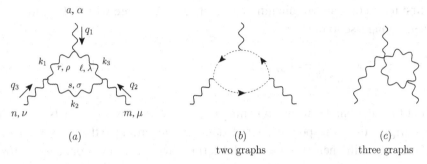

Figure 1.12. R_ξ diagrams contributing to the one-loop three-gluon vertex. Diagrams (c) carry a 1/2 symmetry factor. Fermion diagrams are not shown.

with $\widehat{\Delta}^{-1}(q) = q^2 + i\widehat{\Pi}(q^2)$; thus we find the naive one-loop generalization of the tree-level identity of Eq. (1.44). Notice that (except in ghost-free gauges) the rhs of the preceding equation is not the difference of two inverse gluon propagators because the projection operator P has no inverse; also notice that there is no longer reference to auxiliary ghost Green's functions so that Eq. (1.92) is completely gauge invariant.

The intrinsic construction As an application of the intrinsic PT algorithm described in the previous section, let us see in detail how one can construct the one-loop PT three-gluon vertex. The conventional R_ξ diagrams are shown in Figure 1.12 and read

$$\Lambda_{\alpha\mu\nu}(q_1, q_2, q_3) = -\frac{i}{2}g^2 C_A \int_{k_1} \frac{1}{k_1^2 k_2^2 k_3^2} N_{\alpha\mu\nu} + \frac{9}{2}\widehat{B}_{\alpha\mu\nu}, \qquad (1.93)$$

with

$$N_{\alpha\mu\nu} = \Gamma_{\alpha\lambda\rho}(q_1, k_3, -k_1)\Gamma_{\mu\sigma\lambda}(q_2, k_2, -k_3)\Gamma_{\nu\rho\sigma}(q_3, k_1, -k_2)$$
$$- k_{1\alpha}k_{2\nu}k_{3\mu} - k_{1\alpha}k_{2\nu}k_{3\mu}. \qquad (1.94)$$

Let us then introduce the shorthand notation $\Gamma_1\Gamma_2\Gamma_3$ for the combination of three-level three-gluon vertices appearing in Eq. (1.94). In this notation, all Lorentz indices are suppressed, and the number appearing in each vertex is the number corresponding to the vertex's external momentum q_i. Then, decomposing each of the Γ_i into $\Gamma_i^F + \Gamma_i^P$, we obtain

$$\Gamma_1\Gamma_2\Gamma_3 = \Gamma_1^F\Gamma_2^F\Gamma_3^F + \Gamma_1^P\Gamma_2\Gamma_3 + \Gamma_1\Gamma_2^P\Gamma_3 + \Gamma_1\Gamma_2\Gamma_3^P - \Gamma_1^P\Gamma_2^P\Gamma_3 - \Gamma_1^P\Gamma_2\Gamma_3^P$$
$$-\Gamma_1\Gamma_2^P\Gamma_3^P + \Gamma_1^P\Gamma_2^P\Gamma_3^P. \qquad (1.95)$$

The first term contains no pinching momenta and therefore will be kept in the PT answer, giving rise to the term

$$(\widehat{a}) = -\frac{\mathrm{i}}{2}g^2 C_A \int_{k_1} \frac{1}{k_1^2 k_2^2 k_3^2} \Gamma^{\mathrm{F}}_{\alpha\lambda\rho}(q_1, k_3, -k_1)\Gamma^{\mathrm{F}}_{\mu\sigma\lambda}(q_2, k_2, -k_3)\Gamma^{\mathrm{F}}_{\nu\rho\sigma}(q_3, k_1, -k_2).$$

(1.96)

Each of the other six terms has pinching terms generated when Γ^{P}_i acts on the full Γs that trigger the corresponding Ward identities; according to the rules of intrinsic pinching, we will then discard all the terms that are proportional to $d^{-1}(q_i^2)$. However, these d^{-1} terms can also refer to a virtual momentum k_i, in which case, they give rise to an integral with only two propagators. The last term on the rhs of Eq. (1.95) yields terms of this sort as well as a contribution that adds to the ghost graph

$$\Gamma^{\mathrm{P}}_1\Gamma^{\mathrm{P}}_2\Gamma^{\mathrm{P}}_3 = -d^{-1}(k_1^2)\left(g_{\mu\nu}k_{3\alpha} + g_{\alpha\mu}k_{1\alpha}\right) - d^{-1}(k_2^2)\left(g_{\alpha\mu}k_{1\nu} + g_{\alpha\nu}k_{3\mu}\right)$$
$$- d^{-1}(k_3^2)\left(g_{\alpha\nu}k_{2\mu} + g_{\mu\nu}k_{1\alpha}\right) - k_{1\alpha}k_{2\mu}k_{3\nu} - k_{1\nu}k_{2\mu}k_{3\alpha}. \quad (1.97)$$

The rest of the terms instead have external pinches, which we drop, keeping only the relevant terms:

$$\Gamma^{\mathrm{P}}_1\Gamma_2\Gamma_3 = d^{-1}(k_3^2)\left[\Gamma_{\nu\alpha\mu}(k_1, -k_2) + \Gamma_{\mu\nu\alpha}(k_2, -k_3)\right]$$
$$+ k_{2\mu}\left[d^{-1}(k_1^2)g_{\alpha\nu} - k_{1\alpha}k_{1\nu}\right] + k_{2\nu}\left[d^{-1}(k_3^2)g_{\alpha\mu} - k_{3\alpha}k_{3\mu}\right] \quad (1.98)$$

$$\Gamma^{\mathrm{P}}_1\Gamma^{\mathrm{P}}_2\Gamma_3 = d^{-1}(k_3^2)\Gamma_{\nu\alpha\mu}(k_1, -k_2) - k_{3\alpha}\left[d^{-1}(k_2^2)g_{\mu\nu} - k_{2\mu}k_{2\nu}\right]$$
$$- k_{3\mu}\left[d^{-1}(k_1^2)g_{\nu\alpha} - k_{1\nu}k_{1\alpha}\right]. \quad (1.99)$$

Similar expressions can be found for all the other terms appearing on the rhs of Eq. (1.95). Isolating all the pinching terms that do not pinch out any (internal) propagator and adding them to the conventional ghost graph of Figure 1.12(b), we get the result

$$(\widehat{b}) = 2\frac{\mathrm{i}}{2}g^3 C_A \int_{k_1} \frac{1}{k_1^2 k_2^2 k_3^2} 2(k_1 + k_3)_\alpha (k_2 + k_3)_\mu (k_1 + k_2)_\nu, \quad (1.100)$$

with the remaining pinching contribution giving

$$(c)^{\mathrm{P}} = -\frac{\mathrm{i}}{2}g^2 C_A \int_{k_2} \frac{1}{k_2^2 k_3^2} \left[g_{\alpha\mu}(k_1 - q_3)_\nu + 2g_{\alpha\nu}(q_3 - q_1)_\mu + g_{\mu\nu}(k_1 + q_1)_\alpha\right]$$
$$- \frac{\mathrm{i}}{2}g^2 C_A \int_{k_1} \frac{1}{k_1^2 k_3^2} \left[g_{\alpha\mu}(k_2 + q_3)_\nu + g_{\alpha\nu}(k_2 - q_2)_\mu + 2g_{\mu\nu}(q_2 - q_3)_\alpha\right]$$
$$- \frac{\mathrm{i}}{2}g^2 C_A \int_{k_1} \frac{1}{k_1^2 k_2^2} \left[2g_{\alpha\mu}(q_1 - q_2)_\nu + g_{\alpha\nu}(k_3 + q_2)_\mu + g_{\mu\nu}(k_3 - q_1)_\alpha\right].$$

(1.101)

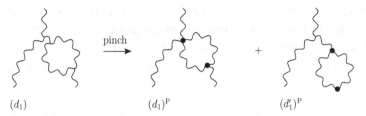

(d_1) $\qquad\qquad$ $(d_1)^{\mathrm{P}}$ $\qquad\qquad$ $(d_1')^{\mathrm{P}}$

Figure 1.13. 1PR diagram giving rise to effectively 1PI pinching contributions (diagram $(d_1)^{\mathrm{P}}$). Two more diagrams (corresponding to having the gluon self-energy correction on the remaining legs) that give rise to similar terms are not shown.

As we have seen, in the presence of longitudinal momenta, the topology of a Feynman diagram is not a well-defined property because longitudinal momenta will pinch out internal propagators, turning t-channel diagrams into s-channel diagrams. This same caveat applies also to the notion of one-particle reducibility. It turns out that by pinching out internal propagators, one can effectively convert 1PR diagrams into 1PI diagrams (see Figure 1.13); of course, the opposite cannot happen. Evidently, one-particle reducibility is a gauge-dependent concept. Thus, when constructing the purely (1PI) gauge-invariant three-gluon vertex at one loop, one has to take into account possible 1PI pinching contributions coming from apparently 1PR diagrams, like those coming from the graphs shown in Figure 1.13. Notice also that not all the pinching terms coming from the diagrams of Figure 1.13 will be producing 1PI terms but only those that will remove the *internal* gluon propagator (diagram $(d_1)^{\mathrm{P}}$). Those removing the *external* gluon propagator (diagram $(d_1')^{\mathrm{P}}$) ought to be discarded, in full accordance with the rules of the intrinsic pinch technique, because they will inevitably cancel against analogous contributions coming (in the S-matrix PT implementation) from non-Abelian vertices attached to the external test-quark.

We show in detail what happens in the case shown in Figure 1.13. One has

$$(d_1) = -\frac{i}{2}g^2 C_A \Gamma_{\alpha\mu'\nu}(q_1, q_2, q_3) d(q_2^2) g^{\mu'\nu'}$$

$$\times \int_k \frac{1}{k^2(k+q_2)^2} \Gamma_{\nu'\rho\sigma}(-q_2, k+q_2, -k) \Gamma_\mu^{\rho\sigma}(-q_2, k+q_2, -k). \quad (1.102)$$

As explained, of all the possible pinching contributions appearing after the splitting of the two self-energy three-gluon vertices, shown in Eqs (1.59) and (1.60), one needs to retain only half of the first term appearing on the rhs of Eq. (1.59); the other half removes instead the external propagator, thus generating diagram $(d_1')^P$ of Figure 1.13. Therefore one has

$$(d_1)^{\mathrm{P}} = ig^2 C_A \Gamma_{\alpha\mu\nu}(q_1, q_2, q_3) I(q_2), \quad (1.103)$$

where we kept only the $g^{\mu\sigma}$ part of the $P^{\mu\sigma}$ appearing in the pinching term because the $q_2^\mu q_2^\sigma$ term will remove the external propagator and thus ought to be discarded. All that is left to do is to add this term to the first term appearing in Eq. (1.101), denoted by $(c_1)^P$; a straightforward (setting $k_2 = k$) calculation shows then that

$$(c_1)^P + (d_1)^P = i\frac{7}{2}g^2 C_A (g_{\mu\nu}q_{2\alpha} - g_{\alpha\mu}q_{2\nu})I(q_2). \tag{1.104}$$

The same procedure can be repeated for the diagrams (d_2) and (d_3), which would show the gluon self-energy on the ν and α leg, respectively; after adding them to the corresponding contributions $(c_2)^P$ and $(c_3)^P$, we get

$$(c_2)^P + (d_2)^P = i\frac{7}{2}g^2 C_A (g_{\alpha\nu}q_{3\mu} - g_{\mu\nu}q_{3\alpha})I(q_3) \tag{1.105}$$

$$(c_3)^P + (d_3)^P = i\frac{7}{2}g^2 C_A (g_{\alpha\nu}q_{1\mu} - g_{\alpha\nu}q_{1\nu})I(q_1). \tag{1.106}$$

Because these terms have exactly the same structure as the conventional (c) diagrams of Figure 1.12, they can be combined with them (viz. with the last term in Eq. (1.93)). Then, putting everything together, we recover exactly the same result found in Eq. (1.84).

1.5.3 The four-gluon vertex

Clearly, there should be a generalization of the three-gluon pinch technique to the four-gluon proper vertex; it has been given at one-loop order in [22] with the by-now standard technique of forming an S-matrix element with, in this case, eight on-shell quark legs and then finding the pinch graphs in the Feynman gauge. We will state here only the Ward identity that this vertex satisfies, which is the naive ghost-free generalization that we have learned to expect. This one-loop ghost-free Ward identity has exactly the structure of the tree-level Ward identity. One of four Ward identities, one for each momentum q_i, reads

$$q_1^\alpha \Gamma_{\alpha\mu\nu\rho}^{amnr}(q_1, q_2, q_3, q_4) = g f^{adm} \widehat{\Gamma}_{\mu\nu\rho}^{dnr}(q_1 + q_2, q_3, q_4)$$

$$+ g f^{adr} \widehat{\Gamma}_{\nu\rho\mu}^{drm}(q_1 + q_3, q_4, q_2)$$

$$+ g f^{adn} \widehat{\Gamma}_{\nu\mu\rho}^{dmr}(q_1 + q_4, q_2, q_3), \tag{1.107}$$

where the $\widehat{\Gamma}$ with three indices are the PT three-gluon vertices we found earlier. Note that a possible new renormalization constant $\widehat{Z}_4$ for the four vertex is, by virtue of the Ward identity, equal to $\widehat{Z}_3$. We now have four ghost-free Ward identities (at least at the one-loop level), as given by Eqs. (1.62), (1.82), (1.92), and (1.107).

1.6 The pinch technique in the light-cone gauge

The light-cone gauge is one of a class of gauges that is ghost free, which simplifies our conceptual tasks in understanding the pinch technique (in fact, it was the first gauge used [1, 2, 3] in the development of the pinch technique). We recount here the pinch technique as explained in [3]. It is not necessarily any easier to compute in this gauge – in fact, in some respects, it is harder. But it is easier, as we will see, than the axial gauge, which is similar but also definable in Euclidean space.

The light-cone gauge can only be defined in Minkowski space, but that will not prevent us from using it in Euclidean space *after* cancellation of all gauge-dependent terms has taken place. In fact, this cancellation takes place *before* any momentum-space integrations are done, so we can convert easily to Euclidean integrals in stating PT results derived from the light cone; this is useful for applications to finite-temperature gauge theory [23].

The light-cone gauge introduces a lightlike vector n_μ, with $n^2 = 0$, for the gauge-fixing:

$$n^\mu A_\mu = 0. \tag{1.108}$$

To fix the gauge completely, some other lower-dimensional constraints are needed that give a precise meaning to operators like $(n \cdot \partial)^{-1}$. But we do not even need to know the constraints because all such inverse operators will disappear from the PT propagator before it is necessary to define them.

Although it is not required to implement light-cone gauge fixing as we do here, it is convenient; the alternative is canonical quantization in a gauge such as $A_0 = A_3$. We replace the gauge-fixing term of the R_ξ gauges by

$$\frac{1}{2\eta} \text{Tr} \left(n^\mu A_\mu \right)^2 ; \tag{1.109}$$

later, we will take the limit $\eta = 0$ to enforce the light-cone gauge. This limit can only be taken after all calculations are done. The free propagator and inverse propagator are as follows:

$$i\Delta_{\mu\nu}^{(0)}(q) = \frac{1}{q^2} Q_{\mu\nu}(q) + \eta \frac{q_\mu q_\nu}{(n \cdot q)^2}, \tag{1.110}$$

$$-i[\Delta^{(0)}]_{\mu\nu}^{-1}(q) = q^2 P_{\mu\nu}(q) + \frac{n_\mu n_\nu}{\eta},$$

where[8]

$$Q_{\mu\nu}(q) = g_{\mu\nu} - \frac{n_\mu q_\nu + n_\nu q_\mu}{n \cdot q}. \tag{1.111}$$

The propagator should be annihilated by n_μ and indeed $n^\mu Q_{\mu\nu} = 0$. The term multiplying η does not vanish on multiplication by n^μ, but this is no surprise because it comes from the gauge-dependent gauge-fixing term. Of course, it vanishes in the physical limit $\eta \to 0$.

The virtue of the light-cone gauge is that (except for the unphysical η-dependent term) the propagator is *homogeneous* of degree zero in the vector n_μ, as a moment's thought shows it must be. Any scalar function of a single momentum constructed from the light-cone gauge propagator can *only* depend on q^2 and not in any way on n_μ itself because of this homogeneity requirement. It would seem to follow that the only way in which the gauge choice can be manifested in the propagator is through kinematic factors such as $n_\mu n_\nu/(n \cdot q)^2$. This is to be contrasted with the explicit dependence of the R_ξ-gauge propagators on ξ.[9] So results for the propagator in the light-cone gauge must be very close to those of the pinch technique, even without taking pinching into account.

The conventional one-loop Feynman propagator calculated in the light-cone gauge is [3]

$$i\Delta_{\mu\nu}(q) = i\Delta_{\mu\nu}^{(0)}(q) + Q_{\mu\nu}(q) \left[\frac{22}{3} I(q) + 8I'(q) \right] + \frac{n_\mu n_\nu}{(n \cdot q)^2} \left[8I(q) + 8I'(q) \right], \tag{1.112}$$

where $I(q)$ has already been defined in Eq. (1.87), while

$$I'(q) = \frac{i}{2} g^2 C_A (n \cdot q) \int_k \frac{1}{k^2 (k-q)^2 (n \cdot k)}. \tag{1.113}$$

The radiative corrections have the most general form allowed in the light-cone gauge, where the propagator must be annihilated (except for the gauge-fixing term) by n_μ or n_ν. The integral $I(q) = 1/(16\pi^2) \ln(-q^2/\Lambda^2) + \cdots$ is the one appearing in the R_ξ pinch technique, and if the first term on the rhs were the only contribution, we would again recover exactly the same PT propagator involving the gauge-invariant running charge, except for the kinematics such as the factor $Q_{\mu\nu}(q)$. As for the integral $I'(q)$, it is, as advertised, homogeneously degree zero in n_μ, but it is not clear how to evaluate it. In fact, many learned papers have been written on how to regulate the $1/(n \cdot k)$ singularity and to evaluate integrals such as $I'(q)$, but we do

---

[8] The notation used here differs slightly from that of [3]; in particular $Q_{\mu\nu}$ is defined with the opposite sign.
[9] Certain integrals arise, such as I', that have n_μ in their definition, and their value is not clear. Fortunately, the pinch technique cancels all such terms before the integrations need to be done.

not need them. The reason is that when we add the pinch terms to the conventional light-cone propagator, now stemming from the longitudinal terms in $Q_{\mu\nu}(q)$, the terms in $I'(q)$ and the terms multiplying $n_\mu n_\nu$ are all canceled *before* one needs to face up to the task of doing the strange integral in $I'(q)$. The only term remaining is the one giving the gauge-invariant running charge. We can therefore write the PT propagator in the light-cone gauge as

$$i\widehat{\Delta}_{\mu\nu}(q) = Q_{\mu\nu}(q)\frac{1}{q^2}\left[1 - bg^2\ln\left(\frac{-q^2}{\Lambda^2}\right) + \dots\right] + \eta\frac{q_\mu q_\nu}{(n\cdot q)^2}, \quad (1.114)$$

where b is the by-now familiar one-loop coefficient in the running charge (see Eq. (1.69)).

The absence of the $n_\mu n_\nu$ term in the PT propagator persists to all orders. Such a term is kinematically allowed because it is annihilated by n_μ or n_ν, and its absence to all orders is not a trivial matter.

The light-cone version of the PT propagator differs from the earlier PT propagator not only through the gauge-dependent term multiplying η but also in the kinematical factor $Q_{\mu\nu}$, which depends on the gauge. The first gauge dependence is expected, but perhaps not the second. We can exhibit a more exact correspondence between PT propagators in various gauges by looking not at the propagator but at its inverse (or more to the point, at the PT proper self-energy).

It is straightforward to calculate the inverse of the pinch propagator given in Eq. (1.114); the renormalized version follows:

$$-i\widehat{\Delta}_{\mu\nu}^{-1}(q) = P_{\mu\nu}(q)\left\{q^2\left[1 + bg^2\ln\left(\frac{-q^2}{\mu^2}\right)\right]\right\} + \frac{n_\mu n_\nu}{\eta}. \quad (1.115)$$

We see that the inverse of the light-cone PT propagator is exactly what we found earlier, except for the η-dependent terms. These gauge-dependent terms never receive radiative corrections, except for a multiplicative renormalization of η. In effect, they are only associated with *free* Green's functions.

The importance of the inverse PT propagator, or equivalently, the proper self-energy as well as proper vertices, is that they are the natural ingredients of Ward identities. No matter what gauge is used, the proper self-energy has the simple transverse form (cf. Eq. (1.55))

$$\widehat{\Pi}_{\mu\nu}(q) = P_{\mu\nu}(q)\widehat{\Pi}(q), \quad (1.116)$$

where $\widehat{\Pi}(q)$ is independent of any gauge choice. There are, as we have seen, PT three-point vertices $\widehat{\Gamma}_{\mu\nu\alpha}$ that obey, in any gauge, certain ghost-free Ward identities (cf. Eq. (1.92)). In the light-cone gauge, this Ward identity actually has true inverse propagators on the rhs:

$$q^\alpha \widehat{\Gamma}_{\mu\nu\alpha}(k, q - k, -q) = \widehat{\Delta}^{-1}_{\mu\nu}(k) - \widehat{\Delta}^{-1}_{\mu\nu}(q - k). \qquad (1.117)$$

Note that although the inverse of the PT propagator (Eq. (1.115)) depends on the gauge parameter η, the difference of two inverse propagators is independent of η so that all terms in the Ward identity are strictly gauge invariant. This is just the naive Ward identity that one would expect if NAGTs behaved like QED, with no ghosts to worry about. Of course, this Ward identity is not the Slavnov–Taylor identity satisfied by the conventional full vertex Γ, which has ghost contributions and is gauge dependent.

Just as in QED, this Ward identity ensures that $\widehat{Z}_1 = \widehat{Z}_{3g}$, as we found earlier in the covariant gauge.

1.7 The absorptive pinch technique construction

Here we show how unitarity is defined for the pinch technique, with one-loop examples. In one respect, unitarity for PT Green's functions is the same as for the S-matrix; in another respect, it differs – and it must differ, or it is impossible to reconcile with asymptotic freedom.

The two aspects of PT unitarity are as follows:

1. Off-shell PT Green's functions obey dispersion relations of conventional type, having only physical threshholds (i.e., Goldstone and ghost masses, which are generically gauge dependent, cannot occur in the set of allowed threshholds). The Feynman (background-field) gauge is singled out here because the ghost and Goldstone masses are the same as the gauge-boson mass in this gauge, and so a threshhold criterion cannot rule out that a propagator ultimately stemmed from a ghost or Goldstone particle.
2. The absorptive parts of PT Green's functions are calculated from the PT Green's functions with the standard (Cutkosky rules) construction. However, because the PT Green's functions differ from the conventional ones by terms that subtract out gauge dependence and ghost lines, it happens for a NAGT – but not for QED – that there can be absorptive parts with a negative sign. This is in no way inconsistent with physical unitarity for the S-matrix, which contains not only PT parts but other parts that restore positivity for the absorptive parts of the S-matrix.

The difference between QED and NAGTs is the following: the PT is empty for QED, which is a situation realized by an exact cancellation of terms that would contribute to the pinch parts. But for QCD, the cancellation is not exact, as we have seen. The study of unitarity properties for the pinch technique reveals a similar situation: unitarity has its familiar form for QED, which means, among other things, that the imaginary part of the photon propagator is positive definite (i.e., essential for the propagator to obey the Källen–Lehmann representation) and the beta function of QED is also positive definite. But as we show here, the construction of the absorptive part of a PT Green's function, such as the propagator, usually involves an incomplete cancellation between positive and negative terms that can leave negative terms in absorptive parts where a positive term would normally be expected. This is essential for a negative beta function in an asymptotically free theory. The appearance of negative absorptive terms in a PT propagator is certainly not an indication that the PT fails to respect normal unitarity for the S-matrix, any more than the need for ghosts means a violation of unitarity.

Such cancellations have appeared in a related context, even before there was a pinch technique. Long ago, people asked whether it was possible to derive the fundamental structure of a renormalizable theory of multiple massive vector mesons from some straightforward assumption such as high-energy unitarity, or whether one simply had to adopt by fiat theories of the NAGT-Higgs type. Several authors [24, 25, 26] gave the answer: starting from the most general Lagrangian of spin 0, 1/2, and 1 particles that is renormalizable by power counting for massless vector bosons, one can, by imposing the requirements of high-energy unitarity, derive uniquely the structure of a NAGT–Higgs theory coupled minimally to spin 0 and 1/2 matter fields. The issue is that the massive vector mesons of a general Lagrangian have longitudinal modes that, if their effects were uncanceled because of some relations among couplings that give the Lagrangian a very specific structure, would lead through the usual optical theorem to unbridled growth in energy E of perturbative amplitudes through a power of E/M for every longitudinal mode. Another way to say this is that, in the unitary gauge, a massive vector propagator has a longitudinal numerator part $\sim k_\mu k_\nu / M^2$ that is unrenormalizable unless the theory has a special form – that of an NAGT. These authors showed that requiring the longitudinal modes to be canceled led uniquely to an NAGT–Higgs theory (at least in perturbation theory). But they did not cancel completely; they simply were tamed to the point where total amplitudes behaved like positive powers of M rather than negative powers. Studies of PT unitarity show similar incomplete cancellations for NAGTs [27, 28], as we review here.

Because positivity is often an important physical constraint on absorptive parts, one might question whether PT unitarity, with some negative terms, can be physically useful. In [29], it was conjectured that the product of the PT propagator and the coupling g^2 factors into two terms, both with positive absorptive parts, and the product fails to have a positive absorptive part only because of asymptotic freedom. This factorization allows a re-organization of terms in the Schwinger–Dyson equations for higher-point PT Green's functions such that ordinary positivity requirements still hold.

1.7.1 The strong version of the optical theorem

The T-matrix element of a reaction $i \rightarrow f$ is defined via the relation

$$\langle f|S|i \rangle = \delta_{fi} + \mathrm{i}(2\pi)^4 \delta^{(4)}(P_f - P_i)\langle f|T|i \rangle, \tag{1.118}$$

where P_i (P_f) is the sum of all initial (final) momenta of the $|i\rangle$ $(|f\rangle)$ state. Furthermore, imposing the unitarity relation $S^\dagger S = 1$ leads to the generalized optical theorem

$$\langle f|T|i \rangle - \langle i|T|f \rangle^* = \mathrm{i} \sum_j (2\pi)^4 \delta^{(4)}(P_j - P_i)\langle j|T|f \rangle^* \langle j|T|i \rangle. \tag{1.119}$$

In Eq. (1.119), the sum $\sum_j$ is over the entire phase space and spins of all possible on-shell intermediate particles j.

An important corollary of this theorem is obtained if $f = i$, corresponding to the case of so-called forward scattering. Then, Eq. (1.119) reduces to

$$\Im m \langle i|T|i \rangle = \frac{1}{2} \sum_j (2\pi)^4 \delta^{(4)}(P_j - P_i)|\langle j|T|i \rangle|^2. \tag{1.120}$$

In what follows, we will refer to the relation given in Eq. (1.120) as the optical theorem.

The rhs of the optical theorem consists of the sum of the (squared) amplitudes, $\mathcal{M}^{ij}$, of all kinematically allowed elementary processes connecting the initial and final states. Note in particular that only physical particles may appear as intermediate $|j\rangle$ states. If the particles involved are fermions, gauge bosons, or both when calculating $\mathcal{M}^{ij}$, one averages over the initial-state polarizations and sums over the final-state polarizations. In addition, the integration over all available phase space, implicit in the sum $\sum_j$, must be carried out. The left-hand side (lhs) of the optical theorem is given by the imaginary part of the entire amplitude, i.e., including all Feynman diagrams contributing to it. For example, in the case of NAGTs, to obtain the lhs of the optical theorem, one must calculate the imaginary part of all diagrams, regardless of whether they contain physical (gluons, quarks) or unphysical (ghosts or would-be Goldstone bosons) fields inside their loops. To do that, one usually uses the Cutkosky rules or cutting rules, whereby in all diagrams, the propagators of physical and unphysical particles are put simultaneously on shell.

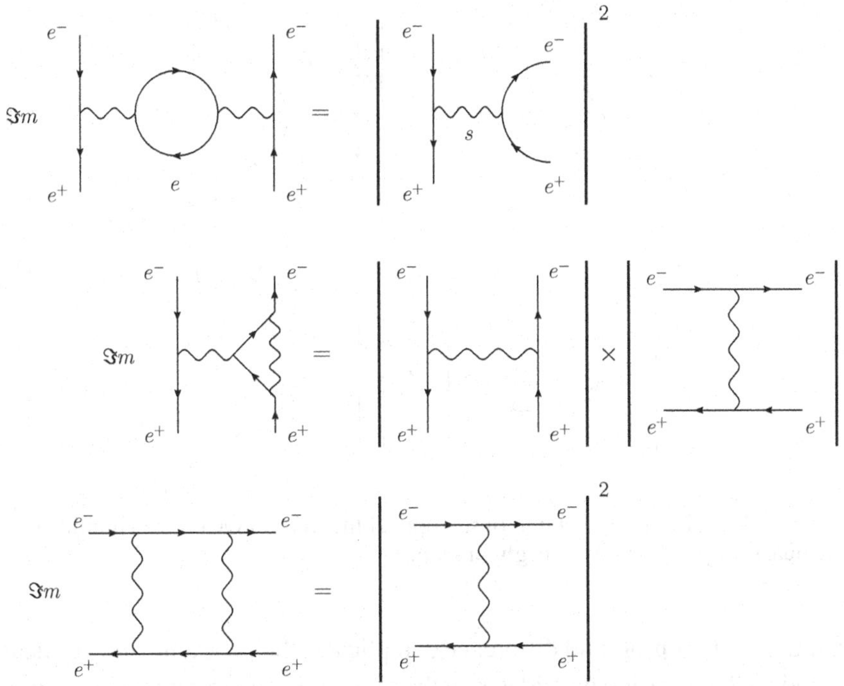

Figure 1.14. The strong version of the optical theorem in QED.

An issue of central importance for what follows is the way the optical theorem is realized at the level of the conventional diagrammatic expansion, or equivalently, at the level of the individual propagator-, vertex-, and boxlike amplitudes. Specifically, in its general formulation of Eq. (1.120), the optical theorem is a statement at the level of entire amplitudes and not individual Feynman graphs or Green's functions. Thus, the imaginary part of a given diagram appearing on the rhs does not necessarily correspond to an easily identifiable diagrammatic (or kinematic) piece on the rhs. Of course, there are theories in which the optical theorem holds also at the level of individual graphs and kinematic subamplitudes. This strong version of the optical theorem is realized in scalar theories and QED (see Figure 1.14) but fails in NAGTs (see Figure 1.15). This is so because, with the exception of certain gauges, in the NAGTs, the propagator-, vertex-, and box-like subamplitudes of each side of the optical theorem are totally different. For example, in the case of the forward QCD process $q(p_1)\bar{q}(p_2) \rightarrow q(p_1)\bar{q}(p_2)$, the propagator-like part of the lhs, computed in the renormalizable gauges, is determined by cutting through one-loop graphs containing ξ-dependent gluon propagators and unphysical ghosts (omit quark loops), whereas the propagator-like part of the rhs contains the polarization tensors corresponding to physical massless particles of spin 1 (two physical

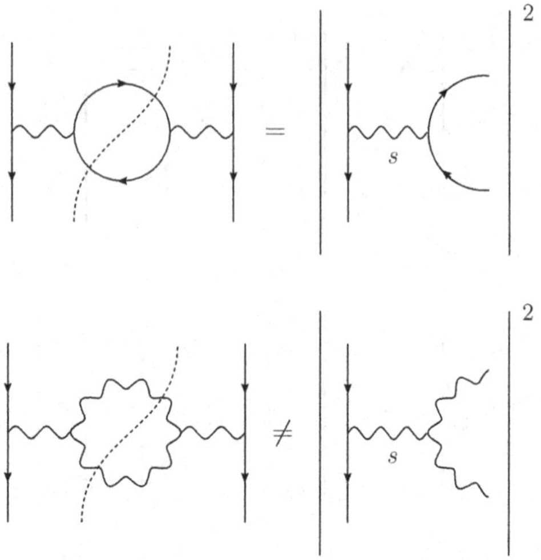

Figure 1.15. The strong version of the optical theorem in QCD, which holds for the quark loop but fails for the gluon loop.

polarizations). This profound difference complicates the diagrammatic verification of the optical theorem and invalidates, at the same time, its strong version. A crucial advantage of the PT is that it permits the realization of the optical theorem at the level of kinematically distinct, well-defined subamplitudes, even in the context of non-Abelian gauge theories; these privileged subamplitudes are, of course, none other than the PT Green's functions. In other words, the strong version of the optical theorem holds if and only if the identification of the subamplitudes on each side occurs after the application of the PT.

1.7.2 The fundamental s-t cancellation

As we demonstrate in this section, the application of the PT on the rhs (the physical side) of the optical theorem is tantamount to the explicit use of an underlying fundamental cancellation between s-channel and t-channel graphs [27, 28]. This cancellation results in a nontrivial reshuffling of terms, which, in turn, allows for the definition of kinematically distinct contributions; interestingly enough, they correspond to the imaginary parts of the one-loop PT subamplitudes constructed in the previous section.

To see all this in detail, we consider the forward-scattering process $q(p_1)\bar{q}(p_2) \rightarrow q(p_1)\bar{q}(p_2)$ and concentrate on the optical theorem to lowest order. Obviously, the intermediate states appearing on the rhs may involve quarks or gluons. The quarks can be treated essentially as in QED and are, in that sense, completely straightforward. We will therefore focus on the part of the optical theorem where

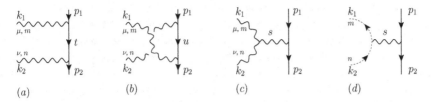

Figure 1.16. Diagrams defining (a and b) the amplitudes $\mathcal{T}_t$ and (c) $\mathcal{T}_s$. Diagram (d) is related to the amplitude $\mathcal{S}$ defined in Eq. (1.126).

the intermediate states are two gluons; we have that

$$\Im m\langle q\bar{q}|T|q\bar{q}\rangle = \frac{1}{2}\times\frac{1}{2}\int_{\mathrm{PS}_{gg}}\langle q\bar{q}|T|gg\rangle\langle gg|T|q\bar{q}\rangle^*, \qquad (1.121)$$

where $\int_{\mathrm{PS}_{gg}}$ is the two gluon phase space measure integral.[10] The extra $\frac{1}{2}$ factor is statistical and arises from the fact that the final on-shell gluons should be considered identical particles in the total rate. Let us now focus on the rhs of Eq. (1.121) and set $T \equiv \langle q\bar{q}|T|gg\rangle$. Diagrammatically, the tree-level amplitude T is the sum of two distinct parts: t- and u-channel graphs that contain an internal quark propagator, $T_{t\,\mu\nu}^{mn}$, as shown in of Figure 1.16(a) and 1.16(b), and an s-channel amplitude, $T_{s\,\mu\nu}^{mn}$, given in Figure 1.16(c). Defining $V_\alpha^a = \bar{v}(p_2)t^a\gamma_\alpha u(p_1)$, we have that

$$T_{s\,\mu\nu}^{mn} = g^2 f^{mnc}V_\alpha^c d(q)\Gamma_{\mu\nu}^\alpha(q,k_1,k_2) \qquad (1.122)$$

$$T_{t\,\mu\nu}^{mn} = ig^2\bar{v}(p_2)\left[t^n\gamma_\nu S^{(0)}(p_1+k_1)t^m\gamma_\mu + t^m\gamma_\mu S^{(0)}(p_1+k_2)\gamma_\nu t^n\right]u(p_1).$$

The subscripts s and t refer, as usual, to the corresponding Mandelstam variables, i.e., $s = q^2 = (p_1 + p_2)^2 = (k_1 + k_2)^2$ and $t = (p_1 - k_1)^2 = (p_2 - k_2)^2$. We then have

$$\mathcal{M} = [\mathcal{T}_s + \mathcal{T}_t]_{\mu\nu}^{mn}\, L^{\mu\mu'}(k_1)L^{\nu\nu'}(k_2)\left[\mathcal{T}_s^* + \mathcal{T}_t^*\right]_{\mu'\nu'}^{mn}, \qquad (1.123)$$

where the polarization tensor $L^{\mu\nu}(k)$ corresponding to a massless spin one particle is given by

$$L_{\mu\nu}(k) = -g_{\mu\nu} + \frac{n_\mu k_\nu + n_\nu k_\mu}{n\cdot k} + \eta^2\frac{k_\mu k_\nu}{(n\cdot k)^2}. \qquad (1.124)$$

For on-shell gluons, i.e., for $k^2 = 0$, $k^\mu L_{\mu\nu}(k) = 0$. By virtue of this last property, we see immediately that if we carry out the PT decomposition of Eq. (1.41) to

---

[10] In general the (4-dimensional) two body phase space integral $\int_{\mathrm{PS}}$ is defined as

$$\int_{\mathrm{PS}} = \frac{1}{(2\pi)^2}\int d^4k_1\int d^4k_2\delta_+(k_1^2 - m_1^2)\delta_+(k_2^2 - m_2^2)\delta^{(4)}(q - k_1 - k_2),$$

where m_1 and m_2 are the masses of the intermediate particles produced.

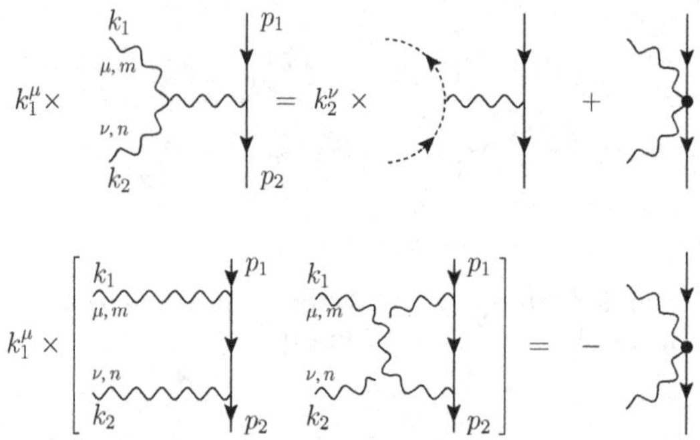

Figure 1.17. The s-t cancellation at tree level.

the three-gluon vertex Γ, the term Γ^{P} vanishes after being contracted with the polarization vectors, and only the Γ^{F} piece of the vertex survives. Thus Eq. (1.123) becomes

$$\mathcal{M} = [\mathcal{T}_s^{\mathrm{F}} + \mathcal{T}_t]_{\mu\nu}^{mn} L^{\mu\mu'}(k_1)\, L^{\nu\nu'}(k_2)[\mathcal{T}_s^{\mathrm{F}} + \mathcal{T}_t]_{\mu'\nu'}^{mn*}, \qquad (1.125)$$

where $\mathcal{T}_s^{\mathrm{F}}$ is given by Figure 1.16(c) after substituting Γ by Γ^{F}.

Let us now define the quantities $\mathcal{S}^{mn}$ and $\mathcal{R}_\mu^{mn}$ (see Figure 1.16(d)):

$$\mathcal{S}^{mn} = \frac{1}{2} g f^{amn} d(q^2)(k_1 - k_2)^\mu \mathcal{V}_\mu^a \qquad \mathcal{R}_\mu^{mn} = g f^{amn} \mathcal{V}_\mu^a, \qquad (1.126)$$

where $\mathcal{V}_\rho^a(p_2, p_1) = \bar{v}(p_2) g t^a \gamma_\rho u(p_1)$; they are related by $k_1^\mu \mathcal{R}_\mu^{mn} = -k_2^\mu \mathcal{R}_\mu^{mn} = q^2 \mathcal{S}^{mn}$. Then, using the conditions $k_1^2 = k_2^2 = 0$, together with current conservation $q^\rho \mathcal{V}_\rho^a = 0$, we obtain the WI

$$k_1^\mu \Gamma_{\alpha\mu\nu}^{\mathrm{F}}(q, -k_1, -k_2) = -q^2 g_{\alpha\nu} + (k_1 - k_2)_\alpha k_{2\nu}. \qquad (1.127)$$

Now the crucial point is that the q^2 term on the rhs of the preceding Ward identity will cancel against the $d(q^2)$ inside $\mathcal{T}_s^{\mathrm{F}}$, allowing the communication of this part with the (contracted) t-channel graph. Specifically,

$$k_1^\mu [\mathcal{T}_s^{\mathrm{F}}]_{\mu\nu}^{mn} = 2k_{2\nu} \mathcal{S}^{mn} - \mathcal{R}_\nu^{mn} \qquad k_1^\mu [\mathcal{T}_t]_{\mu\nu}^{mn} = \mathcal{R}_\nu^{mn} \qquad (1.128)$$

so that

$$k_1^\mu [\mathcal{T}_s^{\mathrm{F}} + \mathcal{T}_t]_{\mu\nu}^{mn} = 2k_{2\nu} \mathcal{S}^{mn}. \qquad (1.129)$$

This is the s-t cancellation: the term $\mathcal{R}$ comes with opposite sign and drops out in the sum (see Figure 1.17). An exactly analogous cancellation takes place if we contract by k_2^ν.

It is now easy to check that, indeed, all dependence on both n_μ and η^2 cancels in Eq. (1.125), as it should, and we are finally left with (omitting the fully contracted color and Lorentz indices)

$$\mathcal{M} = \left(T_s^F T_s^{F*} - 8SS^* \right) + \left(T_s^F T_t^* + T_s^{F*} T_t \right) + T_t T_t^*. \quad (1.130)$$

The reader may wonder what happens if, in Eq. (1.125), we do not eliminate Γ^P using the transversality of the polarization tensors and keep the full Γ instead of just Γ^F. In that case, the tree-level WI to use would be that of Eq. (1.35) instead of Eq. (1.127). This modification leaves the s-t cancellation unaffected but changes the terms proportional to S^{mn}. However, the presence of the longitudinal parts in Γ^P triggers further s-t cancellations, exposed by using the decomposition of Eq. (1.57). As was shown in [27] and [28], the end result of this equivalent but slightly lengthier procedure is exactly that given in Eq. (1.130).

At this point, i.e., after the implementation of the s-t cancellation, we can define the *genuine* propagator-like, vertexlike, and boxlike subamplitudes, corresponding to the first, second, and third terms on the rhs of Eq. (1.130). Thus the propagator-like part of the rhs of the optical theorem, to be denoted by (rhs)$_1$, is given by

$$(\text{rhs})_1 = \frac{1}{2} \times \frac{1}{2} \int_{\text{PS}_{gg}} \left(T_s^F T_s^{F*} - 8SS^* \right). \quad (1.131)$$

It is elementary to verify that

$$T_s^F T_s^{F*} - 8SS^* = g^2 C_A \mathcal{V}_\mu^a d(q^2) \left[8q^2 P^{\mu\nu}(q) + 2(k_1 - k_2)^\mu (k_1 - k_2)^\nu \right] d(q^2) \mathcal{V}_\nu^a. \quad (1.132)$$

For the case of two massless gluons in the final state, the phase-space integrals give

$$\int_{\text{PS}_{gg}} = \frac{1}{8\pi}, \qquad \int_{\text{PS}_{gg}} (k_1 - k_2)_\mu (k_1 - k_2)_\nu = -\frac{1}{24\pi} q^2 P_{\mu\nu}(q), \quad (1.133)$$

and thus Eq. (1.131) becomes

$$(\text{rhs})_1 = \mathcal{V}_\mu^a d(q^2) [\pi b g^2 q^2 P_{\mu\nu}(q) d(q^2)] \mathcal{V}_\nu^a. \quad (1.134)$$

On the other hand, for the propagator-like part of the lhs of the optical theorem, we have

$$(\text{lhs})_1 = \mathcal{V}_\mu^a d(q^2) \Im m \, \widehat{\Pi}^{\mu\nu}(q) d(q^2) \mathcal{V}_\nu^a, \quad (1.135)$$

where the $\Im m \, \widehat{\Pi}^{\mu\nu}(q)$ should be obtained from the one-loop expression for $\widehat{\Pi}^{\mu\nu}(q)$ in Eqs (1.67) and (1.68); it is then obvious that, indeed, (lhs)$_1$ = (rhs)$_1$, namely, that the PT gluon self-energy satisfies the strong version of the optical theorem, as announced.

1.8 Positivity and the pinch technique gluon propagator

The Ward identity of Eq. (1.92) (or Eq. (1.119), in the light-cone gauge) tells us that the renormalization constant $\widehat{Z}_V$ for the PT proper vertex is the same as the wave function renormalization constant $\widehat{Z}_G$ for the PT propagator; just as in QED, the charged vertex renormalization constant Z_1 equals the charged propagator renormalization constant Z_2. In view of this equality, the relations between the radiatively corrected but unrenormalized PT propagator $\widehat{d}_U$ and the bare coupling g_U, and their renormalized counterparts, both renormalized at momentum μ, are as follows:

$$\widehat{d}(\mu; q^2) = \widehat{Z}_G^{-1}\widehat{d}_U(q^2) \qquad g(\mu) = g_U \widehat{Z}_G^{1/2}, \qquad (1.136)$$

from which it follows that the renormalized product $g^2\widehat{d}$ is not only gauge invariant but renormalization group invariant (i.e., independent of μ). The same is true of the running charge $\bar{g}^2(q^2)$, so it is natural to make the factorization

$$g^2\widehat{d}(q^2) = \bar{g}^2(q^2)\widehat{H}(q^2), \qquad (1.137)$$

where $\widehat{H}$ is a propagator of conventional type; for example,

$$-\widehat{H}(q^2) = \frac{1}{m^2(q^2) - q^2 - i\epsilon}, \qquad (1.138)$$

which is also gauge and renormalization group invariant. Both the running charge $\bar{g}^2(q^2)$ and the other factor $-\widehat{H}(q^2)$ should obey the Källen–Lehmann (K-L) representation, for example,

$$\bar{g}^2(q^2) = \frac{1}{\pi} \int_{4m^2}^{\infty} d\sigma \, \frac{\rho(\sigma)}{\sigma - q^2 - i\epsilon}, \qquad (1.139)$$

with a positive spectral function ρ and threshholds determined by the gluon mass m. This ensures that $\bar{g}^2$ is always positive for spacelike (negative) q^2 and has no real zeroes.

Now consider the product $\bar{g}^2\widehat{H}$. Asymptotic freedom tells us that as $|q^2|$ grows, $\bar{g}^2(q^2) \to 1/[b \ln(-q^2/\Lambda^2)]$, and we certainly expect that in the same limit, $\widehat{H}$ behaves like a free propagator so that $-\widehat{H} \sim 1/q^2$. But their product decreases *faster* than $1/q^2$ and therefore cannot obey a K-L representation with a positive spectral function.[11]

It turns out [29] that the Schwinger–Dyson equations for both the PT propagator and gluonic PT proper vertices can be expressed solely in terms of the well-behaved

---

[11] In QED, where e^2 times the photon propagator is gauge and renormalization group invariant, it is possible to write a K-L representation for the photon propagator. But this only holds because QED is asymptotically unstable and has a positive beta function, consistent with positivity of the spectral function, as the renormalization group equation for the propagator shows.

factor $\widehat{H}$ and other pieces that are both gauge and renormalization group invariant, such as given in Eq. (1.84) for the three-gluon PT vertex and Eq. (1.137) for the PT propagator. Thus, as a matter of practice, the nonpositivity of the PT propagator spectral function is not apparent or harmful.

As one might expect, the ansatz of factorization of the PT propagator into parts obeying the K-L represesntation is feasible only if the (zero-momentum) dynamical gluon mass m is large enough; in a simple model of the PT gluon gap equation [29], it was estimated that m/Λ should exceed about 1.2. This corresponds to an upper limit on the strong coupling at zero momentum of roughly $\alpha_s(0) = \bar{g}^2(0)/(4\pi) \lesssim 0.7$.

References

[1] J. M. Cornwall, Confinement and infrared properties of Yang–Mills theory, paper presented at the U.S.–Japan Seminar on Geometric Models of Elementary Particles, Osaka, Japan (1976).

[2] J. M. Cornwall, Nonperturbative mass gap in continuum QCD, in *Proceedings of the French–American Seminar on Theoretical Aspects of Quantum Chromodynamics*, Marseille, France (1981).

[3] J. M. Cornwall, Dynamical mass generation in continuum QCD, *Phys. Rev.* **D26** (1982) 1453.

[4] J. M. Cornwall and J. Papavassiliou, Gauge-invariant three-gluon vertex in QCD, *Phys. Rev.* **D40** (1989) 3474.

[5] D. Binosi and J. Papavassiliou, The pinch technique to all orders, *Phys. Rev.* **D66** (2002) 111901(R).

[6] D. Binosi and J. Papavassiliou, Pinch technique self-energies and vertices to all orders in perturbation theory, *J. Phys.* **G30** (2004) 203.

[7] L. F. Abbott, The background field method beyond one loop, *Nucl. Phys.* **B185** (1981) 189.

[8] D. Binosi, Electroweak pinch technique to all orders, *J. Phys.* **G30** (2004) 1021.

[9] S. Hashimoto, J. Kodaira, Y. Yasui, and K. Sasaki, The background field method: Alternative way of deriving the pinch technique's results, *Phys. Rev.* **D50** (1994) 7066.

[10] A. Denner, G. Weiglein, and S. Dittmaier, Gauge invariance of Green functions: Background field method versus pinch technique, *Phys. Lett.* **B333** (1994) 420.

[11] J. Papavassiliou, On the connection between the pinch technique and the background field method, *Phys. Rev.* **D51** (1995) 856.

[12] P. Di Vecchia, L. Magnea, A. Lerda, R. Russo, and R. Marotta, String techniques for the calculation of renormalization constants in field theory, *Nucl. Phys.* **B469** (1996) 235.

[13] M. Lavelle, Gauge invariant effective gluon mass from the operator product expansion, *Phys. Rev.* **D44**, 26 (1991).

[14] M. E. Peskin and D. V. Schroeder, *An Introduction to Quantum Field Theory* (Addison Wesley; New York, 1994).

[15] C. Becchi, A. Rouet, and R. Stora, Renormalization of gauge theories, *Ann. Phys.* **98** (1976) 287.

[16] I. V. Tyutin, Gauge invariance in field theory and statistical physics in operator formalism, *LEBEDEV-75-39*.

[17] N. Nakanishi, Covariant quantization of the electromagnetic field in the Landau gauge, *Prog. Theor. Phys.* **35** (1966) 1111.

[18] B. Lautrup, Canonical quantum electrodynamics in covariant gauges, *Mat. Fys. Medd. Dan. Vid. Selsk.* **35** (1966) 1.

[19] G. 't Hooft, Renormalization of massless Yang-Mills fields, *Nucl. Phys.* **B33** (1971) 173.

[20] P. Pascual and R. Tarrach, QCD: Renormalization for the practitioner, *Lect. Notes Phys.* **194** (1984) 53.

[21] J. S. Ball and T. W. Chiu, Analytic properties of the vertex function in gauge theories, 2, *Phys. Rev.* **D22** (1980) 2550 [Erratum, ibid. **D23** (1981) 3085].

[22] J. Papavassiliou, The gauge invariant four gluon vertex and its ward identity, *Phys. Rev.* **D47** (1993) 4728.

[23] J. M. Cornwall, W. S. Hou, and J. E. King, Gauge invariant calculations in finite temperature QCD: Landau ghost and magnetic mass, *Phys. Lett.* **B153** (1985) 173.

[24] J. M. Cornwall, D. N. Levin, and G. Tiktopoulos, Uniqueness of spontaneously broken gauge theories, *Phys. Rev. Lett.* **30** (1973) 1268 [Erratum, ibid. **31** (1973) 572].

[25] J. M. Cornwall, D. N. Levin, and G. Tiktopoulos, Derivation of gauge invariance from high-energy unitarity bounds on the *S* matrix, *Phys. Rev.* **D10** (1974) 1145 [Erratum, ibid. **D11** (1975) 972].

[26] C. H. Llewellyn Smith, High-energy behavior and gauge symmetry, *Phys. Lett.* **B46** (1973) 233.

[27] J. Papavassiliou and A. Pilaftsis, Gauge-invariant resummation formalism for two point correlation functions, *Phys. Rev.* **D54** (1996) 5315.

[28] J. Papavassiliou, E. de Rafael, and N. J. Watson, Electroweak effective charges and their relation to physical cross sections, *Nucl. Phys.* **B503** (1997) 79.

[29] J. M. Cornwall, Positivity issues for the pinch-technique gluon propagator and their resolution, *Phys. Rev. D* **80** (2009) 096001.

2

Advanced pinch technique: Still one loop

In this chapter, we study several more advanced aspects of the pinch technique (PT), still sticking to one-loop processes, mostly in perturbation theory but with some discussion of one-dressed-loop effects related to gluon mass generation. One of these applications of the pinch technique also has nonperturbative consequences, coming from the invocation of a gauge-field condensate; it allows us to conclude, as we show in this chapter, that the dynamical gauge-boson mass in QCD vanishes like q^{-2}, modulo logarithms, at large momentum. Finally, we introduce one of the main themes of the rest of the book: the pinch technique is realized to all orders by calculating conventional Feynman graphs in the background-field Feynman gauge. The subjects covered include the following:

1. The pinch technique and the operator product expansion (OPE) at one loop, where we see how only *gauge-invariant* condensates such as $\langle \mathrm{Tr}\, G_{\mu\nu} G^{\mu\nu} \rangle$ arise in PT Green's functions and how this condensate governs the vanishing at large momentum of dynamically generated gauge-boson mass in QCD.
2. Uses of the pinch technique in studying gauge-boson mass generation, both dynamic in QCD (no symmetry breaking, whether by Higgs–Kibble fields or other mechanisms) and with spontaneous symmetry breaking.
3. The background field method and the effective action.
4. The one-loop equivalence between the pinch technique and the background field method in the Feynman gauge.

2.1 The pinch technique and the operator product expansion: Running mass and condensates

As mentioned more than once, the pinch technique is essential to unveiling the nonperturbative effects that are vital in understanding confinement in QCD. One of the oldest and most familiar nonperturbative phenomena of QCD is the gauge-field condensate $\langle \text{Tr}\, G_{\mu\nu} G^{\mu\nu} \rangle$ appearing in the OPE of hadronic or leptonic currents. This condensate is explicitly gauge invariant, so it has physical significance.[1]

If there is to be dynamical mass generation (by which we mean generation of mass where gauge invariance of the usual classical action forbids such a mass), the dynamical mass must be a function of the momentum and must decrease at large momentum. If it did not vanish at infinite momentum, there would be a corresponding bare mass, not allowable in cases of interest to us. From the viewpoint of Schwinger–Dyson equations, there simply would be no massive solution for the gauge propagator unless the mass vanished at infinite momentum. This situation is already familiar for the constituent mass of the light quarks in QCD, which for all practical purposes have zero bare mass protected by a chiral symmetry forbidding a quark mass at any finite order of perturbation theory. Nevertheless, there is a large constituent mass that must also decrease with momentum. The mass is a sign of spontaneous chiral symmetry breaking, and there is another characteristic sign of chiral symmetry breaking: a nonzero value of the quark condensate $\langle \bar{q}(x)q(x) \rangle$. One cannot exist without the other. The OPE tells us how these are related: the running quark mass $M(q)$ decreases at large momentum as

$$M(q) \rightarrow \text{const.} \frac{-\langle \bar{q}q \rangle}{q^2}. \tag{2.1}$$

What is the corresponding relation between gluon mass and gauge-field condensate? As in every use of the OPE, the first step is to pick a matrix element with the right quantum numbers and bring together the space-time arguments of the condensate fields, thereby picking up the appropriate c-number multiplier of the condensate for that particular matrix element.

The OPE was used to find the contribution of the $\langle \text{Tr}\, G_{\mu\nu} G^{\mu\nu} \rangle$ condensate to the conventional gluon propagator, with disappointing but not unexpected results at one-loop order. Not only did this condensate appear, but gauge-dependent condensates involving the ghost fields c and $\bar{c}$ also appeared. It seemed that no physical results could be obtained from the OPE for a gauge-dependent quantity such as the usual gluon propagator. Then Lavelle [1] did the same calculation for the PT propagator in $d = 3, 4$, with very different results. *Only* the gauge-invariant condensate appeared and in just such a way that it could be interpreted as contributing to a running mass. Lavelle's results are equivalent to saying that the scalar

---

[1] The condensate $\langle \text{Tr}\, A^{\mu} A_{\mu} \rangle$, which is explicitly not gauge invariant, can be made gauge invariant by turning it into the gauged nonlinear sigma model, as we indicate in Section 2.2.4. This is equivalent to minimizing the space-time integral of $\text{Tr}\, A_{\mu} A^{\mu}$ over all local gauge transformations.

inverse of the PT propagator, in Euclidean space, behaves in $d = 3, 4$, and at large momentum as

$$\widehat{d}^{-1}(q) \rightarrow q^2 + c_d \frac{g^2 \langle \text{Tr} \, G_{\mu\nu} G^{\mu\nu} \rangle}{q^2}, \tag{2.2}$$

where

$$c_3 = \frac{29N}{30(N^2 - 1)} \qquad c_4 = \frac{17N}{18(N^2 - 1)}. \tag{2.3}$$

(Actually, powers of logarithms of q can also occur, but we ignore them here.) The constants are positive, so this OPE correction has the right sign to represent a running mass because the condensate is also positive. In both cases, quark constituent mass and dynamical gluon mass, the running mass decreases like q^{-2} times a vacuum expectation value (VEV) of a gauge-invariant condensate. The difference is, of course, that there is no symmetry breaking for gluon mass generation, and indeed the gauge-field condensate is in no sense an order parameter for any kind of symmetry breaking. The simple physical reasoning for the connection of the condensate and the gluon mass is that a gluon mass allows for the construction of many different quantum solitons that cannot exist at the classical (zero-mass) level, including center vortices and nexuses. Condensates of these solitons are favored because of their large entropy (large number of possible space-time configurations) relative to their (finite) action and so lead to a gauge-field condensate. Lavelle's finding is the converse: a condensate allows for gluonic mass generation. Ultimately, this connection exists only because the gauge theories of interest show *infrared slavery* – the infrared manifestation of the ultraviolet phenomenon of asymptotic freedom in $d = 4$. Infrared slavery means that the perturbative PT propagator has a one-loop proper self-energy of the wrong sign (opposite to that of QED), with consequent intolerable infrared diseases such as tachyons. These theories must find a cure for infrared slavery, and that cure is dynamical gluon mass generation.

2.2 The pinch technique and gauge-boson mass generation

2.2.1 General remarks

As in Chapter 1, we consider only the one-loop case, postponing the all-orders generalization to later chapters.

There are several closely related ways of endowing a gauge boson with mass. The most straightforward way to generate gauge-boson masses is through

Higgs–Kibble–Goldstone symmetry breaking[2] with elementary scalar fields, as in the Georgi–Glashow model [2] or the electroweak (EW) sector of the standard model. A second way [3, 4] of generating gauge-boson mass is through dynamical effects associated with taming the infrared singularities of strongly coupled gauge theories such as QCD, which has no elementary scalars. Some gauge theories with no scalar fields can show dynamical symmetry breaking, in which the elementary scalar fields are replaced by composites arising from homogeneous solutions of the Schwinger–Dyson equations. A variant of these cases has elementary Higgs–Kibble–Goldstone fields and possibly symmetry breaking, but the VEVs in the scalar sector are too small (perhaps even zero, as in the EW sector at high temperature) to remove the infrared singularities of the underlying NAGT [5]. A third, rather specialized way, is through a Chern–Simons term in three dimensions. It can happen that the perturbative gauge-boson mass coming from the Chern–Simons term is too small to overcome infrared slavery, and dynamical mass generation comes into play. The pinch technique is important in estimating the critical Chern–Simons coupling (which is quantized) separating perturbative behavior of the theory from the need for nonperturbative dynamical mass generation [6].

All these cases have two vital ingredients in common. First, they require massless longitudinally coupled scalars, one for each gauge boson that gets mass. This is a subtle matter because the massless scalars do not appear (at least directly) in the S-matrix, yet they can appear in the pinch technique proper Green's functions. Every Goldstone-like scalar, whether elementary or composite, that is eaten by a gauge boson to give it mass is canceled out of the S-matrix by other massless poles or current conservation.

If the massless scalars are not elementary Goldstone fields, then they arise as composite excitations in a strongly coupled gauge theory.[3] By a composite excitation, we mean a pole in an off-shell Green's function representing a field that does not exist in the classical action but that occurs in the solution of the Schwinger–Dyson equation for that Green's function, as a sort of bound state. Therefore, the second vital ingredient is strong coupling, which, as far as we know, can only come from the infrared instabilities of a NAGT.

The residue, at zero momentum, of these Goldstone-like poles is essentially the square m^2 of a gauge-boson mass. The classical action does not have the

---

[2] If the gauge theory has local gauge symmetry at the classical level, so-called spontaneous symmetry breaking is not actually breaking this local gauge symmetry but simply realizing it in a different way. Without explicit symmetry breaking, such as fermion masses for local axial symmetries, no gauge-dependent object can have a nonzero VEV, as Elitzur's theorem tells us. Gauge-fixing terms break a gauge symmetry explicitly, but the pinch technique effectively removes such breaking.

[3] Other massive composite excitations may have to arise as well to save unitarity at high energies; see Lee et al. [7].

corresponding elementary field because if it did, there would be a symmetry violation, or a violation of perturbative renormalizability, or both. If renormalizability is at issue, the squared mass runs with momentum q and vanishes at large q; when mass generation is an infrared effect, as it will always be for us, a typical decrease is $m^2(q) \sim q^{-2}$ (modulo logarithms), as Lavelle [1] found. Without some such large-momentum falloff, the Schwinger–Dyson equation would have no solutions without extra infinities not corresponding to perturbative renormalization principles.

Here we will introduce two cases of the dynamical gauge-boson mass generation, saving other examples for later chapters. The first case is dynamical mass generation in QCD; the second is the well-known case of elementary Higgs–Kibble–Goldstone scalars. In both cases, the pinch technique is an essential tool for ensuring gauge invariance of the results.

2.2.2 Dynamical gauge-boson mass generation in QCD

In $d = 4$, the necessary strong coupling comes from asymptotic freedom as expressed in the wrong sign of the beta function (see Eq. (1.69)). Unfortunately, it is not helpful to associate the infrared phenomenon of mass generation with the ultraviolet phenomenon of asymptotic freedom. This is only a question of terminology because (perturbative) asymptotic freedom necessarily implies infrared singularities that are more virulent than in QED. The situation is actually more clearly seen in $d = 3$, where NAGTs are superrenormalizable, there is no renormalization group, and the ideas of asymptotic freedom are not relevant. Yet $d = 3$ NAGTs have – in even worse form than in $d = 4$ – serious infrared singularities.[4] We prefer, then, to suggest these low-momentum singularities with the term *infrared slavery* because, ultimately, they lead to confinement. The one-loop PT propagator in $d = 3$ clearly shows infrared slavery through the negative (and, as with asymptotic freedom, wrong) sign of a certain gauge-invariant constant [3, 4]. This sign is absolutely critical because if it were positive, the infrared behavior of radiatively corrected gluon propagators would be less singular than at tree level. But in the physical case of a negative sign, the infrared singularities show up as potential tachyons or ghost particles, that is, unphysical objects with imaginary mass or couplings. No other solution for the wrong sign is known aside from dynamical gauge-boson mass generation, which generates positive terms in the PT proper self-energy that overcomes the negative and singular behavior.

---

[4] This was not fully appreciated until the pinch technique came along because, before that, people had only investigated the standard Feynman propagator, which is gauge dependent. For the PT results, see [3, 4, 8].

We will briefly explore here, at the one-loop level, the nonperturbative dynamics leading to gauge-boson mass generation, deferring a detailed treatment of the PT Schwinger–Dyson equations to Chapter 6. Dressed propagators and vertices come, at least formally, from resumming Feynman graphs and have the power to inform us about phenomena that do not occur even when perturbation theory is summed to all orders in the coupling g. This kind of resummation is familiar from the skeleton graph expansion of Schwinger–Dyson equations and the resummation of the effective potential [9]. To truncate the otherwise infinite series of Schwinger–Dyson equations requires us to understand how to construct approximate forms for three- and higher-point Green's functions that, in spite of their approximate nature, *exactly* satisfy the PT Ward identities and that are expressed in terms of lower-point functions such as the PT propagator itself. If the vertex functions used do not obey exactly the Ward identities, gauge invariance is lost. The method of vertex constructions that satisfies these Ward identities is called the *gauge technique* and is discussed in Chapter 5. Because the $d = 3$ case is so instructive, we begin with it.

2.2.3 The need for dynamical mass in $d = 2 + 1$ QCD

We will easily see the problems of infrared slavery in $d = 2 + 1$ by calculating the one-loop perturbative PT proper self-energy. This goes exactly as in the $d = 3 + 1$ case of Section 1.3.3 except for the values of the integrals. The result [3, 4] for the scalar part of the one-loop PT inverse propagator is

$$\widehat{d}^{-1}(q) = q^2 + \pi b_3 g_3^2 (-q^2)^{1/2} + \mathcal{O}(g_3^4), \qquad (2.4)$$

where

$$b_3 = \frac{15N}{32\pi} \qquad (2.5)$$

and g_3 is the $d = 3$ coupling with dimensions of (mass)$^{1/2}$. Infrared slavery is simply the fact that b_3 is positive, which has the implication that there is a pole in the propagator for a spacelike momentum ($q^2 < 0$). This indicates a *tachyonic* pole – a pole corresponding to an imaginary mass.

There is also a tachyonic pole in $d = 4$, as one can see from the renormalized version of Eq. (1.68): the propagator has a pole at

$$-q^2 = \text{const.} \times \mu^2 e^{-1/bg(\mu)^2}, \qquad (2.6)$$

again satisfied with tachyonic q^2.

What could be the cure for this unphysical behavior? At first glance, it could be easy: Because the coupling g_3^2 has dimensions of mass, the omitted g_3^4 term might well provide a sufficiently positive term to overcome the negative one-loop term. This is indeed what happens nonperturbatively but not to any order of perturbation theory, where the coefficient of g_3^4 is identically zero to all orders. (If it were not zero, we could add a bare mass term to the action, which is not perturbatively renormalizable.)

This is only the beginning of the bad perturbative behavior. At $\mathcal{O}(g_3^{2N})$, each perturbative integral, by simple dimensional reasoning, has the infrared behavior $g_3^4(g_3^2/q)^{N-2}$, with poles of infinitely high order in the inverse propagator. But with nonperturbative generation of a (nontachyonic) mass m, the infrared behavior of every propagator in a loop is $\sim 1/m^2$, and an easy power counting shows that $|q|$ in the perturbative ordering expression is replaced by the dynamical mass $m \sim g_3^2$ so that all terms are of $\mathcal{O}(m^2)$ for order $N \geq 2$.

Of course, a one-loop pinch technique calculation only clearly shows us (i.e., gauge invariantly) the disease, not the cure. We take up the cure in Chapter 9, but for now, it is important to know something about what this cure looks like.

2.2.4 What do vertices and propagators look like when dynamical mass is generated?

The question is how to write PT Schwinger–Dyson equations with the right structure to represent composite massless scalars in both $d = 3$ and $d = 4$. There is a quite simple answer [10]: abstract this structure from an infrared-effective action in which we add to the usual NAGT action a mass term that is a gauged nonlinear sigma model. We add to the classical action,[5] the integral of Eq. (1.5), the term

$$S_m = \int d^d x \, m^2 \, \text{Tr} \left(U^{-1} \mathcal{D}_\mu U U^{-1} \mathcal{D}^\mu U \right) \tag{2.7}$$

to arrive at a total infrared-effective action (without fermions):

$$S\{A, m\} = \int d^d x \left[-\frac{1}{2} \text{Tr} \, G_{\mu\nu} G^{\mu\nu} + m^2 \text{Tr} \left(U^{-1} \mathcal{D}_\mu U U^{-1} \mathcal{D}^\mu U \right) \right]. \tag{2.8}$$

Here U depends on $\mathcal{R}$, the $N \times N$ unitary matrix representative of the group element $\mathcal{R}$. We also add the prescription that the exponential of this effective action is to be integrated over the Haar measure of the local gauge group.

We emphasize that we are not proposing to take S_m seriously as an addition to the true NAGT action, which always consists only of the usual Yang–Mills term. It is only part of an effective action, whose consequences should be studied only at the

---

[5] This action is to be used in $d = 3, 4$; for simplicity, we do not explicitly indicate the dimension as we did in the previous section.

classical level. However, it is legitimate for us to guess, from this classical study, what kind of structure the ingredients of the (PT) Schwinger-Dyson equation must have and then show that the Schwinger-Dyson equation with the assumed structure is actually self-consistent and physical. What we will find is what we spoke of earlier: massless longitudinally coupled Goldstone-like scalars, with couplings proportional to squared gauge-boson masses.

Because S_m is an effective action, we interpret integration over the group as finding the extrema over $\mathcal{R}$ of S_m. The matrix U undergoes gauge transformations along with the potential

$$U \to VU, \quad A_\mu \to V A_\mu V^{-1} + V \frac{i}{g} \partial_\mu V^{-1}. \tag{2.9}$$

It is then elementary that the modified potential

$$C_\mu \equiv U^{-1} \mathcal{D}_\mu U, \tag{2.10}$$

is formally gauge invariant. (Furthermore, it is a gauge transformation by U^{-1} of the gauge potential A_μ.) We define a potential that is gauge covariant by multiplication from the left by U and from the right by U^{-1}:

$$\widetilde{A}_\mu \equiv U C_\mu U^{-1} = A_\mu + \frac{i}{g}(\partial_\mu U) U^{-1} \quad \widetilde{A}_\mu \to V \widetilde{A}_\mu V^{-1}. \tag{2.11}$$

It appears that we have added new degrees of freedom to the NAGT action by introducing U, but we have not, at least perturbatively. The reason is that the classical equations of motion for U are not independent of those for A_μ but follow from them, and U can be solved as a (nonlocal) functional of A_μ. The equations of motion for A_μ are as follows:

$$\left[\mathcal{D}^\mu, G_{\mu\nu}\right] + m^2 \widetilde{A}_\nu = 0. \tag{2.12}$$

Because of the identity

$$\left[\mathcal{D}^\nu, \left[\mathcal{D}^\mu, G_{\mu\nu}\right]\right] \equiv 0, \tag{2.13}$$

it must be that

$$[\mathcal{D}^\nu, \widetilde{A}_\nu] = 0. \tag{2.14}$$

But this equation is precisely the equation of motion found by varying U. After a certain amount of algebra, one can show that this U equation of motion is equivalent to

$$\partial^\nu C_\nu = 0. \tag{2.15}$$

In other chapters, we will find nonperturbative features of these equations and connect them to confinement via center vortices and to the Gribov ambiguity.[6] In particular, there are massless scalar excitations longitudinally coupled to the gauge potential. These massless scalars must be present if a gauge boson is to have mass in a way that preserves local gauge symmetry. We will see that such scalar fields actually represent long-range pure-gauge parts of the gauge potential that carry critical topological information about confinement and topological charge.

For now, we pursue the perturbative solution of the equation of motion for U and find massless scalars there. Write $U = \exp[i\omega]$ and find [10]:

$$i\omega = -\frac{1}{\Box}\partial \cdot A - \frac{1}{2}\left[\frac{1}{\Box}\partial \cdot A, \partial \cdot A\right] + \frac{1}{\Box}\left[A_\mu, \partial^\mu \frac{1}{\Box}\partial \cdot A\right] + \mathcal{O}(A^3). \quad (2.16)$$

Substitution of Eq. (2.16) in the gauged nonlinear sigma model action reveals an infinite set of vertices for the potential A_μ, longitudinally coupled to the massless scalars. This massless scalar is completely analogous to the Goldstone scalar of spontaneous symmetry breaking even though, in the present case, there is no symmetry breaking (and no elementary Higgs field). The lowest-order vertex is quadratic and yields a transverse mass term of the form

$$\int d^d x \, m^2 \, \mathrm{Tr}\left[\left(A_\mu - \partial_\mu \frac{1}{\Box}\partial \cdot A\right)\left(A^\mu - \partial^\mu \frac{1}{\Box}\partial \cdot A\right)\right] \quad (2.17)$$

and a free gauge-boson propagator that has the structure expected from a pinch technique propagator:

$$i\Delta^{(0)}_{\alpha\beta}(q) = P_{\alpha\beta}(q)d(q) + \xi\frac{q_\alpha q_\beta}{q^4}, \quad (2.18)$$

where

$$d(q) = \frac{1}{q^2 - m^2}. \quad (2.19)$$

The next vertex is a three-gluon vertex, to be added to the conventional free vertex. We convert the result to PT form by choosing the free vertex to be Γ^ξ of Eq. (1.37) and find [4, 11] that

$$\widehat{\Gamma}^{m,\xi}_{\mu\alpha\lambda}(q, k, -q - k) = \Gamma^\xi_{\mu\alpha\lambda}(q, k, -q - k) - \left[\frac{m^2}{2}\frac{q_\mu k_\alpha(q - k)_\lambda}{q^2 k^2} + \text{c.p.}\right], \quad (2.20)$$

where c.p. stands for *cyclic permutations*. This vertex obeys the PT Ward identity of Eq. (1.40) with the propagator of Eqs. (2.18) and (2.19). The new vertex $\widehat{\Gamma}^{m,\xi}$ has,

---

[6] The Gribov ambiguity is that setting up a covariant gauge fixing in the usual way does not completely fix the gauge, so the gauge potential is ambiguous for a given field strength. For example, suppose that A_μ is in the Landau gauge; then, by Eqs. (2.10) and (2.15), so is C_μ, which is a gauge transformation by U^{-1} of A_μ.

as we mentioned earlier, terms with longitudinally coupled massless poles whose residue is m^2. This is the only way that the Ward identity can be satisfied if the propagator is transverse and has mass, just as the only way a massive gauge-boson propagator can be transverse is if it has similar poles in the transverse projector $P_{\mu\nu}$. There are infinitely many other vertices coming from the gauged nonlinear sigma model term, all with these poles, but we will go no further in exhibiting any of them explicitly.

Now apply the usual pinch technique to the mass-modified action of Eq. (2.8), in which we substitute the solution for U, as in Eq. (2.16). After much calculation [11] (originally done in the light-cone gauge), we find the one-loop pinch technique proper self-energy. In $d = 4$, it is as follows:

$$\widehat{\Pi}(q^2) = \frac{bg^2}{\pi^2} \int d^4k \left[-q^2 d(k)d(k+q) + \frac{4}{11}d(k) + \frac{m^2}{11}d(k)d(k+q) \right]. \quad (2.21)$$

(We will take up the $d = 3$ analog later.) This reduces, as it must, to the formerly calculated one-loop proper self-energy of Eq. (1.68) in the limit of zero mass. Note that there are no massless poles in $\widehat{\Pi}$. In fact, the only trace of the massless poles comes from the second and third terms on the right-hand side (rhs), which do not vanish at $q = 0$. There are then massless poles in the tensorial proper self-energy coming from the longitudinal term in the transverse projector $P_{\mu\nu}$ that multiplies $\widehat{\Pi}$. Ultimately, these terms cannot appear in the S-matrix because of gauge current conservation.

With Eq. (2.21), we are actually only a few steps away from being able to study nonperturbative dressed-loop effects and analyze gauge-boson mass generation. The next steps are to integrate the rhs of this equation over m^2, with a spectral weight that is used in the Källen–Lehmann[7] representation of the pinch technique propagator; by this means, we construct a simple example of a gauge technique vertex and a PT Schwinger–Dyson equation for the proper self-energy. But we will postpone this study until later and use our work here to draw a few lessons that apply to such a nonperturbative study:

1. However gauge-boson mass is generated, it is accompanied by Goldstone-like longitudinally coupled scalars.
2. These scalar poles appear in vertices of all order, as, for example, in the transverse mass term of Eq. (2.17) and the three-vertex of Eq. (2.20), with couplings proportional to squared gauge-boson masses.

---

[7] The spectral weight is not positive definite, but this is not required for the existence of a spectral representation. See the discussion of Section 1.7.

3. These scalars do not appear in the *S*-matrix and (in our gauged nonlinear sigma model approach) even cancel out in the final expression for the pinch technique proper self-energy, as in Eq. (2.21).

4. (This lesson will be learned in part through a one-loop calculation in the next chapter.) A gauge-boson mass violating a symmetry or perturbative renormalizability must be a running mass, vanishing at large q, or the Schwinger–Dyson equation has no finite solution. Furthermore, a nonrunning mass by itself leads to violations of unitarity at high energy.

5. Dynamical generation of mass does not interfere with satisfaction of ghost-free Ward identities.

Next, we see if these lessons apply to Higgs–Kibble symmetry breaking.

2.2.5 *Mass generation through Higgs–Kibble–Goldstone fields*

There are two ways of describing mass generation in this case, depending on the description of gauge fixing in the theory (which, of course, is ultimately irrelevant). The method we mostly use in this book for the pinch technique was first developed in [12] for the Georgi–Glashow model and was later generalized to all orders of electroweak (EW) theory in [13]. Numerous other authors have used a similar form of the pinch technique for EW processes; Degrassi and Sirlin [14] have pointed out the relation of the pinch process to equal-time commutators of symmetry currents and given explicit expressions for the one-loop PT proper self-energies. All these authors use a *modified R_ξ* gauge analyzed by Fujikawa et al. [15] and originally from 't Hooft [16], which we will term the *FLS gauge*. In the FLS gauge, the Goldstone bosons decouple from the gauge bosons that eat them, at least at tree level (but not beyond). The value of this gauge, as we will see shortly, is that the tree-level gauge propagators have no longitudinal parts in the Feynman version of the FLS gauge (or 't Hooft–Feynman gauge), just as there are none for the free gauge propagator in the Feynman R_ξ gauge with $\xi = 1$ (see Eq. (1.31)). But this is not an unalloyed virtue because the PT calculations in the FLS gauge seem to differ from what we have said so far: the Goldstone bosons in general have a gauge-dependent mass that is the same as that of the ghosts.[8] This is a sign that the Goldstone bosons as well as the ghosts do not appear in the *S*-matrix. Moreover, the proper self-energies and vertices are independent of ξ but satisfy Ward identities ostensibly different from those we have already used, such as transversality of the gauge boson proper self-energy. However, these pieces can be reshuffled [12] to

---

[8] In the FLS–Feynman gauge, $\xi = 1$, the ghosts and Goldstone bosons, and the gauge bosons that eat them, all have the same mass.

yield transverse proper self-energies and vertices that do have the properties listed in our lessons from dynamical mass generation in QCD.

There is another approach that is more similar in spirit to what we have done so far for QCD-like theories; it uses the standard sort of R_ξ gauge that we have used for QCD. In this gauge, the Goldstone bosons and ghosts are massless scalar fields longitudinally coupled to gauge bosons. The massless poles do not appear in the S-matrix but are important in enforcing current conservation (Ward identities). Demanding that these Ward identities be satisfied leads uniquely to a set of PT Green's functions that are independent of ξ and obey other physical requirements. It is worthwhile to understand this approach because similar massless scalar excitations must and do occur in the vertices of QCD-like gauge theories if there is to be dynamical gluon mass generation. Although these Goldstone-like excitations do not contribute to the perturbative S-matrix, they are the carriers of long-range topological information that is responsible for nonperturbative phenomena, including confinement and chiral symmetry breakdown. We will begin with a brief sketch of what happens in this QCD-like description. Needless to say, either description results in the same physics, as described by the pinch technique.

Symmetry breaking in a standard R_ξ gauge   The main new feature with symmetry breaking is that Ward identities used for pinching, though unchanged in basic structure (see Eq. (1.62)), have new vertex terms involving the massless Goldstone bosons even at tree level, similar in structure to those found for dynamical mass generation in QCD. These new terms are essential for current conservation or, more generally, for satisfaction of the Ward identities.

To be explicit, consider the Georgi–Glashow model [2] in $d = 4$, containing an $SU(2)$ gauge field, a Fermion doublet ψ, and a triplet ϕ_a of scalar bosons. With these scalars and fermions, the Georgi–Glashow model is asymptotically free, with the beta function coefficient b of the pure gauge-boson theory (see Eq. (1.69)) changed to $19/(48\pi^2)$. The action is as follows:

$$S_{GG} = \int d^4x \left\{ -\frac{1}{2}\text{Tr}\, G_{\mu\nu}G^{\mu\nu} + i\bar{\psi}\left[\slashed{P}- M_0\right]\psi + \frac{1}{2}\left[(\mathcal{D}_\mu\phi_a)(\mathcal{D}^\mu\phi_a)\right.\right.$$

$$\left.\left. - V(\phi_a^2) - h\bar{\psi}\tau_a\psi\phi_a\right\}, \right. \tag{2.22}$$

to which we will add a gauge-fixing term. We will also use the convenient matrix notation

$$\phi = \frac{1}{2}\tau_a\phi_a, \tag{2.23}$$

where the τ_a are the Pauli matrices. The potential V has the form

$$V(\phi_a^2) = \frac{\lambda}{2}(\phi_a^2 - v^2)^2, \tag{2.24}$$

with the symmetry-breaking minimum taken by convention to be in the 3 direction so that $\phi_a = v\delta_{a3}$ at the classical level. This gives a Higgs mass to the gauge bosons with group indices 1,2, with value $M_g = vg$, and the third gauge boson remains massless. There is also a symmetry-breaking term in the fermion mass matrix $\mathcal{M}$, which becomes

$$\mathcal{M} = M_0 + hv\tau_3 \equiv M_0 + m\tau_3. \tag{2.25}$$

In the presence of symmetry breaking, we write the scalar fields in matrix form:

$$\phi = \frac{1}{2}[v\tau_3 + \chi_a\tau_a], \tag{2.26}$$

where, by definition, the VEV of χ_a vanishes.

In a conventional R_ξ gauge, there is a quadratic coupling term between the original gauge potential and the χ_a for $a = 1, 2$. This is just what tells us that the gauge bosons with $a = 1, 2$ swallow the corresponding Goldstone bosons and become massive (the gauge potential $A_{3\mu}$ remains massless, and χ_3 describes a massive Higgs–Kibble field). From now on, we use the notation W_μ for the massive gauge bosons.

This quadratic coupling is no particular problem. Because the Goldstone fields are eaten by the W-bosons, they become – at least in perturbation theory – dependent fields, expressible entirely in terms of the gauge potential and possibly other fields. Save only the W-bosons and the ϕ fields with indices $a = 1, 2$ and write the quadratic part of the Lagrangian, including the gauge-fixing term:

$$S_2 = \int d^4x \left\{ -\frac{1}{2}\widetilde{\mathrm{Tr}}\left(\partial_\mu W_\nu - \partial_\nu W_\mu\right)\left(\partial^\mu W^\nu - \partial^\nu W^\mu\right) \right. \tag{2.27}$$

$$\left. + \widetilde{\mathrm{Tr}}\left(\partial_\mu \chi + v\left[W_\mu, \frac{1}{2}\tau_3\right]\right)^2 - \frac{1}{\xi}\widetilde{\mathrm{Tr}}\left(\partial_\mu W_\mu\right)^2 \right\},$$

where $\widetilde{\mathrm{Tr}}$ means taking the trace only over terms involving $\tau_{1,2}$. We define an anti-Hermitean Goldstone matrix $G = \frac{1}{2i}\tau_a G_a$, with $a = 1, 2$, by a re-ordering of $\chi_{1,2}$:

$$\chi = \left[G, \frac{1}{2}\tau_3\right] \quad \text{or} \quad \chi_1 = -G_2, \quad \chi_2 = G_1. \tag{2.28}$$

Now couple the fields W_μ and G to currents V_μ and T, respectively. A short calculation using the action S_2, in terms of G rather than χ, yields (in momentum space) the two equations

$$(q^2 g_{\mu\nu} - q_\mu q_\nu) \, W^\nu + \frac{1}{\xi} q_\mu q_\nu W^\nu - M_g^2 \left(W_\mu - \frac{q_\mu G}{M_g} \right) = V_\mu \qquad (2.29)$$

$$q^2 G - M_g q_\nu W^\nu = T.$$

The solution for the G equation is

$$G = \frac{M_g q_\mu W^\mu + T}{q^2}, \qquad (2.30)$$

and this, substituted in the equation for W_μ, yields an equation for this potential with a modified source term:

$$(q^2 - M_g^2) \left(g_{\mu\nu} - \frac{q_\mu q_\nu}{q^2} \right) W^\nu + \frac{1}{\xi} q_\mu q_\nu W^\nu = V_\mu - \frac{q_\mu M_g T}{q^2}. \qquad (2.31)$$

Note that the inverse propagator (coefficient of W^ν on the left-hand side) has just the form of Eq. (1.30) that we expect of a PT propagator, with the proper self-energy given by M_g^2.

We are concerned, as usual, with the gauge dependence of amplitudes, so we note that the ξ-dependent term in the solution of Eq. (2.31) is

$$W^\nu = \frac{\xi q^\nu q^\mu}{q^4} \left(V_\mu - \frac{q_\mu M_g T}{q^2} \right) + \cdots. \qquad (2.32)$$

Without gluon mass generation by symmetry breaking (i.e., if M_g and m were zero), the current V_μ would have to be conserved on shell,[9] or otherwise there would be ξ dependence even in the tree-level S-matrix. The situation would then be just the same as for QCD-like theories. But with symmetry breaking, a different current is conserved on shell; the massless Goldstone pole has modified the massive gauge-boson source and is essential for current conservation and Ward identities, as we now show. This modification of the source term is essential because it must happen that the combined source in Eq. (2.31) must be conserved on shell:

$$q_\mu V^\mu - M_g T = 0 \quad \text{on-shell.} \qquad (2.33)$$

It is this quantity multiplied by ξ that would appear in the tree-level S-matrix and so must vanish.

The Goldstone contribution is necessary because with symmetry breaking, the divergence $q_\mu V^\mu$ is generally not zero, even at tree level. Consider our earlier example $V_\mu = \bar{u}_1(p+q)\gamma_\mu u_2(p)$, where the fermion momenta p, $p+q$ are on

---

[9] For example, $V_\mu \sim \bar{u}_1(p+q)\gamma_\mu u_2(p)$, where the labels 1,2 distinguish different $SU(2)$ fermion eigenstates.

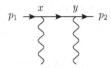

Figure 2.1. Part of a one-loop Feynman graph for on-shell quark-quark scattering. Gauge propagators end at space-time points x, y on the quark line.

shell and the mass of the fermion labeled 1 is $M_0 + m$ and of the fermion labeled 2 is $M_0 - m$. Then $q_\mu V^\mu$ is

$$q_\mu \bar{u}_1(p+q)\gamma^\mu u_2(p) = \bar{u}_1(p+q)\left[S^{-1}(p+q) - S^{-1}(p) - 2m\right]u_2(p)$$
$$= -2m\bar{u}_1 u_2. \tag{2.34}$$

The last equation follows because the inverse propagators annihilate the on-shell spinors. The Goldstone source term of Eq. (2.31), proportional to $M_g T$, turns out to be $2m\bar{u}_1(p+q)u_2(p)q_\mu/q^2$ when one recognizes that $m = (h/g)M_g$. Now we see that the vertex $V_\mu - M_g T q_\mu/q^2$ does obey the expected on-shell Ward identity. One reads off from Eq. (2.34) the off-shell Ward identity of our tree-level example:

$$q_\mu \left[V^\mu - M_g T \frac{q^\mu}{q^2}\right] = S^{-1}(p+q) - S^{-1}(p). \tag{2.35}$$

This is the same as the Ward identity used earlier for QCD-like theories, and the pinch technique proceeds from it as before.

We digress to give the Degrassi–Sirlin explanation of how the symmetry currents coupled to the gauge bosons are related to the pinching out of various propagators. Figure 2.1 shows a part of a Feynman graph that occurs in the S-matrix. This could be a part of several of the complete S-matrix graphs shown in Figure 1.1. This figure by itself is the tree-level version of the matrix element:

$$\langle p_1|T\left[J_\mu^a(x)J_\nu^b(y)\right]|p_2\rangle, \tag{2.36}$$

aside from the two gauge propagators, which we do not exhibit explicitly. The T operation is covariant time ordering, and the currents $J_\alpha^{a,b}$ are symmetry currents coupled to the gauge bosons. A gauge propagator with a longitudinal momentum acts to take the divergence of this time-ordered product, say, with respect to x. The currents themselves are supposed to be conserved, so the only effect is the matrix element of the equal-time commutator:

$$\langle p_1 | \delta (x_4 - y_4) \left[J_0^a(x), J_\nu^b(y) \right] | p_2 \rangle. \tag{2.37}$$

The equal-time commutator here is

$$i f_{abc} \delta(x - y) \langle p_1 | J_\nu^c(y) | p_2 \rangle, \tag{2.38}$$

in which the points x and y are made to coincide and a fermion propagator is missing, just as would happen with our standard PT formalism.

If just the currents V_μ that we introduced in Eq. (2.29) were used for the symmetry currents, our pinch arguments would fail because these are not conserved. We must add the T currents, sources of the Goldstone bosons, to get conserved currents, and conservation arises through the coupling of the massless Goldstone field. It is not hard to check, by resumming graphs, that these Goldstone particles and currents come from the tree-level mixing of Goldstone fields and gauge fields that is no longer eliminated when we use a standard R_ξ gauge.

The Goldstone poles cannot appear in the S-matrix because they are not independent elementary fields; their appearance at one place must be canceled by another appearance elsewhere. The pinch technique, of course, shows this cancellation. The essence of it is that in the pinch technique, the inverse propagator has the form

$$-i\widehat{\Delta}_{\alpha\beta}^{-1}(q) = P_{\mu\nu}(q) \left[q^2 + i\widehat{\Pi}(q) \right] + \frac{1}{\xi} q_\alpha q_\beta. \tag{2.39}$$

For the charged bosons, $\widehat{\Pi}(q = 0) \neq 0$, so the Goldstone bosons appear as the longitudinal massless poles of the transverse projector. In the S-matrix, these terms in the propagator itself strike currents that are conserved and cannot contribute to the S-matrix. On the other hand, the Goldstone poles that allow these currents to be conserved annihilate the physical part of the propagator or inverse propagator, leaving only gauge-dependent kinematic terms. We know that these must cancel in the pinch technique. Now we go on to the more widely used, and equivalent, formulation in the FLS gauge.

Symmetry breaking in the FLS gauge   In the FLS gauge, the gauge-fixing term is chosen to cancel the quadratic coupling of W_μ and $\partial_\mu S$:

$$\mathcal{L}_{\text{GF}} = -\frac{1}{\xi} \widetilde{\text{Tr}} \left(\partial_\mu W_\mu - M_W \xi S \right)^2. \tag{2.40}$$

The quadratic cross-term between W_μ and $\partial_\mu S$ now cancels. However, the S^2 term in the gauge-fixing Lagrangian now makes these Goldstone fields, and also the ghosts, massive. A short calculation gives the following:

$$i\Delta_{\mu\nu} = \frac{g_{\mu\nu}}{q^2 - M_g^2} + \frac{q_\mu q_\nu(\xi - 1)}{(q^2 - M_g^2)(q^2 - \xi M_g^2)} \qquad (2.41)$$

$$i\Delta_S = i\Delta_{gh} = \frac{1}{q^2 - \xi M_g^2},$$

where $\Delta_{\mu\nu}$ is the W propagator, Δ_S is the Goldstone propagator, and Δ_{gh} is the ghost propagator. Clearly, all the ξ dependence in these propagators must cancel in the S-matrix and therefore in the pinch technique, and they do, after some very lengthy calculations [12]. At $\xi = 1$, the tree-level gauge propagator has no longitudinal terms that can pinch, which considerably simplifies the calculations. Moreover, in this Feynman gauge, the gauge bosons, ghosts, and Goldstone particles all have the same mass M_W, and no unphysical masses can appear.

For general ξ, one often decomposes the W propagator as follows [12, 15]:

$$i\Delta_{\mu\nu} = i\Delta_{\mu\nu}^1 + i\Delta_{\mu\nu}^2, \qquad (2.42)$$

$$i\Delta_{\mu\nu}^1 = \left[g_{\mu\nu} - \frac{q_\mu q_\nu}{M_g^2} \right] \frac{1}{q^2 - M_g^2},$$

$$i\Delta_{\mu\nu}^2 = \frac{q_\mu q_\nu}{M_g^2(q^2 - \xi M_g^2)}.$$

This is one way of isolating the gauge dependence of the free propagator into Δ^2. The first term, Δ^1, is the propagator in the so-called unitary gauge $\xi = \infty$. The one-loop pinch technique decomposition has been worked out [12] with this separation of the W-propagator, followed by a demonstration of how to recover the usual results with massless Goldstone particles that we discussed earlier. The idea is simply to recompose the W-propagator by writing the $1/M_g^2$ term in Δ^1 with the identity

$$\frac{1}{M_g^2} = \frac{q^2 - M_g^2}{q^2 M_g^2} + \frac{1}{q^2}. \qquad (2.43)$$

This shows not only the massless Goldstone poles but one of the ways in which they cancel out in the S-matrix.

The pinch technique has been worked out to all orders for the standard electroweak model [13]. We will discuss it in Chapter 10.

2.3 The pinch technique today: Background-field Feynman gauge

It would be very awkward to carry on with the pinch technique in the light-cone gauge (see Section 2.2.4). Even two-loop calculations would be exceptionally difficult, not only because of the intrinsic difficulties of working in a noncovariant gauge but also because what we have done so far at the one-loop level does not really suggest how to generalize the pinch technique. Fortunately, there is a simple way to generalize the pinch technique to all orders of perturbation theory (and to nonperturbative applications, with the help of the gauge technique). It consists of calculating ordinary Feynman graphs in the Feynman gauge of the background-field method. Just as with ordinary Feynman graphs, the sum of such graphs can be reorganized into a dressed-loop expansion from which nonperturbative effects can arise. Much of the rest of this book is devoted to demonstrating these points, which are not at all evident.

The background-field method [17, 18, 19, 20, 21, 22, 23, 24, 25, 26, 27] goes back decades in the study of general relativity and NAGTs. It gives an effective action as a functional of specified background fields (gauge potentials) – an effective gauge action guaranteed to depend on the background fields only through gauge-invariant constructs such as Tr $G_{\mu\nu}^2$. Unfortunately, just as in any other covariant formulation of NAGTs, there has to be gauge fixing and ghosts, and the coefficient functions of the background field constructs depend on the specific gauge chosen (so they depend on ξ in an R_ξ gauge). If these coefficient functions were gauge independent, there would be no need for an independent pinch technique. It is essential that the gauge-fixing term can be chosen to have full local non-Abelian gauge invariance for the background fields in order that this effective potential be gauge invariant for these fields. In other words, gauge fixing is used only for the quantum fields – those integrated out in the functional integral defining the effective potential.

As we said earlier, the first connection was made at the one-loop level [28, 29, 30], where it was shown that the one-loop pinch technique and the one-loop background Feynman gauge method gave precisely the same results. This raises the question of why these two seemingly disparate approaches should give the same answer but by no means answers it. It does not seem plausible that calculations in a specified gauge should actually give gauge-invariant results for Green's functions, as the pinch technique does. In fact, if one were to calculate Green's functions in some other version of the background-field method, for example, the Landau gauge, the results would not be the same as the pinch technique gives. The pinch technique can be used to combine pieces of Feynman graphs in the background Landau gauge, just as in any other gauge, and the usual pinch technique results emerge. Ultimately, as the rest of this book shows, the background Feynman gauge is singled out because of the absence, in this particular gauge, of certain longitudinal numerator parts that give pinches.

We cannot emphasize too strongly that the pinch technique is a way of enforcing gauge invariance and several other physical properties for off-shell Green's functions. The background-field method, in a general gauge, is not. It is a remarkable and extraordinarily useful result that PT Green's functions can also be calculated in the background Feynman gauge, but this needs a very extensive demonstration.

Before presenting the background-field method, we first quickly review the well-known construction of the effective action and the problems encountered with it in gauge theories.

2.3.1 The effective action

A brief review of the scalar field case   Consider first a Euclidean $d = 4$ ϕ^4 field theory. The generating functional is written

$$Z[J] = \int [d\phi] \exp\left\{ iS[\phi] + i \int d^4x\, J(x)\phi(x) \right\} \equiv e^{iW[J]}, \qquad (2.44)$$

where $S[\phi]$ is the scalar field action. Functional derivatives of W with respect to the source J yield connected quantum Green's functions. In particular,

$$\frac{\delta W}{\delta J(x)} = \langle \phi(x) \rangle \equiv \Phi(x), \qquad (2.45)$$

the expectation value of the quantum field in the presence of the source J.

The Legendre transform of W is more useful because it generates the 1PI graphs:

$$\Gamma[\Phi] = W[J] - \int d^4x\, J(x)\Phi(x); \qquad \frac{\delta\Gamma}{\delta\Phi(x)} = -J(x). \qquad (2.46)$$

The functional $\Gamma[\Phi]$ is the effective action, and its functional derivatives with respect to Φ yield 1PI Green's functions. Usually, we are interested in setting the source J to zero at the end of the calculation, and so the preceding equation shows that Γ is stationary in Φ. This condition of stationarity is the Schwinger–Dyson equation for Φ.

The effective action can be found directly by introducing a background field Φ and shifting the argument ϕ of the action $S[\phi]$ by this amount. At the outset, this shift field Φ is arbitrary and independent of J. The shift yields a new generating functional $\tilde{Z}$:

$$Z \to \tilde{Z} = \int [d\phi] \exp\left\{ iS[\phi + \Phi] + i \int d^4x\, J(x)\phi(x) \right\} \equiv e^{i\tilde{W}[J,\Phi]}. \qquad (2.47)$$

We might as well write this by shifting variables back, through $\phi \to \phi - \Phi$, so that

$$\widetilde{W}[J, \Phi] = W[J] - \int d^4x \, J(x)\Phi(x). \tag{2.48}$$

This is precisely the effective action $\Gamma[\Phi]$, provided that J and Φ are related by Eq. (2.45). This equation amounts to $\delta \widetilde{W}/\delta J = 0$.

Let us calculate the effective action $\widetilde{\Gamma}$ corresponding to $\widetilde{Z}$. Do the Legendre transform:

$$\widetilde{\Gamma}[\bar{\phi} + \Phi] = \widetilde{W}[J, \Phi] - \int d^4x \, J(x)\bar{\phi}(x), \tag{2.49}$$

where the Legendre-transform variable $\bar{\phi}$ is defined by

$$\bar{\phi}(x) = \frac{\delta \widetilde{W}}{\delta J(x)}. \tag{2.50}$$

When the equations of motion (2.45) are satisfied, $\bar{\phi}$ vanishes.

We have written $\widetilde{\Gamma}$ as a functional of only one variable. To show this, use

$$\frac{\delta \widetilde{W}}{\delta \Phi(x)} = -J(x), \tag{2.51}$$

to find the total variation of $\widetilde{\Gamma}$:

$$\delta \widetilde{\Gamma} = -J\delta(\bar{\phi} + \Phi). \tag{2.52}$$

The final conclusion is that when the equations of motion in Eq. (2.45) hold,

$$\widetilde{\Gamma}[\bar{\phi} = 0, \Phi] = \Gamma[\Phi]. \tag{2.53}$$

A possible construction of Γ comes from summing all connected 1PI Feynman graphs, using the field-shifted action of Eq. (2.47) to find the 1PI graphs. Or, one can simply integrate the Schwinger–Dyson equation for Φ, written in terms of 1PI skeleton graphs.

All the scalar-field results have analogs for NAGTs but with certain complications from gauge fixing and ghost terms. There is a further generalization [9] of effective-action methods that allows us to construct an effective action that is 2PI, so that no connected graph in Γ can be separated by cutting only two (distinct) lines. We discuss this generalization briefly in the following section.

The two-particle-irreducible effective action By introducing a two-point source $K(x, y)$ and corresponding Legendre transform, we can [9] define a generating functional and its Legendre transform, the effective action $\Gamma[\Phi, G]$ that depends on a propagator function Δ as well as on Φ:

$$Z[J, K] = \int [\mathrm{d}\phi] \exp \left\{ iS[\phi] + i \int \mathrm{d}^4 x \, J(x)\phi(x) \right.$$

$$\left. - \frac{i}{2} \int \mathrm{d}^4 x \int \mathrm{d}^4 y \, \phi(x) K(x, y)\phi(y) \right\}$$

$$= e^{iW[J,K]}$$

$$\Gamma[\Phi, \Delta] = W[J, K] - \int \mathrm{d}^4 x \, J(x)\Phi(x)$$

$$- \frac{1}{2} \int \mathrm{d}^4 x \int \mathrm{d}^4 y \, [\Phi(x)K(x, y)\Phi(y) + \Delta(x, y)K(x, y)]. \quad (2.54)$$

The functional derivatives

$$\frac{\delta W}{\delta J(x)} = \langle \phi(x) \rangle \equiv \Phi(x) \qquad (2.55)$$

$$\frac{\delta W}{\delta K(x, y)} = \frac{1}{2} [\Phi(x)\Phi(y) + \Delta(x, y)],$$

where Δ is the connected two-point function, lead to the functional derivatives for the effective action:

$$\frac{\delta \Gamma}{\delta \Phi(x)} = -J(x) - \int \mathrm{d}^4 y \, K(x, y)\Phi(y)$$

$$\frac{\delta \Gamma}{\delta \Delta(x, y)} = -\frac{1}{2} K(x, y). \qquad (2.56)$$

As before, physical processes correspond to vanishing sources, so now Γ is stationary with respect to both the one- and two-point functions. The vanishing variation of $\delta\Gamma/\delta\Delta$ yields the Schwinger–Dyson equation for Δ. The graphical construction for Γ now involves the sum of connected 2PI graphs, those that cannot be separated by cutting only two (distinct) lines. These graphs necessarily have dressed propagators for their lines. There are, in addition, some one-dressed-loop terms. Even if Γ is approximated by saving only a few terms in the dressed-loop expansion, the equations resulting from requiring stationarity may well reveal nonperturbative effects not visible even in resummed perturbation theory.

One can go further and introduce sources for three- and four-point functions, along with Legendre transforms analogous to those of Eq. (2.54). The resulting Γ is now stationary (at vanishing sources) with respect to N-point functions with $N = 1, 2, 3, 4$, and the stationarity requirements are the corresponding Schwinger–Dyson equations. Or [31], one can simply look at the sum of connected graphs (with correct combinatoric factors!) for W in the absence of sources and resolve these into skeleton graphs for these N-point functions. The derivatives of W with respect to these N-point functions come out to be the Schwinger–Dyson equations.

2.3.2 The background-field method for gauge fields

The goal is to produce an effective action that is gauge invariant in terms of the classical potentials $\widehat{A}_\mu$ that appear in it so that it is a functional of the classical field strengths such as $\operatorname{Tr} \widehat{A}_{\mu\nu}\widehat{A}^{\mu\nu}$. We would like to imitate the principle of shifting the variable of integration (which we also call the quantum potential) by a classical potential and then shift back to produce the effective action, as we did for scalar fields. So we want to split the original quantum variable A_μ in the functional integrals into a classical part $\widehat{A}_\mu$ and a quantum part Q_μ:

$$A_\mu = \widehat{A}_\mu + Q_\mu. \tag{2.57}$$

The original action of Eq. (1.5) is invariant under the inhomogeneous gauge transformation of Eq. (1.10), which is just a change of variable of integration. How do we apportion this gauge transformation,

$$A_\mu \rightarrow V\frac{\mathrm{i}}{g}A_\mu V^{-1} + V\partial_\mu V^{-1}, \tag{2.58}$$

between $\widehat{A}_\mu$ and Q_μ? Furthermore, the NAGT generating functional with no background field given in Eq. (1.5), which we repeat here:

$$
Z[J_\mu] = \int [\mathrm{d}Q_\mu][\mathrm{d}\bar{c}][\mathrm{d}c]\exp\Bigg\{\mathrm{i}S[Q]
$$
$$
+ \mathrm{i}\int \mathrm{d}^4x\left[\frac{1}{2\xi}\operatorname{Tr}\left(\partial_\mu Q^\mu\right)^2 + (\bar{c}\partial_\nu \mathcal{D}^\nu c) + J_\mu(x)Q^\mu(x)\right]\Bigg\}
$$
$$
\equiv \mathrm{e}^{\mathrm{i}W[J_\mu]}, \tag{2.59}
$$

is not gauge invariant because of the ghost-antighost and gauge-fixing terms, as well as the term involving $J\cdot A$. So how do we get overall gauge invariance of some sort in $W[J]$ and its Legendre transformation, the effective action?

Gauge invariance of the effective action as a functional of the classical potential $\widehat{A}_\mu$ means that it is invariant under a standard gauge transformation of this potential in which the full inhomogeneous term goes with $\widehat{A}_\mu$ (and it would seem that no inhomogeneous term goes with the quantum potential Q_μ):

$$\widehat{A}_\mu \rightarrow V\frac{\mathrm{i}}{g}\widehat{A}_\mu V^{-1} + V\partial_\mu V^{-1}; \qquad Q_\mu \rightarrow V\frac{\mathrm{i}}{g}Q_\mu V^{-1}. \tag{2.60}$$

The combined transformations preserve gauge invariance of the NAGT action $S[\widehat{A}_\mu + Q_\mu]$. Following the scalar field construction, we proceed by coupling only the quantum potential to the current so that the term $J \cdot Q$ remains unchanged.

Next is the gauge-fixing term. If it involves derivatives (and therefore has ghosts), we can replace the ordinary derivatives with the covariant derivative with respect to the classical potential. For example, in R_ξ gauges,

$$\frac{1}{2\xi}\mathrm{Tr}\left(\partial_\mu Q_\mu\right)^2 \rightarrow \frac{1}{2\xi}\mathrm{Tr}\left(\left[\mathcal{D}_\mu(\widehat{A}), Q_\mu\right]^2\right), \tag{2.61}$$

where $\mathcal{D}_\mu(\widehat{A}) = \partial_\mu - ig\widehat{A}_\mu$ is the covariant derivative with respect to the classical potential.[10] The Faddeev–Popov determinant undergoes a corresponding change. At this point, the new generating functional $\widetilde{Z}$, given by

$$\widetilde{Z}[J_\mu, \widehat{A}_\nu] = \int [dQ_\mu][d\bar{c}][dc]\exp\left\{iS[C+Q]\right.$$

$$\left. + \int d^4x\, i\left[\frac{1}{2\xi}\mathrm{Tr}\left([\mathcal{D}_\mu(\widehat{A}), Q_\mu]\right)^2 + \mathcal{L}_{\bar{c}c}(x) + J_\mu(x)Q_\mu(x)\right]\right\}$$

$$\equiv e^{i\widetilde{W}[J_\mu, C_\nu]}, \tag{2.62}$$

(where $\mathcal{L}_{\bar{c}c}$ is the Lagrangian expressing the Faddeev–Popov determinant), is invariant under the combined transformations of Eq. (2.60) plus a homogeneous rotation of the current under which

$$J_\mu \rightarrow VJ_\mu V^{-1}. \tag{2.63}$$

The corresponding Legendre transform

$$\widetilde{\Gamma}[\mathcal{A}_\mu, \widehat{A}_\mu] = \widetilde{W}[J_\mu, \widehat{A}_\mu] - \int d^4x\, J^\mu(x)\mathcal{A}_\mu(x)$$

$$\frac{\delta\widetilde{W}}{\delta J_\mu(x)} = \langle Q_\mu(x)\rangle \equiv \mathcal{A}_\mu(x), \tag{2.64}$$

is also invariant.

Just as with the scalar field, the next step is to change the variable of integration back so that in Eq. (2.62), $Q_\mu \rightarrow Q_\mu - \widehat{A}_\mu$. This changes the argument of the action back to a conventional form. Then the generating functional becomes

$$e^{i\widetilde{W}[J_\mu, \widehat{A}_\mu]} = e^{iW[J_\mu] + i\int d^4x\, J_\mu(x)\mathcal{A}_\mu(x)}. \tag{2.65}$$

Here W is the conventional exponent, except that it is calculated in a special gauge. Making the shift $Q_\mu \rightarrow Q_\mu - \widehat{A}_\mu$ in the gauge-fixing term of Eq. (2.61) yields the

---

[10] The unadorned covariant derivative $\mathcal{D}_\mu$ is always with respect to the quantum potential.

gauge-fixing term

$$\frac{1}{2\xi}\text{Tr}\,G^2; \qquad G = \left[\mathcal{D}^\mu(C), Q_\mu - C_\mu\right]. \tag{2.66}$$

This is somewhat unconventional because this term depends on the classical potential. One must also calculate the Faddeev–Popov determinant,

$$\det\frac{\delta G}{\delta\theta}, \tag{2.67}$$

for an infinitesimal gauge transformation $V \approx \mathbb{I} - i\theta$, under which the new variable of integration Q_μ transforms as

$$\delta Q_\mu = -\frac{1}{g}[\mathcal{D}_\mu, \theta]. \tag{2.68}$$

Note that this corresponds to transforming Q_μ inhomogeneously, as in Eq. (2.58), which is necessary because the action is only invariant under such a gauge transformation.

For the scalar field case, we were finished at this point because it was trivial to show that $\widetilde{W} = \Gamma$. It is only slightly more elaborate [23] to show that

$$\widetilde{W}[\mathcal{A}_\mu = 0, \widehat{A}_\mu] = \Gamma_C[\widehat{A}_\mu], \tag{2.69}$$

where Γ_C is the conventional effective action with the gauge-fixing term of Eq. (2.66). The complication is that the gauge-fixing term in the conventional effective action also depends on the external potential.

The Feynman rules for the effective action, consisting, as usual, of the sum of 1PI connected graphs in the presence of the external potential C_μ, are given by Abbott [23] and in the appendix. They are different from the usual rules because this external potential appears in the gauge-fixing term and in the ghost-antighost action. As far as we are concerned right now, two of the main differences are that the three-gluon vertex with one external potential leg is precisely the same as the vertex Γ^ξ of Eq. (1.38) and that the ghost-external potential vertex, unlike the conventional (asymmetric) ghost-gluon vertex, is conserved.

One can, of course, use any value for ξ in the gauge choice for the background-field method without affecting the fact that the effective action is a gauge-invariant functional of C_μ. Unfortunately, the coefficient functions found from the functional integrals over Q_μ still lead to ξ-dependent quantities, and so we are only part way along the path to true quantum gauge invariance. It is nevertheless true that simply by setting $\xi = 1$, we do get the gauge-invariant Green's functions of the pinch technique. In fact, Chapter 1 and our subsequent remarks already give us what amounts to a proof of this for one-loop quantities. The integrands in Eq. (1.61)

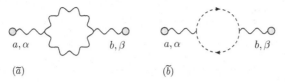

Figure 2.2. Feynman diagrams contributing to the one-loop background gluon self-energy. Shaded circles on external lines represent background fields.

for the PT proper self-energy and Eq. (1.84) for the three-gluon vertex are precisely those that would be found using the background-field method at $\xi = 1$. The background-field method at $\xi \neq 1$ does not give the PT results, but these can be recovered by applying the pinch technique as usual, thereby coming back to the background-field method results at $\xi = 1$.

2.3.3 Pinch technique and background Feynman gauge correspondence

Let us have a closer look at the announced connection between the pinch technique and the background Feynman gauge. The key observation [28, 29] is that at $\xi_Q = 1$, the tree-level vertex that occurs in the action multiplying $\widehat{A}_\alpha(q)A_\mu(k_1)A_\nu(k_2)$, to be denoted by $\widetilde{\Gamma}^{\xi_Q}_{\alpha\mu\nu}(q, k_1, k_2)$ (see Feynman rules of the appendix), collapses to the expression for $\Gamma^F_{\alpha\mu\nu}(q, k_1, k_2)$, given in Eq. (1.42). Because in addition, at $\xi_Q = 1$, the longitudinal parts of the gluon propagator vanish, one realizes that at this point, there is nothing there that could pinch. Thus, ultimately, the background Feynman gauge is singled out because of the total absence, in this particular gauge, of any pinching momenta.

It is relatively straightforward to verify at the one-loop level the correspondence between the PT two- and three-point functions and those of the background Feynman gauge [28, 29]. For example, the two Feynman diagrams contributing to the background Feynman gauge gluon self-energy are shown in Figure 2.2. Using the background Feynman gauge Feynman rules, we obtain

$$(\tilde{a})_{\alpha\beta} = \frac{1}{2}g^2 C_A \int_k \frac{1}{k^2(k+q)^2} \widetilde{\Gamma}_{\alpha\mu\nu}(q, -k-q, k)\widetilde{\Gamma}^{\mu\nu}_\beta(q, -k-q, k)$$

$$(\tilde{b})_{\alpha\beta} = -g^2 C_A \int_k \frac{1}{k^2(k+q)^2}(2k+q)_\alpha(2k+q)_\beta. \tag{2.70}$$

We can simply compare the two terms on the rhs of Eq. (1.63) with the two terms given in Eqs. (2.70). Evidently, the PT and background Feynman gauge gluon self-energies are identical at one loop.

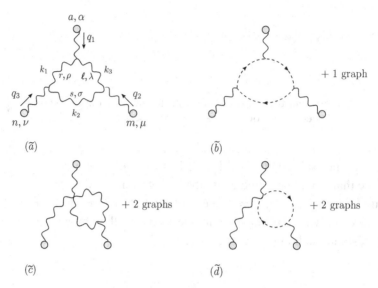

Figure 2.3. One-loop diagrams contributing to the three-gluon vertex in the BFM. Diagrams $(\tilde{c})$ carries a $1/2$ symmetry factor.

Similarly, the one-loop diagrams contributing to the background Feynman gauge three-gluon vertex are shown in Figure 2.3; it is easy to see that the sum $(\tilde{a}) + (\tilde{b})$ coincides with the term $\widehat{N}_{\alpha\mu\nu}$ of Eq. (1.85), while diagrams $(\tilde{c})$ give exactly the term $\widehat{B}_{\alpha\mu\nu}$ of Eq. (1.86). Finally, diagrams $(\tilde{d})$ vanish by virtue of elementary group-theoretical identities.

Although it is a remarkable and extremely useful fact that the one-loop PT Green's functions can be calculated in the background Feynman gauge, particular care is needed for the correct interpretation of this correspondence. First, the pinch technique enforces gauge independence (and several other physical properties, such as unitarity and analyticity) on off-shell Green's functions, whereas the BFM, in a general gauge, does not. This is reflected in the gauge invariance of the BFM n-point functions in the sense that they satisfy (by construction) QED-like Ward identities, but are not gauge independent, i.e., they depend explicitly on ξ_Q. For example, the BFM gluon self-energy at one loop is given by [30]

$$\widetilde{\Pi}_{\alpha\beta}^{(\xi_Q)}(q) = \widetilde{\Pi}_{\alpha\beta}^{(\xi_Q=1)}(q) + \frac{i}{4(4\pi)^2}g^2 C_A(1-\xi_Q)(7+\xi_Q)q^2 P^{\alpha\beta}(q). \quad (2.71)$$

Had the BFM n-point functions been ξ_Q independent, in addition to being gauge invariant, there would be no need to introduce the pinch technique independently.

We emphasize that the objective of the PT construction is not to derive diagrammatically the background Feynman gauge but rather to exploit the underlying BRST symmetry to expose a large number of cancellations and eventually define gauge-independent Green's functions satisfying Abelian Ward identities. Thus, that the PT Green's functions can also be calculated in the background Feynman gauge always needs a very extensive demonstration. Therefore, the correspondence must be verified at the end of the PT construction and should not be assumed beforehand. Moreover, the ξ_Q-dependent BFM Green's functions are not physically equivalent. This is best seen in theories with spontaneous symmetry breaking: the dependence of the BFM Green's functions on ξ_Q gives rise to *unphysical* thresholds inside these Green's functions for $\xi_Q \neq 1$, which limits their usefulness for resummation purposes (this point will be studied in detail in Chapter 11). Only the case of the background Feynman gauge is free from unphysical poles because then (and only then) do the BFM results collapse to the physical PT Green's functions.

It is also important to realize that the PT construction goes through unaltered under circumstances in which the BFM Feynman rules cannot even be applied. Specifically, if instead of an S-matrix element, one were to consider a different observable, such as a current correlation function or a Wilson loop (as was in fact done in the original formulation [4]), one could not start out using the background Feynman rules because *all* fields appearing inside the first nontrivial loop are quantum ones. Instead, by following the PT rearrangement inside these physical amplitudes, the unique PT answer emerges again.

Perhaps the most compelling fact that demonstrates that the PT and BFM are intrinsically two completely disparate methods is that one can apply the PT within the BFM. Operationally, this is easy to understand: away from $\xi_Q = 1$, even in the BFM, there are longitudinal (pinching) momenta that will initiate the pinching procedure. Thus, one starts out with the S-matrix written with the BFM Feynman rules using a general ξ_Q and applies the PT algorithm as in any other gauge-fixing scheme; one will recover again the unique PT answer for all Green's functions involved (i.e., the Green's functions will be projected to $\xi_Q = 1$).

2.3.4 The generalized pinch technique

As we have seen in detail, the PT projects us dynamically to the background Feynman gauge, regardless of the gauge-fixing scheme from which we may start. A question that arises naturally at this point is the following: could we devise a PT-like procedure that would project us to some other value of the background gauge-fixing parameter ξ_Q? As was shown by Pilaftsis [32], such a construction is indeed possible; the systematic algorithm that accomplishes this is known as the *generalized pinch technique*.

The starting point of the generalized PT is precisely the decomposition given in Eqs. (1.37), (1.38), and (1.39). However, unlike the pinch technique where *all* longitudinal momenta are allowed to pinch, in the generalized pinch technique, the Γ^ξ of Eq. (1.38) does not trigger any pinching (even though it contains longitudinal momenta), playing essentially the role of Γ^F; the pinching momenta of the generalized pinch technique come from the $\Gamma^{P\xi}$ of Eq. (1.39) and, of course, the tree-level gluon propagators. At the end of this procedure, one recovers diagrammatically the background Green's functions calculated at the desired value of $\xi \to \xi_Q$.

To be sure, the generalized pinch technique represents a fundamental departure from the primary aim of the pinch technique, which is to construct gauge-fixing, parameter–independent, off-shell Green's functions. The generalized pinch technique, instead, deals exclusively with gauge-fixing, parameter–dependent Green's functions, with all the pathologies that this dependence entails. Nonetheless, it is certainly useful to have a method that allows us to move systematically from one gauge-fixing scheme to another at the level of individual Green's functions. In addition to the possible applications mentioned by Pilaftsis [32], we would like to emphasize the usefulness of the generalized pinch technique in truncating gauge-invariant (i.e., maintaining transversality) sets of Schwinger–Dyson equations written in gauges other than the Feynman gauge (see Chapter 6). This possibility becomes particularly relevant, for example, in attempts to compare SDE predictions with lattice simulations, which are carried out usually in the Landau gauge.

The method can be systematically generalized to more complicated situations [32]. For instance, a method may be projected from the R_ξ gauges to one of the generalized BFM gauges, such as the BFM axial gauge. This, of course, leads to a proliferation of pinching momenta; the resulting construction is therefore more cumbersome but remains conceptually rather straightforward.

2.4 What to expect beyond one loop

Everything in Chapters 1 and 2 illustrates the pinch technique at the one-loop level. The pinch technique would be of little interest unless everything in these chapters had an all-order generalization. A good part of the rest of the book is devoted to showing that the PT propagator has the following indispensable properties to all orders, and even nonperturbatively:

1. It truly is gauge independent.
2. It is independent of what group representation or spin the external particles used to construct the S-matrix have. (The reader should check this for the one-loop pinch technique.)
3. It has only physical threshholds even if (or especially if) the gluons get a mass through the strong interactions. There are no unphysical ghost contributions.

4. Its Green's functions have conventional analytic properties and spectral representations, except that in certain cases, conventional positivity requirements do not hold.
5. It, along with similarly defined pinch technique vertices, participates in ghost-free Ward identities that are analogous to those of QED and with similar consequences such as generalizations of the familiar QED identity $Z_1 = Z_2$.
6. The PT propagator defines a running charge that is gauge invariant and scheme independent.
7. Once the ghost-free Ward identities are imposed, it is unique, which can be understood because (as we will show) PT propagators and vertices are simply those of the background Feynman gauge, which is a uniquely defined graphical prescription with the same Ward identities.

As a result of the preceding requirements – and we emphasize once again not the other way around – we show that the PT Green's functions to all orders are identical to those of the background-field method in the Feynman gauge.

After the technical developments that establish these points come the applications. They range from perturbative effects, such as a physical and gauge-invariant definition of the neutrino charge radius, to nonperturbative effects, such as the all-order resummation needed to define the widths of unstable gauge bosons beyond tree level, to setting up the tools necessary for calculating the dynamical mass of gauge bosons in the magnetic sector of QCD or of high-temperature electroweak theory.

References

[1] M. Lavelle, Gauge invariant effective gluon mass from the operator product expansion, *Phys. Rev.* **D44** (1991) R26.
[2] H. Georgi and S. L. Glashow, Unified weak and electromagnetic interactions without neutral currents, *Phys. Rev. Lett.* **28** (1972) 1494.
[3] J. M. Cornwall, Nonperturbative mass gap in continuum QCD, in *Proceedings of the French–American Seminar on Theoretical Aspects of Quantum Chromodynamics*, Marseille, France (1981).
[4] J. M. Cornwall, Dynamical mass generation in continuum QCD, *Phys. Rev.* **D26** (1982) 1453.
[5] J. M. Cornwall, Center vortices, nexuses, and the Georgi-Glashow model, *Phys. Rev.* **D59** (1999) 125015.
[6] J. M. Cornwall, On the phase transition in D = 3 Yang-Mills Chern-Simons gauge theory, *Phys. Rev.* **D54** (1996) 1814.
[7] B. W. Lee, C. Quigg, and H. B. Thacker, Weak interactions at very high-energies: The role of the Higgs boson mass, *Phys. Rev.* **D16** (1977) 1519.
[8] J. M. Cornwall, W. S. Hou, and J. E. King, Gauge invariant calculations in finite temperature QCD: Landau ghost and magnetic mass, *Phys. Lett.* **B153** (1985) 173.

[9] J. M. Cornwall, R. Jackiw, and E. Tomboulis, Effective action for composite operators, *Phys. Rev.* **D10** (1974) 2428.

[10] J. M. Cornwall, Spontaneous symmetry breaking without scalar mesons: 2, *Phys. Rev.* **D10** (1974) 500.

[11] J. M. Cornwall and W. S. Hou, Extension of the gauge technique to broken symmetry and finite temperature, *Phys. Rev.* **D34** (1986) 585.

[12] J. Papavassiliou, Gauge invariant proper self-energies and vertices in gauge theories with broken symmetry, *Phys. Rev.* **D41** (1990) 3179.

[13] D. Binosi, Electroweak pinch technique to all orders, *J. Phys.* **G30** (2004) 1021.

[14] G. Degrassi and A. Sirlin, Gauge invariant self-energies and vertex parts of the standard model in the pinch technique framework, *Phys. Rev.* **D46** (1992) 3104.

[15] K. Fujikawa, B. W. Lee, and A. I. Sanda, Generalized renormalizable gauge formulation of spontaneously broken gauge theories, *Phys. Rev.* **D6** (1972) 2923.

[16] G. 't Hooft, Renormalization of massless Yang-Mills fields, *Nucl. Phys.* **B33** (1971) 173.

[17] B. S. DeWitt, Quantum theory of gravity. II. The manifestly covariant theory, *Phys. Rev.* **162** (1967) 1195.

[18] J. Honerkamp, The question of invariant renormalizability of the massless Yang-Mills theory in a manifest covariant approach, *Nucl. Phys.* **B48** (1972) 269.

[19] R. E. Kallosh, The renormalization in nonabelian gauge theories, *Nucl. Phys.* **B78** (1974) 293.

[20] H. Kluberg-Stern and J. B. Zuber, Renormalization of nonabelian gauge theories in a background field gauge. 1. Green functions, *Phys. Rev.* **D12** (1975) 482.

[21] I. Y. Arefeva, L. D. Faddeev, and A. A. Slavnov, Generating functional for the *S*-Matrix in gauge theories, *Theor. Math. Phys.* **21** (1975) 1165.

[22] G. 't Hooft, The background field method in gauge field theories, in *Karpacz 1975, Proceedings, Acta Universitatis Wratislaviensis No. 368, Vol. 1*, Wroclaw, Poland (1976).

[23] L. F. Abbott, The background field method beyond one loop, *Nucl. Phys.* **B185** (1981) 189.

[24] S. Weinberg, Effective gauge theories, *Phys. Lett.* **B91** (1980) 51.

[25] G. M. Shore, Symmetry restoration and the background field method in gauge theories, *Ann. Phys.* **137** (1981) 262.

[26] L. F. Abbott, M. T. Grisaru, and R. K. Schaefer, The background field method and the *S* matrix, *Nucl. Phys.* **B229** (1983) 372.

[27] C. F. Hart, Theory and renormalization of the gauge invariant effective action, *Phys. Rev.* **D28** (1983) 1993.

[28] S. Hashimoto, J. Kodaira, Y. Yasui, and K. Sasaki, The background field method: Alternative way of deriving the pinch technique's results, *Phys. Rev.* **D50** (1994) 7066.

[29] A. Denner, G. Weiglein, and S. Dittmaier, Gauge invariance of Green functions: Background field method versus pinch technique, *Phys. Lett.* **B333** (1994) 420.

[30] J. Papavassiliou, On the connection between the pinch technique and the background field method, *Phys. Rev.* **D51** (1995) 856.

[31] R. E. Norton and J. M. Cornwall, On the formalism of relativistic many body theory, *Ann. Phys.* **91** (1975) 106.

[32] A. Pilaftsis, Generalized pinch technique and the background field method in general gauges, *Nucl. Phys.* **B487** (1997) 467.

3

Pinch technique to all orders

In this chapter, we present the generalization of the pinch technique (PT) beyond one loop. The key observation is that the one-loop PT rearrangements described in Chapter 1 constitute the lowest-order manifestation of a fundamental cancellation taking place between graphs of distinct kinematic nature. This cancellation is encoded in the Slavnov–Taylor identity satisfied by a special Green's function, which serves as a common kernel to all higher-order self-energy and vertex diagrams. This allows for the collective treatment of entire sets of diagrams, providing a compact way of extending the PT construction to higher orders. In addition, we will show that, quite remarkably, the correspondence between the pinch technique and the background Feynman gauge established in Chapter 1 is not accidental but persists to all orders.

3.1 The s-t cancellation to all orders

The generalization of the pinch technique to all orders relies on the following basic observations. The vast PT cancellations between one-loop Feynman diagrams, studied in Chapter 1, are in fact encoded in the Slavnov–Taylor identity obeyed by the kernel $A_\mu A_\nu q\bar{q}$ (with the gluons off shell and the quarks on shell). In the Feynman gauge, this Slavnov–Taylor identity is triggered by the longitudinal momenta k_1^μ and k_2^ν contained in $\Gamma_{\alpha\mu\nu}^{\rm P}(q, k_1, k_2)$. The tree-level version of this Slavnov-Taylor identity gives rise precisely to the s-t cancellation discussed in Section 1.7.2 (but with the gluons on shell) for the tree-level process $gg \rightarrow q\bar{q}$, namely, the lowest-order contribution to the aforementioned amplitude $A_\mu A_\nu q\bar{q}$.

Indeed, as explained in Section 1.7.2, at tree-level, the preceding amplitude, denoted by $T_{\mu\nu}^{mn}$ is the sum of two distinct parts: an s-channel subamplitude, $T_{s,\mu\nu}^{mn}$, given in Figure 1.16(c) and t- and u-channel subamplitudes containing an internal quark propagator, $T_{t,\mu\nu}^{mn}$, shown in diagrams (a) and (b) of the same figure.

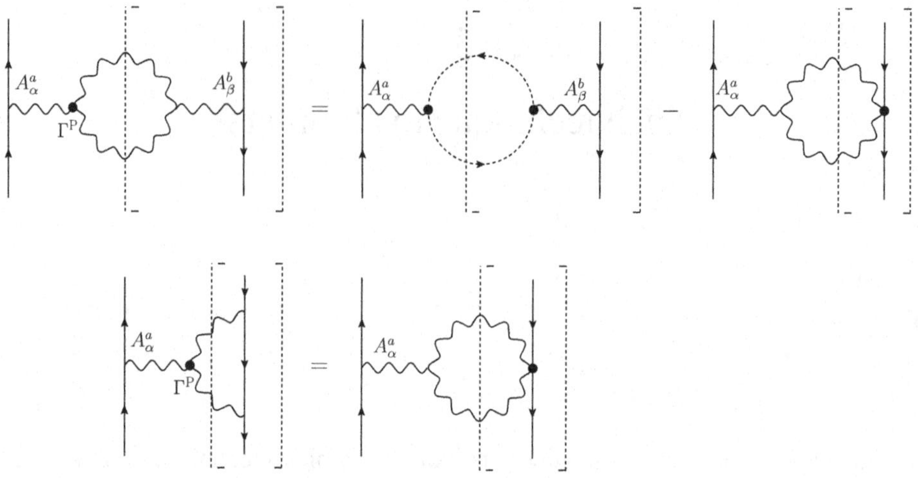

Figure 3.1. The one-loop pinch technique seen in terms of the fundamental *s-t* cancellation. The self-energy-like contribution coming from the vertex cancels exactly against the contribution coming from the propagator. Notice that none of the effective vertices induced after the cancellation is contained in the original Lagrangian of the theory; their field-theoretic interpretation will be presented in Chapter 4.

When $T^{mn}_{\mu\nu}$ is contracted by k_1^μ or k_2^ν, a characteristic cancellation takes place between $T^{mn}_{s\,\mu\nu}$ and $T^{mn}_{t\,\mu\nu}$. To see this, use the elementary Ward identity satisfied by $\Gamma^\alpha_{\mu\nu}(q, k_1, k_2)$ and note that the term proportional to q^2 cancels the $d(q)$, thus allowing communication with the *t*-channel graphs (in Section 1.7.2, we used Γ^F instead, but this makes no difference; see the comment following Eq. (1.130)). Using, in addition, current conservation, $q^\alpha \mathcal{V}^c_\alpha = 0$, and keeping the gluons off shell (i.e., not setting $k_1^2 = k_2^2 = 0$, as we did in Section 1.7.2), we have

$$k_1^\mu T^{mn}_{s\,\mu\nu} = g^2 f^{mnc} \mathcal{V}^c_\alpha (k_2^2 g^\alpha_\nu - k_{2\nu} k_2^\alpha) d(q^2) - g^2 f^{mnc} \mathcal{V}^c_\nu$$

$$k_1^\mu T^{mn}_{t\,\mu\nu} = g^2 f^{mnc} \mathcal{V}^c_\nu, \tag{3.1}$$

so that

$$k_1^\mu T^{mn}_{\mu\nu} = g^2 f^{mnc} \mathcal{V}^c_\alpha (k_2^2 g^\alpha_\nu - k_{2\nu} k_2^\alpha) d(q^2). \tag{3.2}$$

The important point to realize is that one can recast the entire one-loop PT construction in terms of the *s-t* cancellation. The precise way in which the preceding cancellation is realized inside the one-loop self-energy and vertex graphs, giving rise to the PT rearrangements described in Chapter 1, is shown schematically in Figure 3.1.

It turns out that the pinch technique may be extended to higher orders simply by pursuing the preceding cancellations beyond tree level [1, 2]. Specifically, the

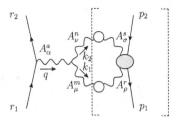

Figure 3.2. The fundamental amplitude receiving the action of the longitudinal momenta stemming from Γ^P. The shaded blob represents the (connected) kernel corresponding to the process $AA \rightarrow q\bar{q}$.

all-order version of the Slavnov–Taylor identity satisfied by $T_{\mu\nu}^{mn}$, appropriately interpreted, allows the generalization of the PT construction to all orders.

The subset of all graphs that receive the action of the longitudinal momenta contained in $\Gamma^P_{\alpha\mu\nu}(q, k_1, k_2)$ is shown in Figure 3.2: it comprises precisely the kernel $A_\mu^m(k_1)A_\nu^n(k_2) \rightarrow q(p_1)\bar{q}(p_2)$, i.e., the all-order version of $T_{\mu\nu}^{mn}$. In terms of Green's functions,

$$T_{\mu\nu}^{mn} = \bar{u}(p_1)\big[C_{\rho\sigma}^{mn}(k_1, k_2, p_1, p_2)\Delta_\mu^\rho(k_1)\Delta_\nu^\sigma(k_2)\big]u(p_2). \tag{3.3}$$

Clearly, the two internal gluons are off shell, whereas the two external quarks are on-shell, satisfying $\bar{u}(p_1)S^{-1}(p_1)\big|_{\not{p}_1=m} = S^{-1}(p_2)u(p_2)\big|_{\not{p}_2=m} = 0$, where $S(p)$ is the full-quark propagator.

Let us focus on the Slavnov–Taylor identity satisfied by the amplitude $T_{\mu\nu}^{mn}$. Following standard techniques [3], one exploits ghost charge conservation to write the trivial position space identity:

$$\langle T[\bar{c}^m(x)A_\nu^n(y)q(z)\bar{q}(w)]\rangle = 0, \tag{3.4}$$

with T denoting the time-ordered product of fields. Rewriting the fields in terms of their BRST-transformed counterparts, using their equations of motion and equal-time commutation relations, and Fourier transforming the final result to momentum space, we find

$$k_1^\mu C_{\mu\nu}^{mn} - k_{2\nu}G_1^{mn} + ig f^{nrs} Q_{1\nu}^{mrs} - gX_{1\nu}^{mn} - g\bar{X}_{1\nu}^{mn} = 0, \tag{3.5}$$

where the various Green's functions appearing on the right-hand side (rhs) are defined in Figure 3.3. Note that the terms $X_{1\nu}$ and $\bar{X}_{1\nu}$ vanish on shell because they are missing one fermion propagator; at lowest order, they are simply the terms proportional to the inverse tree-level propagators $(\not{p}_1 - m)$ and $(\not{p}_2 - m)$ first encountered in the one-loop PT calculations of Chapter 1. After multiplying Eq. (3.5) by the two inverse propagators $S^{-1}(p_1)S^{-1}(p_2)$, we thus arrive at the

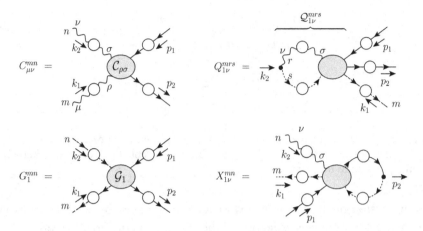

Figure 3.3. Diagrammatic representation of the Green's functions appearing in the Slavnov–Taylor identity (Eq. (3.5)). Ghost Green's functions receive a contribution from similar terms with the ghost arrows reversed (not shown).

on-shell Slavnov–Taylor identity

$$k_1^\mu T_{\mu\nu}^{mn} = S_{1\nu}^{mn},\qquad(3.6)$$

with[1]

$$S_{1\nu}^{mn} = \bar{u}(p_1)\big[gf^{nrs}\mathcal{Q}_{1\nu}^{mrs}(k_1,k_2) - k_{2\nu}\mathcal{G}_1^{mn}(k_1,k_2)D(k_2)\big]D(k_1)u(p_2),\qquad(3.7)$$

with $\mathcal{G}_1^{mn}$ and $\mathcal{Q}_{1\nu}^{ars}$ defined in Figure 3.3.

In perturbation theory, both $T_{\mu\nu}^{mn}$ and $S_{1\nu}^{mn}$ are given by Feynman diagrams, which can be separated into distinct classes according to their kinematic dependence and topological properties (Figure 3.4). Graphs that do not contain information about the external test quarks are self-energy graphs, whereas those depending on the quantum numbers of the test quarks are vertex graphs. The former depend only on the variable s, the latter on both s and the mass m of the test quarks; equivalently, we will refer to them as s- or t-channel graphs, respectively. In addition to the s-t classification, Feynman diagrams can be separated into 1PI and 1PR graphs. The crucial point is that the action of the momentum k_1^μ or k_2^ν on $T_{\mu\nu}^{mn}$ does not respect, in general, the original s-t and 1PI-1PR separations furnished by the Feynman diagrams. In other words, even though Eq. (3.6) holds for the entire amplitude, it is not true for the individual subamplitudes, i.e.,

$$k_1^\mu\big[T_{\mu\nu}^{mn}\big]_{x,\mathrm{Y}} \ne \big[S_{1\nu}^{mn}\big]_{x,\mathrm{Y}}\qquad x=s,t;\ \mathrm{Y}=\mathrm{I},\mathrm{R},\qquad(3.8)$$

---

[1] In what follows, the only momenta we indicate in the Green's functions are the ones corresponding to the gluons (k_i); the quark momenta (p_i) will instead be omitted.

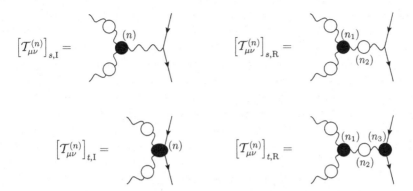

$$\left[\mathcal{T}_{\mu\nu}^{(n)} \right]_{s,\mathrm{I}} = \qquad\qquad \left[\mathcal{T}_{\mu\nu}^{(n)} \right]_{s,\mathrm{R}} =$$

$$\left[\mathcal{T}_{\mu\nu}^{(n)} \right]_{t,\mathrm{I}} = \qquad\qquad \left[\mathcal{T}_{\mu\nu}^{(n)} \right]_{t,\mathrm{R}} =$$

Figure 3.4. Decomposition at an arbitrary perturbative level n of the fundamental amplitude $\mathcal{T}_{\mu\nu}^{mn}$ in terms of s and t channels and 1PI and 1PR components.

where I (R) indicates the one-particle irreducible (reducible) parts of the amplitude involved. Evidently, whereas the characterization of graphs as propagator- and vertex-like is unambiguous in the absence of longitudinal momenta (e.g., in a scalar theory or in QED), their presence in non-Abelian gauge theories tends to mix propagator- and vertex-like graphs. Similarly, 1PR graphs can be effectively converted into 1PI graphs (the opposite cannot happen). The inequality between the two sides of Eq. (3.8) is precisely due to propagator-like terms, such as those encountered in the one-loop PT calculations; they have the characteristic feature that, when depicted by means of Feynman diagrams, contains unphysical vertices, i.e., vertices that do not exist in the original Lagrangian (Figure 3.5). All such terms cancel *exactly* against each other. Thus, after the PT cancellations have been enforced, we have

$$\left[k_1^\mu \mathcal{T}_{\mu\nu}^{mn} \right]_{t,\mathrm{I}}^{\mathrm{PT}} \equiv \left[\mathcal{S}_{1\nu}^{mn} \right]_{t,\mathrm{I}}. \tag{3.9}$$

The nontrivial step for generalizing the pinch technique to all orders is then the following: instead of going through the arduous task of manipulating the left-hand side of Eq. (3.9) to determine the pinching parts and explicitly enforce their cancellation, use directly the rhs, which already contains the answer! Indeed, the rhs involves only conventional (ghost) Green's functions expressed in terms of standard Feynman rules with no reference to unphysical vertices. Thus, its separation into propagator- and vertexlike graphs can be carried out unambiguously because all possibility for mixing has been eliminated.

3.2 Quark-gluon vertex and gluon propagator to all orders

The considerations just presented can be used to generalize the PT construction to all orders. In what follows, we will denote with a caret superscript the PT boxes,

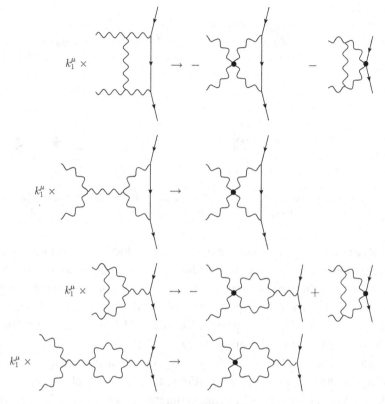

Figure 3.5. Some schematic two-loop examples of PT terms containing unphysical vertices, together with the Feynman diagrams from which they originate. Notice that the sum of all these terms is zero.

self-energies, and vertices and with a tilde the corresponding background Feynman gauge objects; the conventional renormalizable Feynman gauge terms will not carry any superscript.

To begin, it is immediate to recognize that in the renormalizable Feynman gauge, box diagrams of arbitrary order n, $B^{(n)}$, coincide with the PT boxes $\widehat{B}^{(n)}$ because all three-gluon vertices are internal; that is, they do not provide longitudinal momenta because inside the loops there is no preferred direction. Thus, they coincide with the background Feynman gauge boxes, $\tilde{B}^{(n)}$, i.e.,

$$\widehat{B}^{(n)} = B^{(n)} = \tilde{B}^{(n)} \tag{3.10}$$

for every n. The same is true for the PT quark self-energies; for exactly the same reason, they coincide with their renormalizable Feynman gauge (and background Feynman gauge) counterparts, i.e.,

$$\widehat{\Sigma}^{ij\,(n)} = \Sigma^{ij\,(n)} = \tilde{\Sigma}^{ij\,(n)}. \tag{3.11}$$

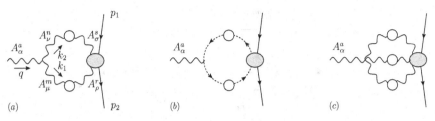

Figure 3.6. The Feynman diagrams contributing to the quark-gluon vertex Γ_α^a in the R_ξ gauge. Diagram (b) has a similar contribution (b′) with the ghost arrow reversed. Kernels appearing in these diagrams are t-channel and 1PI with respect to s-channel cuts.

For the construction of the quark-gluon 1PI vertex $\widehat{\Gamma}_\alpha^a$, start by noting that of all diagrams contributing to this vertex in the renormalizable Feynman gauge (shown in Figure 3.6), the only one receiving the action of the pinching momenta is diagram (a). Thus, we carry out the PT vertex decomposition of Eq. (1.41) in diagram (a) and concentrate on the Γ^P part only; specifically,

$$(a)^P = g f^{amn} \int_{k_1} (g_\alpha^\nu k_1^\mu - g_\alpha^\mu k_2^\nu) \left[T_{\mu\nu}^{mn}(k_1, k_2) \right]_{t,\mathrm{I}}. \tag{3.12}$$

Following the discussion presented in the previous subsection, the pinching action amounts to the replacements

$$k_1^\mu \left[T_{\mu\nu}^{mn} \right]_{t,\mathrm{I}} \rightarrow \left[k_1^\mu T_{\mu\nu}^{mn} \right]_{t,\mathrm{I}}^{\mathrm{PT}} = \left[S_{1\nu}^{mn} \right]_{t,\mathrm{I}} \tag{3.13}$$

$$k_2^\nu \left[T_{\mu\nu}^{mn} \right]_{t,\mathrm{I}} \rightarrow \left[k_2^\nu T_{\mu\nu}^{mn} \right]_{t,\mathrm{I}}^{\mathrm{PT}} = \left[S_{2\mu}^{mn} \right]_{t,\mathrm{I}}, \tag{3.14}$$

or equivalently,

$$(a)^P \rightarrow g f^{amn} \int_{k_1} \left\{ \left[S_{1\alpha}^{mn}(k_1, k_2) \right]_{t,\mathrm{I}} - \left[S_{2\alpha}^{mn}(k_1, k_2) \right]_{t,\mathrm{I}} \right\}. \tag{3.15}$$

At this point, the construction of the effective PT quark-gluon vertex has been completed, and we have

$$\widehat{\Gamma}_\alpha^a(q, p_2, -p_1) = (a)^F + (b) + (b') + (c)$$

$$+ g f^{amn} \int_{k_1} \left\{ \left[S_{1\alpha}^{mn}(k_1, k_2) \right]_{t,\mathrm{I}} - \left[S_{2\alpha}^{mn}(k_1, k_2) \right]_{t,\mathrm{I}} \right\}. \tag{3.16}$$

We emphasize that in the construction presented thus far, we have never resorted to the BFM formalism but have only used the BRST identity of Eq. (3.6) and the replacements (3.13) and (3.14).

The next important question is whether the one-loop correspondence between the pinch technique and the background Feynman gauge persists to all orders. This is

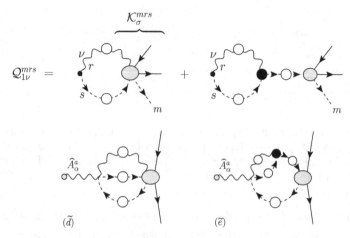

Figure 3.7. (*Top*) The decomposition of the auxiliary function $\mathcal{Q}_{1\nu}^{mrs}$ in terms of its 1PI and 1PR components. Notice that the kernel $\mathcal{K}_{\sigma}^{mrs}$ is 1PI with respect to s-channel cuts. (*Bottom*) Additional topologies present in the BFM quark-gluon vertex and dynamically generated in the PT procedure. Both diagrams have similar contributions (d') and (e') with the ghost arrows reversed.

indeed so, as can be seen by comparing directly the PT vertex $\widehat{\Gamma}_{\alpha}^{a}$ just constructed and the quark-gluon vertex $\widetilde{\Gamma}_{\alpha}^{a}$ written in the background Feynman gauge. We start by observing that all inert terms contained in the original renormalizable Feynman gauge Γ_{α}^{a} vertex carry over to the same subgroups of background Feynman gauge graphs. To facilitate this identification, we recall (see also the Feynman rules reported in the appendix) that to lowest order, one has the identities $\Gamma^{F} = \Gamma_{\widehat{A}AA}$ and $\Gamma_{AAAA} = \Gamma_{\widehat{A}AAA}$, so that

$$(a)^{F} = (\widetilde{a}) \qquad (c) = (\widetilde{c}), \tag{3.17}$$

where a tilde means that the (external) gluon A_{α}^{a} has been effectively converted into a background gluon $\widehat{A}_{\alpha}^{a}$.

As should be familiar by now, the only exception to this rule are the ghost diagrams (d) and (d'): they must be combined with the remaining terms from the PT construction to arrive at the characteristic ghost sector of the background Feynman gauge (see Figure 3.7), namely, the symmetric ghost-gluon vertex $\Gamma_{\widehat{A}c\bar{c}}$ and the four-particle ghost vertex $\Gamma_{\widehat{A}Ac\bar{c}}$, absent in the conventional R_{ξ} gauge fixing. Indeed, using Eq. (3.7), we find (omitting the spinors)

$$g f^{amn} \int_{k_1} \left[S_{1\alpha}^{mn}(k_1, k_2) \right]_{t,I} = -g f^{amn} \int_{k_1} k_{2\alpha} D(k_1) D(k_2) \left[\mathcal{G}_{1}^{mn}(k_1, k_2) \right]_{t,I}$$

$$+ g^2 f^{amn} f^{nrs} \int_{k_1} D(k_1) \left[\mathcal{Q}_{1\alpha}^{mrs}(k_1, k_2) \right]_{t,I}, \tag{3.18}$$

with a similar relation holding for the $\mathcal{S}_2$ term. Then we find

$$(b) - gf^{amn} \int_{k_1} k_{2\alpha} D(k_1)D(k_2) \left[\mathcal{G}_1^{mn}(k_1, k_2)\right]_{t,\mathrm{I}}$$

$$= gf^{amn} \int_{k_1} (k_1 - k_2)_\alpha D(k_1)D(k_2) \left[\mathcal{G}_1^{mn}(k_1, k_2)\right]_{t,\mathrm{I}} = \widehat{(b)}, \qquad (3.19)$$

and using the decomposition for the $\mathcal{Q}_{1\nu}^{mrs}$ shown in Figure 3.2,

$$g^2 f^{amn} f^{nrs} \int_{k_1} D(k_1) \left[\mathcal{Q}_{1\alpha}^{mrs}(k_1, k_2)\right]_{t,\mathrm{I}}$$

$$= ig^2 f^{amn} f^{nrs} \int_{k_1} \int_{k_3} D(k_1)D(k_3)\Delta_\alpha^\sigma(k_4) \left\{\left[\mathcal{K}_\sigma^{mrs}(k_1, k_3, k_4)\right]_{t,\mathrm{I}}\right.$$

$$\left. - i\Gamma_\sigma^{grs}(-k_2, k_3, k_4)D(k_2) \left[\mathcal{G}_1^{mg}(k_1, k_2)\right]_{t,\mathrm{I}}\right\} = \widetilde{(d)} + \widetilde{(e)}, \qquad (3.20)$$

with $\mathcal{K}_\sigma^{mrs}$ representing the 1PI five-particle kernel shown in Figure 3.2, whereas Γ_σ^{grs} is the usual ghost-gluon vertex.

In exactly the same way, the remaining $\mathcal{S}_2$ will generate in $\widetilde{(b')}$ (when added to the R_ξ ghost diagram (b')) as well as $\widetilde{(d')}$ and $\widetilde{(e')}$ so that finally we get the relation

$$\widehat{\Gamma}_\alpha^a(q, p_2, -p_1) = \widetilde{\Gamma}_\alpha^a(q, p_2, -p_1). \qquad (3.21)$$

The final step is to construct the all-order PT gluon self-energy $\widehat{\Pi}_{\alpha\beta}^{ab}(q)$. Notice that at this point, one would expect that it, too, coincides with the background Feynman gauge gluon self-energy $\widetilde{\Pi}_{\alpha\beta}^{ab}(q)$ because the boxes $\widehat{B}$ and the vertices $\widehat{\Gamma}_\alpha^a$ do coincide with the corresponding background Feynman gauge quantities, and the S-matrix is unique.

In what follows, we outline an indirect inductive proof of this result; the gluon self-energy will not be constructed explicitly here but rather in Chapter 6, in the more general context of the Schwinger-Dyson equations. We will use the *strong induction principle*, which states that a given predicate $P(n)$ on $\mathbb{N}$ is true $\forall\, n \in \mathbb{N}$, if $P(k)$ is true whenever $P(j)$ is true $\forall\, j \in \mathbb{N}$ with $j < k$.[2]

To avoid notational clutter, we suppress all color, Lorentz, and momentum labels. At one [4] and two loops (i.e., $n = 1, 2$) [5,6], we know from explicit calculations that the PT and background Feynman gauge Green's functions coincide. Let us then assume that the PT-BFM correspondence

$$\widehat{\Pi}^{(\ell)} = \widetilde{\Pi}^{(\ell)}, \qquad \widehat{\Gamma}^{(\ell)} = \widetilde{\Gamma}^{(\ell)}, \qquad \widehat{B}^{(\ell)} = \widetilde{B}^{(\ell)} \qquad (3.22)$$

---

[2] In simple terms, whereas in the normal induction, one assumes the validity of a property at order $n - 1$ (only) and then demonstrates that it is true also at order n, the strong induction requires the property to be valid at all previous orders $1, 2, \ldots n - 1$.

holds for every $\ell = 1, \ldots, n-1$ (*strong induction hypothesis*). We will then show that the PT gluon self-energy is equal to the background Feynman gauge gluon self-energy at order n, i.e., $\widehat{\Pi}^{(n)} \equiv \widetilde{\Pi}^{(n)}$.

The S-matrix element of order n assumes the form

$$S^{(n)} = \{\Gamma \Delta \Gamma\}^{(n)} + B^{(n)}. \tag{3.23}$$

Moreover, because it is unique, whether written in the renormalizable Feynman gauge or the background Feynman gauge, as well as before and after the PT rearrangement, we have that $S^{(n)} \equiv \widehat{S}^{(n)} \equiv \widetilde{S}^{(n)}$. Using then Eq. (3.10) (which is valid to all orders, implying that Eq. (3.22) holds also when $\ell = n$), we find that

$$\{\Gamma \Delta \Gamma\}^{(n)} = \{\widehat{\Gamma} \widehat{\Delta} \widehat{\Gamma}\}^{(n)} = \{\widetilde{\Gamma} \widetilde{\Delta} \widetilde{\Gamma}\}^{(n)}. \tag{3.24}$$

The preceding amplitudes can then be split into 1PR and 1PI parts; in particular, because of the strong inductive hypothesis (3.22), the 1PR part after the PT rearrangement coincides with the 1PR part written in the background Feynman gauge because

$$\{\Gamma \Delta \Gamma\}^{(n)}_{\mathrm{R}} = \Gamma^{(n_1)} \Delta^{(n_2)} \Gamma^{(n_3)} \qquad \begin{cases} n_1, n_2, n_3 < n \\ n_1 + n_2 + n_3 = n. \end{cases} \tag{3.25}$$

Then Eq. (3.24) states the equivalence of the 1PI parts, i.e.,

$$\{\widehat{\Gamma} \widehat{\Delta} \widehat{\Gamma}\}^{(n)}_{\mathrm{I}} = \{\widetilde{\Gamma} \widetilde{\Delta} \widetilde{\Gamma}\}^{(n)}_{\mathrm{I}}, \tag{3.26}$$

which implies

$$0 = \left[\widehat{\Gamma}^{(n)} - \widetilde{\Gamma}^{(n)}\right] \Delta^{(0)} \Gamma^{(0)} + \Gamma^{(0)} \Delta^{(0)} \left[\widehat{\Gamma}^{(n)} - \widetilde{\Gamma}^{(n)}\right]$$
$$+ \Gamma^{(0)} \Delta^{(0)} \left[\widehat{\Pi}^{(n)} - \widetilde{\Pi}^{(n)}\right] \Delta^{(0)} \Gamma^{(0)}. \tag{3.27}$$

At this point, we do not have the equality we want yet but have only that

$$\widehat{\Gamma}^{(n)} = \widetilde{\Gamma}^{(n)} + f^{(n)} \Gamma^{(0)} \tag{3.28}$$

$$\widehat{\Pi}^{(n)} = \widetilde{\Pi}^{(n)} - 2iq^2 f^{(n)}, \tag{3.29}$$

with $f^{(n)}$ being an arbitrary function of q^2. However, from the *explicit* construction of the PT quark-gluon vertex of the previous section, we have the all-order identity (3.21) so that the second of Eqs. (3.22) actually holds true even when $\ell = n$, i.e., $\widehat{\Gamma}^{(n)} \equiv \widetilde{\Gamma}^{(n)}$. Therefore $f = 0$, and one immediately concludes that

$$\widehat{\Pi}^{(n)} = \widetilde{\Pi}^{(n)}. \tag{3.30}$$

Hence, by strong induction, the preceding relation is true for any given order n.

Reinstating the Lorentz and gauge group structures, we arrive at the announced result[3]:

$$\widehat{\Pi}_{\alpha\beta}^{ab}(q) \equiv \widetilde{\Pi}_{\alpha\beta}^{ab}(q). \tag{3.31}$$

Similar techniques have been used in [7] to generalize to all orders the PT algorithm in the electroweak sector of the standard model (we will briefly touch on this in Chapter 9).

References

[1] D. Binosi and J. Papavassiliou, The pinch technique to all orders, *Phys. Rev.* **D66**(R) (2002) 111901.

[2] D. Binosi and J. Papavassiliou, Pinch technique self-energies and vertices to all orders in perturbation theory, *J. Phys.* **G30** (2004) 203.

[3] P. Pascual and R. Tarrach, *QCD: Renormalization for the Practitioner*, Lect. Notes Phys. **194** (Springer, Heidelberg, 1984).

[4] J. M. Cornwall and J. Papavassiliou, Gauge-invariant three-gluon vertex in QCD, *Phys. Rev.* **D40** (1989) 3474.

[5] J. Papavassiliou, The pinch technique at two loops, *Phys. Rev. Lett.* **84** (2000) 2782.

[6] J. Papavassiliou, The pinch technique at two loops: the case of massless Yang-Mills theories, *Phys. Rev.* **D62** (2000) 045006.

[7] D. Binosi, Electroweak pinch technique to all orders, *J. Phys.* **G30** (2004) 1021.

---

[3] Owing to the validity to all orders of the PT background Feynman gauge correspondence, from now on, we will not make any distinction between PT and background Feynman gauge Green's functions and will indicate both with a caret.

4

The pinch technique in the Batalin–Vilkovisky framework

It is clear from the analysis presented until now that even though the PT Green's functions satisfy naive QED-like Ward identities, their actual derivation relies heavily on Slavnov–Taylor identities obeyed by certain subamplitudes appearing in the ordinary diagrammatic expansion such as the kernel $A_\mu A_\nu q\bar{q}$ considered in the previous chapter. Unlike QED, because of the nonlinearity of the BRST transformations, these Slavnov–Taylor identities are realized through ghost Green's functions involving composite operators such as $\langle 0|T[s\Phi(x)\cdots|0\rangle$, where s is the BRST operator and Φ is a generic QCD field. It turns out that the most efficient framework for dealing with these types of objects is the so-called Batalin-Vilkovisky formalism [1, 2, 3, 4]. In this framework, one adds to the original gauge-invariant action $\Gamma_I^{(0)}$ the term $\mathcal{L}_{BRST} = \sum_\Phi \Phi^* s\Phi$, coupling the composite operators $s\Phi$ to the BRST-invariant external sources (usually called antifields) Φ^* to obtain the new action $\Gamma^{(0)} = \Gamma_I^{(0)} + \sum_\Phi \Phi^* s\Phi$.

From the point of view of the pinch technique, there are considerable conceptual and operational advantages to be gained from employing this formalism [5]. To begin with, the use of antifields [6], which represent a core ingredient of the BV formalism itself, streamlines the derivation of Slavnov–Taylor identities, expressing them in terms of auxiliary functions that can be constructed using a well-defined set of Feynman rules (derived from $\mathcal{L}_{BRST}$). In addition, the formulation of the BFM within the BV formalism gives rise to important all-order identities, to be called *background-quantum identities* [5, 7], relating the BFM n-point functions to the corresponding conventional n-point functions in the R_ξ gauges. These identities are realized by means of unphysical Green's functions involving antifields and background sources. The prime example of such an identity is given in Eq. (4.35), the most important equation in this chapter: the conventional and PT gluon propagators, $\Delta(q)$ and $\widehat{\Delta}(q)$, respectively, are related by means of the auxiliary two-point function $G(q)$.

The basic observation that makes the background-quantum identities so useful is that the unphysical Green's functions appearing in them (such as the $G(q)$ in Eq. (4.35)) are related to the auxiliary Green's functions appearing in the Slavnov–Taylor identities by simple expressions. Put simply, the parts of the diagrams exchanged during the pinching process, namely, the terms containing the unphysical vertices, are involved in relations connecting conventional and BFM n-point functions. In Eq. (4.44), for example, the function $G(q)$ is fully determined in terms of quantities that are defined in the context of the conventional formalism, without recourse to antifields or to the Feynman rules stemming from $\mathcal{L}_{\mathrm{BRST}}$. This, in turn, allows for a direct comparison of the PT and BFM Green's functions: a PT Green's function is obtained from the conventional one by removing the pinching parts; but in doing so, one is practically generating the corresponding background-quantum identity, which carries over to the BFM Green's function.

The background-quantum identities play a central role in the entire PT program for one additional reason. As is already evident at the two-loop level, the two-loop PT gluon self-energy is composed of Feynman diagrams involving the conventional one-loop gluon self-energy and not the one-loop PT self-energy. This might suggest at first that one cannot arrive (eventually) at a genuine Schwinger–Dyson equation involving the same unknown quantity on both sides, i.e., either $\Pi_{\mu\nu}$ or $\widehat{\Pi}_{\mu\nu}$. There is, however, a way around this: the nonperturbative version of the background-quantum identities, and most important, that of Eq. (4.35), allows one to convert the new SD series into a dynamical equation involving either the conventional or the BFM gluon self-energy only. As we will see in Chapter 6, this is instrumental for the success of the entire approach.

In addition to Eq. (4.35), the second identity of Eq. (4.50) captures another important result of this chapter. It turns out that in the Landau gauge (only), the function $G(q)$ coincides with the so-called Kugo-Ojima function [8], $u(q)$, defined in Eq. (4.51). The latter function, and in particular its value in the deep infrared, is intimately connected with the Kugo-Ojima confinement criterion [8], which requires that $u(0) = -1$. The identity of Eq. (4.50) relates the Kugo-Ojima function with the inverse of the ghost-dressing function, $F(q)$, and an auxiliary function, $L(q)$; the latter can be shown to vanish in the deep infrared. The power of Eq. (4.50) is in that it relates the value $u(0)$ and, hence, the fulfillment or nonfulfillment of the corresponding confinement criterion, with the value of $F(0)$: the Kugo–Ojima criterion is satisfied provided that $F(0)$ diverges. However, as we will discuss briefly in Chapter 6, this is not how QCD really works. Both lattice simulations and Schwinger–Dyson equations reveal that $F(0)$ is actually finite – a fact that can ultimately be traced back to the dynamical generation of a gluon mass.

4.1 An overview of the Batalin–Vilkovisky formalism

4.1.1 Green's functions: Conventions

The 1PI Green's functions of any theory are defined in terms of the time-ordered product of interacting fields[1] as

$$\Gamma_{\Phi_1\cdots\Phi_n}(x_1,\ldots,x_n) = \langle T[\Phi_1(x_1)\cdots\Phi_n(x_n)]\rangle^{1\text{PI}} \qquad (4.1)$$

and can be efficiently constructed through a generating functional, which in Fourier space reads

$$\Gamma[\Phi] = \sum_{n=0}^{\infty}\frac{(-i)^n}{n!}\int\prod_{i=0}^{n}d^4p_i\,\delta^4\left(\sum_{j=1}^{n}p_j\right)\Phi_1(p_1)\cdots\Phi_n(p_n)\Gamma_{\Phi_1\cdots\Phi_n}(p_1,\ldots,p_n).$$

$$(4.2)$$

In the preceding formula, the field $\Phi_i(p_i)$ represents the Fourier transform of the field $\Phi_i(x_i)$, with p_i its (in-going) momentum. Then, in terms of the generating functional $\Gamma[\Phi]$, all the (momentum-space) 1PI Green's functions can be obtained by means of functional differentiation:

$$\Gamma_{\Phi_1\cdots\Phi_n}(p_1,\ldots,p_n) = i^n\left.\frac{\delta^n\Gamma}{\delta\Phi_1(p_1)\delta\Phi_2(p_2)\cdots\delta\Phi_n(p_n)}\right|_{\Phi_i=0}. \qquad (4.3)$$

Our convention on the external momenta is summarized in Figure 4.1. From the definition given in Eq. (4.3), it follows that the Green's functions $i^{-n}\Gamma_{\Phi_1\cdots\Phi_n}$ are simply given by the corresponding Feynman diagrams in Minkowski space.

The Green's functions generated by $\Gamma[\Phi]$ can be joined together by full propagators to construct higher-point connected amplitudes, ultimately giving rise to the S-matrix elements of the theory. However, they are by no means a complete set, for the nonlinearity of the BRST transformation of NAGTs implies that auxiliary Green's functions involving ghost fields will appear in the Slavnov–Taylor identities. The latter are precisely the Green's functions with which we have always been working when applying the PT algorithm and constitute the functions we will thoroughly study in the rest of this chapter.

4.1.2 The Batalin–Vilkovisky formalism

The Batalin-Vilkovisky formalism [1, 2, 3, 4] is a powerful quantization scheme that allows us to address in an effective way several aspects of very general gauge

---

[1] We let Φ run over all the fields A, $\psi,\bar\psi$, c, $\bar c$, and B. Sometimes the fields appearing in the gauge-invariant Lagrangian will be collectively indicated as ϕ.

Table 4.1. *Ghost charge, statistics (B for Bose, F for Fermi), and mass dimension of the QCD fields and antifields*

	A_μ^m	ψ_f^i	$\bar\psi_f^i$	c^m	$\bar c^m$	B^m	A_μ^{*m}	ψ_f^{*i}	$\bar\psi_f^{*i}$	c^{*m}	$\bar c^{*m}$
Ghost charge	0	0	0	1	−1	0	−1	−1	−1	−2	0
Statistics	B	F	F	F	F	B	F	B	B	B	B
Dimension	1	$\frac{3}{2}$	$\frac{3}{2}$	0	2	2	3	$\frac{5}{2}$	$\frac{5}{2}$	4	2

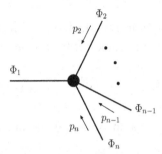

Figure 4.1. Our conventions for the (1PI) Green's functions $\Gamma_{\Phi_1\cdots\Phi_n}(p_1,\ldots,p_n)$. All momenta $p_2,\ldots,p_n$ are assumed to be incoming and are assigned to the corresponding fields starting from the rightmost one. The momentum of the leftmost field Φ_1 is determined through momentum conservation ($\sum_i p_i = 0$) and will be suppressed.

theories (e.g., their quantization, renormalization, and symmetry violation due to quantum effects), including those with open or reducible gauge symmetry algebras.

The Batalin-Vilkovisky formalism starts by introducing for each field Φ a corresponding antifield, to be denoted by Φ^*. The antifield Φ^* has opposite statistics with respect to Φ as well as a ghost charge gh(Φ^*), which is related to the ghost charge gh(Φ) of the corresponding field Φ by gh(Φ^*) = $-1 -$ gh(Φ). The ghost charges, statistics, and mass dimension of the various QCD fields and antifields are summarized in Table 4.1.

Next, one adds to the original gauge-invariant action $\Gamma_I^{(0)}[\phi]$ (with ϕ representing the physical QCD fields A, ψ, and $\bar\psi$), a term coupling the antifields with the BRST variation of the corresponding fields; then one obtains the new action

$$\Gamma^{(0)}[\Phi,\Phi^*] = \Gamma_I^{(0)}[\phi] + \sum_\Phi \Phi^* s\Phi, \qquad (4.4)$$

where

$$\sum_\Phi \Phi^* s\Phi = \int d^4x \left[A_\mu^{*m}(\partial^\mu c^m + gf^{mnr}A_n^\mu c^r) - \frac{1}{2}gf^{mnr}c^{*m}c^nc^r + \bar{c}^{*m}B^m \right.$$

$$\left. + ig\bar{\psi}_f^{*i}c^mt_{ij}^m\psi_f^j - igc^m\bar{\psi}_f^i t_{ij}^m\psi_f^{*j} \right], \tag{4.5}$$

(f is the quark flavor index).[2]

The action (4.4) satisfies the master equation[3]

$$\int d^4x \sum_\Phi \frac{\delta\Gamma^{(0)}}{\delta\Phi^*}\frac{\delta\Gamma^{(0)}}{\delta\Phi} = 0. \tag{4.6}$$

In fact, on one hand, the terms in Eq. (4.4) that are independent of the antifields are zero because of the gauge invariance of the action

$$\sum_\phi s\phi\frac{\delta\Gamma_I^{(0)}}{\delta\phi} = \int d^4x(s\Gamma_I^{(0)}[\phi]) = 0. \tag{4.7}$$

On the other hand, the terms linear in the antifields vanish because of the nilpotency of the BRST operator:

$$\sum_{\Phi'} s\Phi'\frac{\delta(s\Phi)}{\delta\Phi'} = \int d^4x \sum_\Phi s^2\Phi = 0. \tag{4.8}$$

The BRST symmetry is crucial for endowing a (gauge) theory with a unitary S-matrix and gauge-independent physical observables; therefore, it must be implemented to all orders. For achieving this, we establish the quantum corrected version of the master equation (4.6) in the form of the Slavnov–Taylor identity functional

$$\mathcal{S}(\Gamma)[\Phi] = \int d^4x \sum_\Phi \frac{\delta\Gamma}{\delta\Phi^*}\frac{\delta\Gamma}{\delta\Phi} = 0, \tag{4.9}$$

where $\Gamma[\Phi, \Phi^*]$ is now the effective action. In the pure gluodynamics sector, the Slavnov–Taylor functional is given by[4]

$$\mathcal{S}(\Gamma)[\Phi] = \int d^4x \left[\frac{\delta\Gamma}{\delta A_m^{*\mu}}\frac{\delta\Gamma}{\delta A_\mu^m} + \frac{\delta\Gamma}{\delta c^{*m}}\frac{\delta\Gamma}{\delta c^m} + B^m\frac{\delta\Gamma}{\delta\bar{c}^m} \right]. \tag{4.10}$$

---

[2] It can be easily shown that this new action is physically equivalent to the gauge-fixed QCD action because the two are related by a canonical transformation [9].

[3] Our derivatives are all left derivatives, e.g., $\delta(ab) = (\delta a)b + (-1)^{\epsilon_a}a\delta b$, with ϵ_a being the Grassmann parity of a.

[4] Quarks can be easily taken into account by adding to the Slavnov–Taylor functional (4.10) the term

$$\int d^4x \left[\frac{\delta\Gamma}{\delta\psi_f^{*i}}\frac{\delta\Gamma}{\delta\bar{\psi}_f^i} + \frac{\delta\Gamma}{\delta\psi_f^i}\frac{\delta\Gamma}{\delta\bar{\psi}_f^{*i}} \right].$$

The structure of the preceding master equation can be simplified by noticing that the antighost $\bar{c}^a$ and the multiplier B^a have *linear* BRST transformations and therefore do not present with the usual complications of the other QCD fields. Together with their corresponding antifields, they enter bilinearly in the action, which can be then decomposed in the sum of a *minimal* and *nonminimal* sector:

$$\Gamma_C^{(0)}[\Phi, \Phi^*] = \Gamma^{(0)}[A_\mu^m, A_\mu^{*m}, \psi, \psi_f^{*i}, \bar{\psi}_f^i, \bar{\psi}_f^{*i}, c^m, c^{*m}] + \bar{c}^{*m} B^m. \quad (4.11)$$

The last term has no effect on the master equation (4.6), which in fact is satisfied by $\Gamma^{(0)}$ alone. The fields $\{A_\mu^m, A_\mu^{*m}, \psi, \psi_f^{*i}, \bar{\psi}_f^i, \bar{\psi}_f^{*i}, c^m, c^{*m}\}$ are then often called *minimal variables*, whereas $\{\bar{c}^m, B^m\}$ are referred to as *trivial* or *contractible pairs*.[5] Then, in the minimal sector, the *reduced* Slavnov–Taylor functional is given by the complete functional of Eq. (4.10) once the last term $B^m \delta\Gamma/\delta\bar{c}^m$ is left out.

Taking functional derivatives of $\mathcal{S}(\Gamma)[\Phi]$ and setting afterward all fields and anti-fields to zero will generate the complete set of the all-order Slavnov–Taylor identities of the theory.[6] This is an exact analogy (see Eq. (4.3)) to what happens with the generating functional, where taking functional derivatives of $\Gamma[\Phi]$ and setting afterward all fields to zero generates the Green's functions of the theory. However, to reach meaningful expressions, one needs to keep in mind that (1) $\mathcal{S}(\Gamma)$ has ghost charge $+1$ and (2) functions with nonzero ghost charge vanish, for the ghost charge is a conserved quantity. Thus, to extract nonzero identities from Eq. (4.10), one needs to differentiate the latter with respect to a combination of fields containing either one ghost field or two ghost fields and one antifield. The only exception to this rule is when differentiating with respect to a ghost antifield, which needs to be compensated by three ghost fields. Specifically, identities involving one or more gauge fields are obtained by differentiating Eq. (4.10) with respect to the set of fields in which one gauge boson has been replaced by the corresponding ghost field. This is because the linear part of the BRST transformation of the gauge field is proportional to the ghost field: $s A_\mu^m|_{\text{linear}} = \partial_\mu c^m$. Finally, for obtaining Slavnov–Taylor identities involving Green's functions that contain ghost fields, one ghost field must be replaced by two ghost fields because of the quadratic nature of the BRST ghost field transformation ($s c^m \propto f^{mnr} c^n c^r$).

The last technical point to be clarified is the dependence of the Slavnov–Taylor identities on the (external) momenta. One should notice that the integral over $\mathrm{d}^4 x$ present in Eq. (4.10), together with the conservation of momentum flow of the

---

[5] Equivalently, the minimal sector action can be obtained by subtracting from the complete action the local term corresponding to the gauge–fixing Lagrangian $\mathcal{L}_{\text{GF}}$.

[6] In practice, the Slavnov–Taylor identities obtained from the reduced functional coincide with the ones obtained by the complete functional (1) after implementing the Faddeev–Popov equation described in the next section [10] and (2) taking into account that Green's functions involving unphysical fields coincide only up to constant terms proportional to the gauge-fixing parameter, e.g., $\Gamma_{A_\mu^m A_\nu^n}(q) = \Gamma_{A_\mu A_\nu}^C(q) - i\delta^{mn}\xi^{-1} q_\mu q_\nu$.

Table 4.2. *Ghost charge, statistics (B for Bose,*
F for Fermi), and mass dimension of the QCD
background fields and sources

	$\widehat{A}^m_\mu$	Ω^{*m}_μ
Ghost charge	0	1
Statistics	B	F
Dimension	1	1

Green's functions, implies that no momentum integration is left over. As a result, the Slavnov–Taylor identities will be expressed as a sum of products of (at most two) Green's functions.

The (complete) Slavnov–Taylor functional is absolutely general and need not be modified if one changes the way of gauge fixing the Lagrangian, e.g., switching from a general R_ξ gauge to the BFM. In this latter case, however, to control the dependence of the Green's functions on the background fields, some new terms, implementing the equation of motion of the background fields at the quantum level, are conventionally added to the Slavnov-Taylor functional. Specifically, one extends the BRST symmetry to the background gluon field through the relations

$$s\,\widehat{A}^m_\mu = \Omega^m_\mu \qquad s\Omega^m_\mu = 0. \tag{4.12}$$

The expression Ω^m_μ represents a (classical) vector field with the same quantum numbers as the gluon but with ghost charge $+1$ and Fermi statistics (see also Table 4.2). The dependence of the Green's functions on the background fields is then controlled by the modified Slavnov–Taylor functional

$$\mathcal{S}'(\Gamma')[\Phi] = \mathcal{S}(\Gamma')[\Phi] + \int d^4x \; \Omega^\mu_m \left[\frac{\delta\Gamma'}{\delta\widehat{A}^m_\mu} - \frac{\delta\Gamma'}{\delta A^m_\mu} \right] = 0, \tag{4.13}$$

where Γ' denotes the effective action that depends on the background sources Ω^m_μ (with $\Gamma \equiv \Gamma'|_{\Omega=0}$), and $\mathcal{S}(\Gamma')[\Phi]$ is the Slavnov–Taylor identity functional of Eq. (4.10). Differentiation of the Slavnov–Taylor functional (4.13) with respect to the background source and background or quantum fields will then provide the background-quantum identities relating 1PI Green's functions involving background fields to the ones involving quantum fields and already briefly discussed.[7]

Finally, the background gauge invariance of the BFM effective action implies that Green's functions involving background fields satisfy linear Ward identities when

---

[7] As it happens, for Slavnov–Taylor identities, background-quantum identities are not deformed by the renormalization procedure. The new background variables enter, in fact, as BRST doublets, and they cannot change the cohomology of the linearized Slavnov–Taylor operator [11].

contracted with the momentum corresponding to a background leg. These Ward identities are generated by taking functional differentiations of the Ward identity functional

$$\mathcal{W}_\vartheta[\Gamma'] = \int d^4x \sum_\Phi (\delta_\vartheta \Phi) \frac{\delta\Gamma'}{\delta\Phi} = 0, \qquad (4.14)$$

where $\delta_\vartheta \Phi$ are given by the BRST transformation of the corresponding fields when replacing ghosts with the local infinitesimal parameters $\vartheta^a(x)$ corresponding to the $SU(3)$ generators t^a; the background transformations of the antifields $\delta_\vartheta \Phi^*$ coincide with the gauge transformations of the corresponding quantum fields according to their specific representation. To obtain the Ward identity satisfied by the Green's functions involving background gluons $\widehat{A}$, one has then to differentiate the functional (4.14) with respect to the corresponding parameter ϑ.

4.2 Examples

4.2.1 Slavnov–Taylor identities

One of the most useful Slavnov–Taylor identities in a PT context is definitely the one satisfied by the three-gluon vertex. The textbook derivation of this identity has been sketched in Chapter 1 (see Section 1.5.1); here we derive the same identity within the Batalin-Vilkovisky formalism.

According to the rules stated in the previous section, the three-gluon Slavnov–Taylor identity can be obtained by considering the following functional differentiation:

$$\frac{\delta^3 S(\Gamma)}{\delta c^a(q)\delta A_\mu^m(k_1)\delta A_\nu^n(k_2)}\bigg|_{\Phi,\Phi^*=0} = 0, \qquad (4.15)$$

which gives the result

$$-\Gamma_{c^a A_{a'}^{*\alpha}}(-q)\Gamma_{A_\alpha^{a'} A_\mu^m A_\nu^n}(k_1, k_2) = \Gamma_{c^a A_\nu^n A_d^{*\gamma}}(k_2, k_1)\Gamma_{A_\gamma^d A_\mu^m}(k_1)$$

$$+ \Gamma_{c^a A_\mu^m A_d^{*\gamma}}(k_1, k_2)\Gamma_{A_\gamma^d A_\nu^n}(k_2). \qquad (4.16)$$

To further simplify the preceding identity, we need to resort to the so-called Faddeev–Popov (or ghost) equation, which describes the action of longitudinal momenta when acting on auxiliary Green's functions. To derive this equation in the R_ξ gauges, one observes that in the QCD action, the only term proportional to the antighost fields comes from the Faddeev–Popov Lagrangian density, which can be rewritten as

$$\mathcal{L}_{\text{FPG}}^{R_\xi} = -\bar{c}^m \partial^\mu (s A_\mu^m) = -\bar{c}^m \partial^\mu \frac{\delta\Gamma}{\delta A_\mu^{*m}}. \qquad (4.17)$$

Differentiation of the action with respect to $\bar{c}^m$ then yields the Faddeev–Popov equation in the form of the identity

$$\frac{\delta\Gamma}{\delta\bar{c}^m} + \partial^\mu \frac{\delta\Gamma}{\delta A^{*m}_\mu} = 0, \tag{4.18}$$

so that, taking the Fourier transform, we arrive at

$$\frac{\delta\Gamma}{\delta\bar{c}^m} + iq^\mu \frac{\delta\Gamma}{\delta A^{*m}_\mu} = 0. \tag{4.19}$$

Thus, in the R_ξ case, Eq. (4.19) amounts to the simple statement that the contraction of a leg corresponding to a gluon antifield (A^{*m}_μ) by its own momentum (q^μ) converts it to an antighost leg ($\bar{c}^m$). Notice that the Faddeev–Popov equation depends crucially on the form of the ghost Lagrangian, which in turn depends on the gauge-fixing function. In the presence of background gluons and sources, the presence of extra terms in the BFM gauge-fixing function will modify Eq. (4.18), which will read, in this case,

$$\frac{\delta\Gamma'}{\delta\bar{c}^m} + \left(\widehat{\mathcal{D}}^\mu \frac{\delta\Gamma'}{\delta A^*_\mu}\right)^m - (\mathcal{D}^\mu \Omega_\mu)^m = 0. \tag{4.20}$$

Notice that by setting the background field and source to zero, one correctly recovers the R_ξ equation (4.18).

Let us now differentiate Eq. (4.19) with respect to a ghost field c; after setting the fields and antifields to zero, we get

$$\Gamma_{c^m \bar{c}^n}(q) + iq^\nu \Gamma_{c^m A^{*n}_\nu}(q) = 0, \tag{4.21}$$

which can be used to relate the auxiliary function $\Gamma_{c^m A^{*n}_\nu}(q)$ with the full ghost propagator $D^{mn}(q)$. Owing to Lorentz invariance, we can in fact write $\Gamma_{c^m A^{*n}_\nu}(q) = q_\nu \Gamma_{c^m A^{*n}}(q)$, and therefore

$$\Gamma_{c^m \bar{c}^n}(q) = -iq^\nu \Gamma_{c^m A^{*n}_\nu}(q) = -iq^2 \Gamma_{c^m A^{*n}}(q). \tag{4.22}$$

On the other hand, observing that $iD^{mr}(q)\Gamma_{c^r \bar{c}^n}(q) = \delta^{mn}$, we get the announced relation

$$\Gamma_{c^m A^{*n}_\nu}(q) = q_\nu \Gamma_{c^m A^{*n}}(q) = q_\nu [q^2 D^{mn}(q)]^{-1}, \tag{4.23}$$

which, inserted back into Eq. (4.16), gives

$$q^\alpha \Gamma_{A^a_\alpha A^m_\mu A^n_\nu}(k_1, k_2) = [q^2 D^{aa'}(q)] \left\{ \Gamma_{c^{a'} A^n_\nu A^{*\gamma}_d}(k_2, k_1) \Gamma_{A^d_\gamma A^m_\mu}(k_1) \right.$$

$$\left. + \Gamma_{c^{a'} A^m_\mu A^{*\gamma}_d}(k_1, k_2) \Gamma_{A^d_\gamma A^n_\nu}(k_2) \right\}. \tag{4.24}$$

To get the Slavnov–Taylor identity in the same form of Eq. (4.16), one factors out the color structure $ig f^{amn}$, uses the relation

$$\Gamma_{A^a_\alpha A^b_\beta}(q) = (\Delta^{-1})^{ab}_{\alpha\beta}(q) - i\delta^{ab} q_\alpha q_\beta = i\delta^{ab} P_{\alpha\beta}(q)\Delta^{-1}(q^2), \qquad (4.25)$$

and identifies $H_{\mu\gamma}(k_1, k_2)$ with $\Gamma_{cA_\mu A^*_\gamma}(k_1, k_2)$. Notice then that the relation between H and the gluon-ghost vertex is automatic, being a manifestation of the Faddeev-Popov equation; in fact, by differentiating Eq. (4.19) with respect to a gluon and a ghost field, we get the identity

$$\Gamma_{c^r A^n_\nu \bar{c}^m}(k, q) + iq^\mu \Gamma_{c^r A^n_\nu A^{*m}_\mu}(k, q) = 0. \qquad (4.26)$$

We conclude by observing that within the Batalin–Vilkovisky formalism, one can also obtain Slavnov–Taylor identities for kernels appearing e.g., in the usual skeleton expansion of QCD Green's functions. To do so, one decomposes the kernel under scrutiny in terms of 1PI Green's functions, calculates the corresponding Slavnov–Taylor identities by taking functional differentiation of the functional (4.10) with respect to suitable fields' combinations, and then puts together all the pieces. For example, the Batalin-Vilkovisky formalism version of the Slavnov–Taylor identity satisfied by the fundamental PT kernel $\mathcal{K}_{AA\psi\bar{\psi}}$, identified in the previous chapter, reads (suppressing the quark flavor and color indices)

$$k_1^\mu \mathcal{K}_{A^m_\mu A^n_\nu \psi\bar{\psi}}(k_2, p_2, -p_1) = [k_1^2 D^{mm'}(k_1)]$$

$$\times \left\{ \Gamma_{c^{m'} A^n_\nu A^{*\gamma}_d}(k_2, -k_1 - k_2)\Gamma_{A^d_\gamma \psi\bar{\psi}}(p_2, -p_1) \right.$$

$$+ \Gamma_{\psi\bar{\psi}}(p_1)\mathcal{K}_{A^n_\nu \psi c^{m'} \bar{\psi}^*}(p_2, k_1, -p_1)$$

$$+ \mathcal{K}_{A^n_\nu \psi^* \bar{\psi} c^{m'}}(p_2, -p_1, k_1)\Gamma_{\psi\bar{\psi}}(p_2)$$

$$\left. + \Gamma_{c^{m'} A^{*\gamma}_d \psi\bar{\psi}}(k_2, p_2, -p_1)\Gamma_{A^d_\gamma A^n_\nu}(k_2) \right\}, \qquad (4.27)$$

where we have defined the auxiliary kernels

$$\mathcal{K}_{A^n_\nu \psi c^{m'} \bar{\psi}^*}(p_2, k_1, -p_1) = \Gamma_{A^n_\nu \psi c^{m'} \bar{\psi}^*}(p_2, k_1, -p_1) \qquad (4.28)$$

$$- i\Gamma_{\psi c^{m'} \bar{\psi}^*}(k_1, -p_1)S(\ell)\Gamma_{A^n_\nu \psi\bar{\psi}}(p_2, -\ell)$$

$$\mathcal{K}_{A^n_\nu \psi^* \bar{\psi} c^{m'}}(p_2, -p_1, k_1) = \Gamma_{A^n_\nu \psi^* \bar{\psi} c^{m'}}(p_2, -p_1, k_1) \qquad (4.29)$$

$$- i\Gamma_{A^n_\nu \psi\bar{\psi}}(\ell, -p_1)S(\ell)\Gamma_{\psi^* \bar{\psi} c^{m'}}(-\ell, k_1).$$

4.2.2 Background-quantum identities

The first background-quantum identity we can construct is the one relating the conventional with the BFM gluon self-energies. To this end, consider the following functional differentiations ($q + p = 0$):

$$\frac{\delta^2 S'\left(\Gamma'\right)}{\delta \Omega_\alpha^a(p)\delta A_\beta^b(q)}\bigg|_{\Phi,\Phi^*,\Omega=0} = 0; \qquad \frac{\delta^2 S'\left(\Gamma'\right)}{\delta \Omega_\alpha^a(p)\delta \widehat{A}_\beta^b(q)}\bigg|_{\Phi,\Phi^*,\Omega=0} = 0, \quad (4.30)$$

which give the relations

$$i\Gamma_{\widehat{A}_\alpha^a A_\beta^b}(q) = \left[ig_\alpha^\gamma \delta^{ad} + \Gamma_{\Omega_\alpha^a A_d^{*\gamma}}(q)\right]\Gamma_{A_\gamma^d A_\beta^b}(q) \qquad (4.31)$$

$$i\Gamma_{\widehat{A}_\alpha^a \widehat{A}_\beta^b}(q) = \left[ig_\alpha^\gamma \delta^{ad} + \Gamma_{\Omega_\alpha^a A_d^{*\gamma}}(q)\right]\Gamma_{A_\gamma^d \widehat{A}_\beta^b}(q). \qquad (4.32)$$

We can now combine Eqs. (4.31) and (4.32) such that the two-point function mixing background and quantum fields drop out; then, using the transversality of the gluon two-point function Γ_{AA}, we get the background-quantum identity

$$i\Gamma_{\widehat{A}_\alpha^a \widehat{A}_\beta^b}(q) = i\Gamma_{A_\alpha^a A_\beta^b}(q) + 2\Gamma_{\Omega_\alpha^a A_d^{*\gamma}}(q)\Gamma_{A_\gamma^d A_\beta^b}(q)$$

$$- i\Gamma_{\Omega_\alpha^a A_d^{*\gamma}}(q)\Gamma_{A_\gamma^d A_\epsilon^e}(q)\Gamma_{\Omega_\beta^b A_\epsilon^{*\epsilon}}(q). \qquad (4.33)$$

This identity can be rewritten in a more suggestive form by trading the two-point functions Γ_{AA} and $\Gamma_{\widehat{A}\widehat{A}}$ for the corresponding (inverse) propagators and setting[8]

$$\Gamma_{\Omega_\alpha^a A_\gamma^{*d}}(q) = i\delta^{ad}\left[g_{\alpha\gamma} G(q^2) + \frac{q_\alpha q_\gamma}{q^2} L(q^2)\right]. \qquad (4.34)$$

One then gets

$$\widehat{\Delta}^{-1}(q^2) = \left[1 + G(q^2)\right]^2 \Delta^{-1}(q^2). \qquad (4.35)$$

As we know from Chapter 1, the quantity $\widehat{\Delta}(q^2)$ appearing on the left-hand side (lhs) of the preceding equation captures the running of the QCD beta function, exactly as happens with the QED vacuum polarization.[9] For example, to lowest order, one can use the closed expression (4.44) to get (in the Landau gauge)

$$1 + G(q^2) = 1 + \frac{9}{4}\frac{C_A g^2}{48\pi^2} \ln\left(\frac{q^2}{\mu^2}\right)$$

$$\Delta^{-1}(q^2) = q^2\left[1 + \frac{13}{2}\frac{C_A g^2}{48\pi^2} \ln\left(\frac{q^2}{\mu^2}\right)\right], \qquad (4.36)$$

thus recovering the well-known result

$$\widehat{\Delta}^{-1}(q^2) = q^2\left[1 + bg^2 \ln\left(\frac{q^2}{\mu^2}\right)\right], \qquad (4.37)$$

---

[8] From Tables 4.1 and 4.2, one sees that the dimensions of the gluon antifield A^* and background source Ω are, respectively, 3 and 1; then simple power counting shows that the (logarithmically) divergent part of $\Gamma_{\Omega_\alpha^a A_\gamma^{*d}}(q)$ can be proportional to $g_{\alpha\gamma}$ only, whereas the longitudinal form factor $L(q^2)$ is ultraviolet finite.

[9] Recall that this is a fundamental property of the BFM gluon self-energy, valid for every value of the (quantum) gauge-fixing parameter [12].

where b is the usual one-loop beta function coefficient of Eq. (1.69). As we will see in Chapter 6, Eq. (4.35) plays a pivotal role in the derivation of a new set of QCD Schwinger–Dyson equations [13, 14] that can be truncated in a manifestly gauge-invariant way [15].

Other background-quantum identities involving, e.g., the quark-gluon vertex and the three-gluon vertex read

$$i\Gamma_{\widehat{A}^a_\alpha \psi \bar{\psi}}(p_2, -p_1) = [ig^\gamma_\alpha \delta^{ad} + \Gamma_{\Omega^a_\alpha A^{*\gamma}_d}(q)]\Gamma_{A^d_\gamma \psi \bar{\psi}}(p_2, -p_1)$$
$$+ \Gamma_{\psi^* \bar{\psi} \Omega^a_\alpha}(-p_1, q)\Gamma_{\psi \bar{\psi}}(p_2) \qquad (4.38)$$
$$+ \Gamma_{\psi \bar{\psi}}(p_1)\Gamma_{\psi \Omega^a_\alpha \bar{\psi}^*}(q, -p_1)$$

$$i\Gamma_{\widehat{A}^a_\alpha A^r_\rho A^s_\sigma}(p_2, -p_1) = [ig^\gamma_\alpha \delta^{ad} + \Gamma_{\Omega^a_\alpha A^{*\gamma}_d}(q)]\Gamma_{A^d_\gamma A^r_\rho A^s_\sigma}(p_2, -p_1)$$
$$+ \Gamma_{\Omega^a_\alpha A^s_\sigma A^{*\gamma}_d}(-p_1, p_2)\Gamma_{A^d_\gamma A^r_\rho}(p_2) \qquad (4.39)$$
$$+ \Gamma_{\Omega^a_\alpha A^r_\rho A^{*\gamma}_d}(p_2, -p_1)\Gamma_{A^d_\gamma A^s_\sigma}(p_1).$$

Notice first that the auxiliary function appearing in square brackets on the right-hand side (rhs) of the preceding identities is always $\Gamma_{\Omega A^*}$: this is at the root of the process independence of the PT algorithm. Second, observe that in more general identities other than the two-point one, the form factor $L(q^2)$ is also relevant.

4.2.3 Closed expressions for auxiliary functions

From the PT point of view, it would be not enough to be able to derive the Slavnov–Taylor identities and the background-quantum identities in the form given earlier. In fact, one is really striving for a formal link between the Slavnov–Taylor identities, which are triggered by the action of the longitudinal momenta, and the background-quantum identities, which relate Green's functions written in the conventional (R_ξ) and BFM gauges.

The key observation that makes this link possible is that one can always replace an antifield or BFM source with the corresponding BRST composite operator to which it is coupled. This means that we can use the replacements[10] (see Figure 4.2)

$$A^{*a}_\alpha(q) \rightarrow -i\Gamma^{(0)}_{c^{e'} A^{n'}_{\nu'} A^{*a}_\alpha} \int_{k_1} \Delta^{\nu'\nu}_{n'n}(k_2)D^{e'e}(k_1)\cdots \qquad (4.40)$$

$$\Omega^a_\alpha(q) \rightarrow -i\Gamma^{(0)}_{\Omega^a_\alpha A^{n'}_{\nu'} \bar{c}^{e'}} \int_{k_1} \Delta^{\nu'\nu}_{n'n}(k_2)D^{e'e}(k_1)\cdots, \qquad (4.41)$$

---

[10] For consistency with the definition (4.3), we use here (and later in Chapter 6) a definition of the full gluon propagator in which the rhs of Eq. (1.25) corresponds to $-\Delta_{\alpha\beta}$; this will not affect the inverse propagator, which will now be determined by the equation $i\Delta_{\alpha\mu}(\Delta^{-1})^{\mu\beta} = g^\beta_\alpha$. Full gluon lines will then contribute a factor of $i\Delta_{\alpha\beta}$ to the corresponding amplitude (ghost lines will contribute an iD factor).

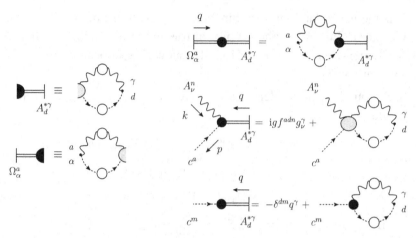

Figure 4.2. (*Left*) Expansion of the gluon antifield and BFM source in terms of the corresponding composite operators. Notice that if the antifield or the BFM sources are attached to a 1PI vertex, such an expansion will in general convert the 1PI vertex into a (connected) Schwinger–Dyson kernel. (*Right*) The corresponding expansion of the two-point function $\Gamma_{\Omega A^*}$ and the three-point function Γ_{cAA^*}.

to write, e.g.,[11]

$$-\Gamma_{c^m A_d^{*\gamma}}(q) = -\delta^{dm} q_\gamma + g f^{dne} \int_{k_1} D(k_1) \Delta^{\gamma v}(k_2) \Gamma_{c^m A_v^n \bar{c}^e}(k_2, k_1), \qquad (4.42)$$

$$i\Gamma_{c^a A_v^n A_d^{*\gamma}}(k, q) = ig f^{adn} g_v^\gamma - ig f^{edr} \int_{k_1} D(k_1) \Delta^{\gamma\rho}(k_2) \mathcal{K}_{c^a A_v^n A_\rho^r \bar{c}^e}(k, k_2, k_1), \qquad (4.43)$$

$$-\Gamma_{\Omega_\alpha^a A_d^{*\gamma}}(q) = g f^{aen} \int_{k_1} D(k_1) \Delta_\alpha^v(k_2) \Gamma_{c^e A_v^n A_d^{*\gamma}}(k_2, -q). \qquad (4.44)$$

Equation (4.43) then shows explicitly the equivalence between Γ_{cAA^*} and the function H introduced earlier (compare also Figures 1.11 and 4.2).

The systematic use of this expansion to write closed expressions for the auxiliary functions appearing in the Slavnov–Taylor identities as well as in the background-quantum identities allows one to unveil a pattern that will be exploited when applying the pinch technique to the Schwinger–Dyson equations of QCD: the auxiliary functions appearing in the background-quantum identity satisfied by a particular Green's function can be written in terms of kernels appearing in the Slavnov–Taylor identities triggered when the PT procedure is applied to that same Green's function.

---

[11] The expansion of Eqs. (4.40) and (4.41) is, of course, not valid at tree level, which must be explicitly accounted for when present.

4.2.4 A special case: The (background) Landau gauge

When choosing to quantize the theory in the background Landau gauge $(\widehat{\mathcal{D}}^\mu A_\mu)^m = 0$, a new local equation (called the *antighost equation*) appears [16]:

$$\frac{\delta\Gamma}{\delta c^m} - \left(\widehat{\mathcal{D}}^\mu \frac{\delta\Gamma}{\delta\Omega_\mu}\right)^m - (\mathcal{D}^\mu A_\mu^*)^m - f^{mnr}c^{*n}c^r + f^{mnr}\frac{\delta\Gamma}{\delta B^n}\bar{c}^r = 0. \quad (4.45)$$

This equation fully constrains the dynamics of the ghost field c and implies that the latter will not get an independent renormalization constant. To see this, let us differentiate Eq. (4.20) with respect to a ghost field and a background source to get (after a Fourier transform)

$$\Gamma_{c^m\bar{c}^n}(q) = -iq^\nu\Gamma_{c^m A_\nu^{*n}}(q)$$
$$\Gamma_{\bar{c}^n\Omega_\mu^m}(q) = q_\mu\delta^{mn} - iq^\nu\Gamma_{\Omega_\mu^m A_\nu^{*n}}(q). \quad (4.46)$$

On the other hand, differentiating the antighost equation (4.45) with respect to a gluon antifield and an antighost, one gets

$$\Gamma_{c^m A_\nu^{*n}}(q) = q_\nu\delta^{mn} - iq^\mu\Gamma_{\Omega_\mu^m A_\nu^{*n}}(q)$$
$$\Gamma_{c^m\bar{c}^n}(q) = -iq^\mu\Gamma_{\bar{c}^a\Omega_\mu^m}(q). \quad (4.47)$$

Next, contracting the first equation in Eq. (4.47) with q^ν, and making use of the first equation in Eq. (4.46), we see that the dynamics of the ghost sector are entirely captured by the $\Gamma_{\Omega_\mu A_\nu^*}$ auxiliary function because

$$\Gamma_{c\bar{c}}(q) = -iq^2 - q^\mu q^\nu\Gamma_{\Omega_\mu A_\nu^*}(q). \quad (4.48)$$

Introducing the Lorentz decompositions

$$\Gamma_{cA_\mu^*}(q) = q_\mu C(q^2); \qquad \Gamma_{\bar{c}\Omega_\mu}(q) = q_\mu E(q^2), \quad (4.49)$$

we find that Eq. (4.48), together with the last equations of Eqs. (4.46) and (4.47), gives the identities [16, 17]

$$C(q^2) = E(q^2) = F^{-1}(q^2)$$
$$F^{-1}(q^2) = 1 + G(q^2) + L(q^2), \quad (4.50)$$

where $F(q^2)$ is the so-called ghost dressing function (with $D(q^2) = iF(q^2)/q^2$ being the ghost propagator).

In addition, in this gauge, one can prove that the form factor G coincides with the well-known Kugo–Ojima function $u(q^2)$ [8], defined (in Euclidean space) through the two-point composite operator function

$$\int d^4x\, e^{-iq\cdot(x-y)}\langle T[(\mathcal{D}_\mu c)_x^m (\mathcal{D}_\mu\bar{c})_y^n]\rangle = -\frac{q_\mu q_\nu}{q^2}\delta^{mn} + P_{\mu\nu}(q)\delta^{mn}u(q^2). \quad (4.51)$$

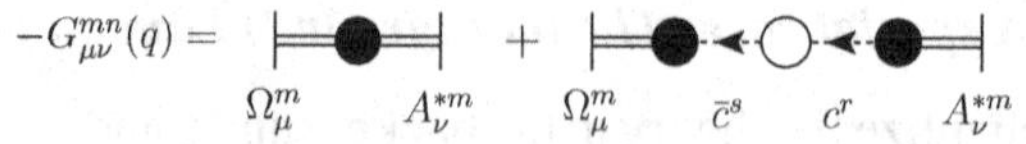

$$-G_{\mu\nu}^{mn}(q) = \quad\underset{\Omega_{\mu}^{m}\qquad A_{\nu}^{*m}}{\rule{0pt}{0pt}} \quad + \quad \underset{\Omega_{\mu}^{m}\qquad \bar{c}^{s}\quad c^{r}\qquad A_{\nu}^{*m}}{\rule{0pt}{0pt}}$$

Figure 4.3. Connected components contributing to the function $\mathcal{G}_{\mu\nu}^{mn}(q)$.

In fact, in the background Landau gauge, the function appearing on the lhs of the preceding equation is precisely given by

$$-\mathcal{G}_{\mu\nu}^{mn}(q) = \frac{\delta^2 W}{\delta\Omega_\mu^m \delta A_\nu^{*n}}, \tag{4.52}$$

where W is the generator of the connected Green's functions and the two connected diagrams contributing to $\mathcal{G}_{\mu\nu}$ are shown in Figure 4.3. Factoring out the color structure and making use of the identities (4.50), one has

$$\mathcal{G}_{\mu\nu}(q) = \Gamma_{\Omega_\mu A_\nu^*}(q) + i\Gamma_{\Omega_\mu\bar{c}}(q)D(q^2)\Gamma_{A_\nu^* c}(q)$$

$$= -i\frac{q_\mu q_\nu}{q^2} + iP_{\mu\nu}(q)G(q^2). \tag{4.53}$$

Passing to the Euclidean formulation, and comparing with Eq. (4.51), we then arrive at the announced equality[12]

$$u(q^2) = G(q^2). \tag{4.54}$$

4.3 Pinching in the Batalin–Vilkovisky framework

It is now important to make contact between the PT algorithm and the Batalin-Vilkovisky formalism. This is, of course, best done at the one-loop level, where all calculations are straightforward and it is relatively easy to compare the standard diagrammatic results with those we will be finding. Not only will this comparison help us in identifying the pieces that will be generated when applying the PT algorithm but it will also be useful for establishing the rules to distribute them among the different Green's functions appearing in the calculation.

The starting point is the embedding of the (one-loop) gluon propagator into an S-matrix element (Figure 4.4), exactly as done in Chapter 1. Then, carrying out the PT decomposition $\Gamma = \Gamma^P + \Gamma^F$ on the tree-level three-gluon vertex of diagram (b), we get for the pinching part

$$(b)^P = -gf^{amn}g_\alpha^\nu \int_{k_1} \frac{1}{k_1^2}\frac{1}{k_2^2}k_1^\mu \mathcal{K}_{A_\mu^m A_\nu^n \psi\bar\psi}^{(0)}(k_2, p_2, -p_1). \tag{4.55}$$

---

[12] Many of the results of this section turn out to be valid also in the conventional R_ξ Landau gauge [17, 18], where, however, only an integrated version of the ghost equation (4.45) is available [19].

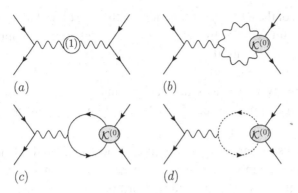

Figure 4.4. The S-matrix one-loop PT setting for constructing the gluon propagator.

On the other hand, observing that $\Gamma_{cA^*\psi\bar{\psi}}$ is zero at tree level, we find that the Slavnov–Taylor identity of Eq. (4.27) reduces to

$$k_1^\mu \mathcal{K}^{(0)}_{A_\mu^m A_\nu^n \psi\bar{\psi}}(k_2, p_2, -p_1) = -gg_\nu^\gamma f^{dmn} \Gamma^{(0)}_{A_\gamma^d \psi\bar{\psi}}(p_2, -p_1)$$

$$+ \Gamma^{(0)}_{\psi\bar{\psi}}(p_1)\mathcal{K}^{(0)}_{A_\nu^n \psi c^m \bar{\psi}^*}(p_2, k_1, -p_1)$$

$$+ \mathcal{K}^{(0)}_{A_\nu^n \psi^* \bar{\psi} c^m}(p_2, -p_1, k_1)\Gamma^{(0)}_{\psi\bar{\psi}}(p_2). \quad (4.56)$$

Now, notice that when the external legs are on shell, the last two terms of the preceding Slavnov–Taylor identity drop out by virtue of the quark equations of motion; thus, making use of Eq. (4.44), we are left with the final result

$$(b)^{\mathrm{P}} = g^2 C_A \delta^{ad} g_\alpha^\gamma \int_{k_1} \frac{1}{k_1^2} \frac{1}{k_2^2} \Gamma^{(0)}_{A_\gamma^d \psi\bar{\psi}}(p_2, -p_1)$$

$$= -\Gamma^{(1)}_{\Omega_\alpha^a A_d^{*\gamma}}(-q)\Gamma^{(0)}_{A_\gamma^d \psi\bar{\psi}}(p_2, -p_1). \quad (4.57)$$

At this point, the calculation is over, and one needs to reshuffle the pieces generated. On one hand, to get the PT (on-shell) quark-gluon vertex, one adds to the Abelian diagram (c) the Γ^{F} part of diagram (b); thus one is left with the combination $(b) + (c) - (b)^{\mathrm{P}}$ or

$$i\widehat{\Gamma}^{(1)}_{A_\alpha^a \psi\bar{\psi}}(p_2, -p_1) = i\Gamma^{(1)}_{A_\alpha^a \psi\bar{\psi}}(p_2, -p_1) + \Gamma^{(1)}_{\Omega_\alpha^a A_d^{*\gamma}}(-q)\Gamma^{(0)}_{A_\gamma^d \psi\bar{\psi}}(p_2, -p_1). \quad (4.58)$$

On the other hand, the PT self-energy will be given by adding to the diagram (a) twice the pinching contribution $(b)^{\mathrm{P}}$ (one for each vertex), i.e.,

$$\widehat{\Pi}^{(1)}_{\alpha\beta}(q) = \Pi^{(1)}_{\alpha\beta}(q) + 2i\Gamma^{(1)}_{\Omega_\alpha^a A_d^{*\gamma}}(q)\Gamma^{(0)}_{A_\gamma^d A_\beta^b}(q). \quad (4.59)$$

The comparison of the PT Green's functions with those of the background Feynman gauge is now immediate by virtue of the background-quantum identities. Equation (4.58) represents the one-loop (on-shell) version of the background-quantum identity (4.38), and, recalling that $-\Gamma_{A_\mu^m A_\nu^n} = \delta^{mn} \Pi_{\mu\nu}$, we find that Eq. (4.59) correctly reproduces the background-quantum identity (4.33). Thus we have (once again) proved the PT background Feynman gauge correspondence at one loop.

The procedure just described goes through almost unaltered when choosing the external legs of the embedding process to be gluons, rendering the (one-loop) proof of the PT process's independence effortless.

References

[1] I. A. Batalin and G. A. Vilkovisky, Feynman rules for reducible gauge theories, *Phys. Lett.* **B120** (1983) 166.
[2] I. A. Batalin and G. A. Vilkovisky, Quantization of gauge theories with linearly dependent generators, *Phys. Rev.* **D28** (1983) 2567 [Erratum, ibid. **D30** (1984) 508].
[3] I. A. Batalin and G. A. Vilkovisky, Closure of the gauge algebra, generalized Lie equations and Feynman rules, *Nucl. Phys.* **B234** (1984) 106.
[4] I. A. Batalin and G. A. Vilkovisky, Existence theorem for gauge algebra, *J. Math. Phys.* **26** (1985) 172.
[5] D. Binosi and J. Papavassiliou, Pinch technique and the Batalin-Vilkovisky formalism, *Phys. Rev.* **D66** (2002) 025024.
[6] J. Zinn-Justin, Renormalization of gauge theories, lecture given at the International Summer Institute for Theoretical Physics, Bonn, West Germany (1974).
[7] P. A. Grassi, T. Hurth, and M. Steinhauser, Practical algebraic renormalization, *Ann. Phys.* **288** (2001) 197.
[8] T. Kugo and I. Ojima, Local covariant operator formalism of non-Abelian gauge theories and quark confinement problem, *Prog. Theor. Phys. Suppl.* **66** (1979) 1.
[9] S. Weinberg, *The Quantum Theory of Fields. Vol. 2: Modern Applications* (Cambridge: Cambridge University Press, 1996).
[10] C. Itzykson and J. B. Zuber, *Quantum Field Theory* (McGraw-Hill, New York, 1980).
[11] P. A. Grassi, Stability and renormalization of Yang-Mills theory with background field method: A regularization independent proof, *Nucl. Phys.* **B462** (1996) 524.
[12] L. F. Abbott, The background field method beyond one loop, *Nucl. Phys.* **B185** (1981) 189.
[13] D. Binosi and J. Papavassiliou, Gauge-invariant truncation scheme for the Schwinger-Dyson equations of QCD, *Phys. Rev.* **D77**(R) (2008) 061702.
[14] D. Binosi and J. Papavassiliou, New Schwinger-Dyson equations for non-Abelian gauge theories, *JHEP* **0811** (2008) 063.
[15] A. C. Aguilar and J. Papavassiliou, Gluon mass generation in the PT-BFM scheme, *JHEP* **0612** (2006) 012.
[16] P. A. Grassi, T. Hurth, and A. Quadri, On the Landau background gauge fixing and the IR properties of YM Green functions, *Phys. Rev.* **D70** (2004) 105014.

[17] T. Kugo, The universal renormalization factors Z(1)/Z(3) and color confinement condition in non-Abelian gauge theory, talk given at the International Symposium on BRS Symmetry on the Occasion of Its 20th Anniversary, Kyoto, Japan (1995).

[18] K. I. Kondo, Kugo-Ojima color confinement criterion and Gribov-Zwanziger horizon condition, *Phys. Lett.* **B678** (2009) 322.

[19] O. Piguet and S. P. Sorella, Algebraic renormalization: Perturbative renormalization, symmetries and anomalies, *Lect. Notes Phys.* **M28** (1995) 1.

5

The gauge technique

The gauge technique goes back a long way [1, 2], having been introduced to deal with the Schwinger–Dyson equations of scalar electrodynamics. Its fundamental idea is to find an approximate electron-photon proper vertex expressed in terms of the electron propagator in such a way that the Ward identity is exactly satisfied. These methods can be extended to NAGTs for constructing approximate three- and four-point vertices that are gauge invariant and exactly obey the ghost-free Ward identities of the pinch technique or background-field method. In particular, we will give here several examples of the three-point proper PT vertex approximately but gauge invariantly expressed in terms of the gauge-invariant PT proper self-energy. We do not discuss the four-point vertex, which has been studied elsewhere [3].

Implementing gauge-invariant studies of NAGTs absolutely requires the gauge technique or something like it because, unless the Ward identities are satisfied, gauge invariance is an impossible goal. Outside of perturbation theory, which is not important for us, or exactly solving the Schwinger–Dyson equations, which is not possible for us, there is no other method known for systematically and usefully constructing Green's functions obeying the right Ward identities.

The gauge technique has two related potential drawbacks. The first is that there is no such thing as a unique gauge technique vertex. A gauge technique three-point vertex $\Gamma^{\mathrm{GT}}_{\mu}$ is one that depends only on two-point functions Δ and identically satisfies a Ward identity of the form $k^{\mu}\Gamma^{\mathrm{GT}}_{\mu}(k,\, p) = \Delta^{-1}(p) - \Delta^{-1}(p - k)$. To any proposed gauge technique vertex, we can always add a term $\widetilde{\Gamma}_{\mu}$ obeying $k^{\mu}\widetilde{\Gamma}_{\mu} \equiv 0$. The second drawback is that the gauge technique vertex is quantitatively but not qualitatively wrong for ultraviolet momenta because it is just the omitted, exactly conserved terms that become as important as the gauge technique terms at large momentum. How then can the gauge technique be justified? The answer is that in a theory with a mass gap,[1] these identically conserved vertex forms, such as $\tilde{\Gamma}_\mu$, vanish more rapidly by at least one power of k near zero momentum compared to the gauge technique vertex. Both these problems suggest that in gauge theories with a mass gap, which include QED and all the NAGTs of interest to us, the identically conserved terms are unimportant in the infrared and can be dropped. So the gauge technique is meant to be used strictly for small momenta. Fortunately, this is just the region of interest to us, where infrared slavery needs to be cured by nonperturbative phenomena. In practice, we are forced to use it out to momenta that are not small compared to a mass scale but comparable to it, so some quantitative error is inevitable. But we are more interested in the *qualitative* behavior of QCD with its infrared slavery, and in particular, we want to know qualitatively how it is forced to generate a dynamical mass.

Usually, the failure in the ultraviolet can be described as a (partial) failure of the gauge technique to satisfy the renormalization group; for example, the gauge technique vertex may have a renormalization group of standard form, but the beta function coefficients may be wrong. It is possible, in principle at least, to correct these gauge technique errors systematically both in QED [4] and in asymptotically free theories [5], but we will not go into such matters here. For a critique of the gauge technique, see the review [6] by one of the early workers.

The simple QED gauge technique is given first, as a warmup to more complex problems involving spontaneous symmetry breaking and to the ultimate challenge of NAGTs. For both Abelian and non-Abelian gauge theories, there are two basic approaches: via dispersion relations for the propagator, which express the vertex as a spectral integral involving the spectral weight for the propagator, and via purely algebraic methods, expressing the vertex directly in terms of the propagators.

5.1 The original gauge technique for QED

5.1.1 Scalar QED

Consider first the proper vertex $\Gamma_\mu(p_1, p_2)$ coupling a photon of momentum q to a charged scalar field. We take all momenta as coming into the vertex and normalize so that the bare vertex is $\Gamma_\mu(p_1, p_2) = i(p_1 - p_2)_\mu$, with $p_1 + p_2 + q = 0$. This vertex obeys the Ward identity

$$q^\mu \Gamma_\mu(p_1, p_2) = \Delta^{-1}(p_1) - \Delta^{-1}(p_2). \tag{5.1}$$

---

[1] By a mass gap, we mean that there are no massless particles carrying gauge-symmetry charge that appear in the S-matrix. For NAGTs, as we already know, generation of a dynamical gluon mass requires longitudinally coupled massless particles akin to Goldstone particles, and such particles are indeed present in the gauge technique vertex. But these particles are absent from the S-matrix because they are eaten by the gauge bosons.

Although throughout this book, *proper* Green's functions have been at the forefront for the pinch technique, for the gauge technique, it is sometimes useful to emphasize the improper vertex. Multiplying Γ_μ by the two charged-scalar propagators gives the improper vertex or form factor

$$F_\mu(p_1, p_2) = \Delta(p_1)\Gamma_\mu(p_1, p_2)\Delta(p_2), \qquad (5.2)$$

with a corresponding change in the Ward identity:

$$q^\mu F_\mu(p, -q - p) = \Delta(p_2) - \Delta(p_1). \qquad (5.3)$$

At tree level, this is satisfied with the usual expressions

$$iF_\mu(p, -p - q) = \frac{1}{p_1^2 - m^2}(p_1 - p_2)_\mu \frac{1}{p_2^2 - m^2} \qquad \Delta(p) = \frac{i}{p^2 - m^2}. \qquad (5.4)$$

The Ward identity is true no matter what the charged-particle mass m is, provided that it is the same for both charged lines in the form factor. This is the basis for one useful form of the gauge technique.

The charged-field propagators have a Källen–Lehmann representation:

$$- i\Delta(p) = \int d\sigma \, \frac{\rho(\sigma)}{p^2 - \sigma}. \qquad (5.5)$$

If, in Eq. (5.4), we replace m^2 by σ and integrate with weight function $\rho(\sigma)$, the Ward identity is still satisfied. So we define the *gauge technique improper vertex* as

$$iF_\mu^{\mathrm{GT}}(p_1, p_2) = \int d\sigma \, \rho(\sigma) \frac{1}{p_1^2 - \sigma}(p_1 - p_2)^\mu \frac{1}{p_2^2 - \sigma}. \qquad (5.6)$$

Clearly, this is not a unique solution because we can add any identically conserved function $[G_\mu(p, q), q^\mu G_\mu \equiv 0]$ to the gauge technique vertex and still solve the Ward identity. Nevertheless, the gauge technique form factor F_μ^{GT} is still useful in the region of infrared photon momentum $q_\mu \sim 0$, provided that there are no massless charged particles in the S-matrix.[2] The reason, for scalar charged particles, is a simple kinematic one: an identically conserved function without massless poles must vanish at least quadratically in q_μ for small q_μ. This is proved by exhaustion of a finite number of cases, of which we give only one example:

$$G_\mu = (q^2 p_\mu - q_\mu p \cdot q)H(p, q). \qquad (5.7)$$

---

[2] The only massless charged particles that gauge theories can tolerate are Goldstone-like bosons that get eaten by the gauge particles.

Without some special condition on the theory, there is no reason that H should vanish at $q_\mu = 0$. So in the case at hand, corrections to the gauge technique form factor are $\mathcal{O}(q^2)$ at small q.

Equation (5.6) for the gauge technique vertex can easily be transcribed with simple algebra into a form that does not use the Källen–Lehmann representation. We give the result for the proper vertex:

$$\Gamma_\mu^{\text{GT}}(p, -p - q) = \frac{(2p + q)_\mu}{2p \cdot q + q^2}[\Delta^{-1}(p + q) - \Delta^{-1}(p)]. \tag{5.8}$$

Note that there are no singularities at $q_\mu = 0$.

The great virtue of the gauge technique, and the reason for its existence, is that one can express an otherwise very complicated three-particle vertex entirely in terms of a propagator, always maintaining exact local gauge invariance. In this way, the Schwinger–Dyson equation for this propagator becomes self-contained.

5.1.2 Fermionic QED

The principles are exactly the same; most of the difference is in notation. In particular, to conform to the usual conventions, we take one momentum to be ingoing and one to be outgoing. There is a proper and an improper vertex, related by fermionic propagators:

$$F_\mu(p, p + q) = S(p)\Gamma_\mu(p, p + q)S(p + q), \tag{5.9}$$

and the fermion propagator obeys the Källen–Lehmann representation:

$$S(p) = \int_{-\infty}^{\infty} dW\, \frac{\rho(W)}{\slashed{p} - W}. \tag{5.10}$$

The Ward identity

$$q^\mu F_\mu(p, p + q) = S(p) - S(p + q) \tag{5.11}$$

is solved with the gauge technique form

$$F_\mu^{\text{GT}}(p, p + q) = \int_{-\infty}^{\infty} dW\, \rho(W) \frac{1}{\slashed{p} - W} \gamma_\mu \frac{1}{\slashed{p} + \slashed{q} - W}. \tag{5.12}$$

Remarkably, using this gauge-technique vertex in the Schwinger–Dyson equation linearizes it, as King [4] shows.

In this simple case, it is also straightforward to construct the algebraic version of the gauge technique. Define the electron proper self-energy Σ by

$$S^{-1}(p) = \slashed{p} - M + \Sigma(p). \tag{5.13}$$

Then [4] the *proper* gauge technique vertex $\tilde{\Gamma}_\mu^{\text{GT}}$ that follows from the spectral form of Eq. (5.12) is

$$\tilde{\Gamma}_\mu^{\text{GT}} = \gamma_\mu + \frac{1}{p^2 - p'^2} \left[\Sigma(p)(\not{p}\gamma_\mu + \gamma_\mu \not{p}') - (\not{p}\gamma_\mu + \gamma_\mu \not{p}')\Sigma(p') \right]. \quad (5.14)$$

There is no singularity at $p = p'$. As before, one could add an identically conserved term, such as $i\sigma_{\mu\nu}q^\nu/M$, but it is one power of q higher, not two, as in the scalar case, at small momentum compared to the mass.

5.2 Massless longitudinal poles

QED has an exact $U(1)$ gauge symmetry, but it is certainly possible to find gauge technique vertices for gauge theories with dynamically broken gauge symmetry, as a simple $O(2) \times U(1)$ gauge model illustrates [7] (Jackiw and Johnson [8] give an entirely equivalent illustration). There are no scalar fields of any sort in the model, just the fermions and gauge potentials, so the conventional Higgs mechanism cannot apply. Nonetheless, a gauge symmetry can be broken dynamically, with Higgs and associated Goldstone bosons arising as elements of the solution of the Schwinger–Dyson equations of the original model. Only the $O(2)$ symmetry is relevant for us (the $U(1)$ gauge field furnishes a critical attractive force that permits nontrivial symmetry-breaking solutions of the Schwinger–Dyson equations). The fermions form a two-vector in the $O(2)$ space of the form

$$\psi(x) = \begin{pmatrix} \psi_1 \\ \psi_2 \end{pmatrix}, \quad (5.15)$$

and they interact with a gauge potential B_μ through the interaction $\bar{\psi}\gamma^\mu B_\mu \tau_2 \psi$, where τ_i are the usual Pauli matrices. The idea is to look for symmetry-breaking solutions of the Schwinger–Dyson equations where the fermion proper self-energy has the form

$$\Sigma(p) = \Sigma_s(p) + \tau_3 \Sigma_v(p). \quad (5.16)$$

Thus, Σ_s preserves the gauge symmetry, and Σ_v violates it. In particular, the symmetry-violating self-energy can split the fermion masses.

The Ward identity for the proper fermion B_μ vertex is

$$(p - p')^\mu \Gamma_\mu(p, p') = S^{-1}(p)\tau_2 - \tau_2 S^{-1}(p'). \quad (5.17)$$

Without the τ_3 term in the fermion self-energy, the vertex, just like the bare vertex, would point solely in the τ_2 direction, and the model would be like two copies of QED. But with the τ_3 term, there must be a term in the vertex behaving like τ_1. Moreover, this part must be singular at small $q \equiv p - p'$ because at $q_\mu = 0$, the right-hand side of the preceding equation does not vanish. This singularity turns out to be a massless, longitudinally coupled scalar – in other words, a dynamical Goldstone boson. This pole part of the proper vertex must have the following singularity for $q \simeq 0$:

$$\Gamma_\mu(p, p') = \frac{2i\tau_1 \Sigma_v(p)q_\mu}{q^2} + \cdots, \tag{5.18}$$

as one can readily check; the omitted terms are regular at $q = 0$. It is this Goldstone boson, now appearing in the completely dressed vertex and not classically, that gives the B-field a mass because its pole in the vertex, behaving like $1/q^2$, cancels the usual kinematical factor of q^2 in the B-field proper self-energy that would otherwise prevent mass generation in a gauge theory. This cancellation of the pole is what we mean when we say that a Goldstone particle is eaten by a gauge boson.

We will not pursue this model further, except to say that it is essential that the gauge-boson forces acting on the fermions be attractive (which is why there is a second gauge potential). If they are, the Schwinger–Dyson equations indeed have a symmetry-violating fermion self-energy that vanishes at large momentum, fermionic mass splitting, a dynamical Higgs boson, and a nonzero B-boson mass, all of whose properties are calculable in terms of the parameters of the original model, whose Lagrangian had none of these effects. However, if the forces are not attractive, the Schwinger–Dyson equations are inconsistent and nonrenormalizable and can only be made consistent[3] by introducing *bare* fermion and B-boson masses.

For us, the point of considering this model is that much the same properties will turn up in non-Abelian gauge theories (with no matter fields of any sort): gauge-boson mass generation necessarily accompanied by longitudinally coupled massless scalars (really long-range, pure-gauge parts of the gauge potential). And the gauge-boson mass will vanish roughly as $1/q^2$ at large momentum, making the Schwinger–Dyson equations renormalizable and self-consistent.

5.3 The gauge technique for NAGTs

It took many years after the QED gauge technique to develop similar tools for non-Abelian gauge theories. The first construction [9] used a spectral form analogous to the original QED gauge technique, where the spectral integral is the Lehmann representation of the electron propagator. Later, a very general nonspectral construction was given [10] that expressed the gauge technique three-gluon vertex algebraically in terms of the PT proper self-energy. This construction was general

---

[3] In actuality, no Abelian gauge theory is really consistent at asymptotically high energies because of the Landau singularity induced by a positive beta function, but this is not of interest to us.

enough to use at finite temperature or for situations involving dynamical symmetry breaking (which requires other fields in appropriate representations; QCD cannot undergo dynamical symmetry breaking because the quarks are in the fundamental representation [11]).

We repeat here the notation and structure used in Chapter 1 for the PT propagator, both in a covariant R_ξ gauge and in the light-cone gauge. In both cases, $\widehat{d}(q)$ is the gauge-invariant scalar part of the PT propagator; these two propagators differ only in gauge terms that receive no corrections and play no essential role. The Ward identity for the (inverse) propagator is simply that it is transverse, aside from the irrelevant gauge-fixing terms:

• Covariant gauge

$$i\widehat{\Delta}_{\alpha\beta}(q) = P_{\alpha\beta}(q)\widehat{d}(q) + \xi\frac{q_\alpha q_\beta}{q^4} \tag{5.19}$$

$$-i\widehat{\Delta}_{\alpha\beta}^{-1}(q) = P_{\alpha\beta}(q)\widehat{d}(q)^{-1} + \frac{1}{\xi}q_\alpha q_\beta. \tag{5.20}$$

• Light-cone gauge

$$i\widehat{\Delta}_{\alpha\beta}(q) = Q_{\alpha\beta}\widehat{d}(q) + \eta\frac{q_\alpha q_\beta}{(n\cdot q)^2} \tag{5.21}$$

$$-i\widehat{\Delta}_{\alpha\beta}^{-1}(q) = P_{\alpha\beta}(q)\widehat{d}(q)^{-1} + \frac{1}{\eta}n_\alpha n_\beta, \tag{5.22}$$

where the gauge-fixing parameter η (which has dimensions of (mass)^2) is set to zero at the end of calculations.

We repeat the definition of $P_{\mu\nu}$ and $Q_{\mu\nu}$ given in Chapter 1:

$$P_{\mu\nu}(q) = g_{\mu\nu} - \frac{q_\mu q_\nu}{q^2} \tag{5.23}$$

$$Q_{\alpha\beta} = g_{\alpha\beta} - \frac{n_\alpha q_\beta + n_\beta q_\alpha}{n\cdot q}.$$

In both cases, the PT self-energy is defined by

$$\widehat{d}^{-1}(q) = q^2 + i\widehat{\Pi}(q). \tag{5.24}$$

The general Ward identity for the PT vertex relates the corrections to the free vertex to the proper self-energy and has already been found at the one-loop level in Chapter 1:

$$q_1^\mu\widehat{\Lambda}_{\mu\nu\alpha}(q_1, q_2, q_3) = \widehat{\Pi}_{\nu\alpha}(q_2) - \widehat{\Pi}_{\nu\alpha}(q_3), \tag{5.25}$$

where the full PT proper vertex $\widehat{\Gamma}_{\mu\nu\alpha}$ is the sum of the free vertex and $\widehat{\Lambda}_{\mu\nu\alpha}$. The free vertex could be either the usual bare vertex or the Γ^ξ vertex of Chapter 1. Suppose that we use the usual free vertex; then another form of this Ward identity is as follows:

$$q_1^\mu \widehat{\Gamma}_{\mu\nu\alpha}(q_1, q_2, q_3) = \widehat{d}^{-1}(q_2)P_{\mu\nu}(q_2) - \widehat{d}^{-1}(q_3)P_{\mu\nu}(q_3). \qquad (5.26)$$

In the light-cone gauge, the right-hand side is really the difference between two inverse propagators:

$$q_1^\mu \widehat{\Gamma}_{\mu\nu\alpha}(q_1, q_2, q_3) = \widehat{\Delta}_{\alpha\beta}^{-1}(q_2) - \widehat{\Delta}_{\alpha\beta}^{-1}(q_3). \qquad (5.27)$$

This is not so in covariant gauges, unless the vertex used is the Γ^ξ vertex.

5.3.1 The gauge technique in the light-cone gauge

Just as for scalar charged particles, there is one special case where Eq. (5.27) is an identity, and that is the case of free massive gauge bosons. We generate the massive propagator by keeping only the quadratic terms of the kinetic energy plus the gauged nonlinear sigma (GNLS) model (Eq. (2.7)), and the vertex by keeping only the free cubic vertex plus the cubic term of the GNLS model. For the propagator, we need only replace $\widehat{\Pi}$ by m^2 in Eq. (5.24), and the cubic vertex is

$$\widehat{\Gamma}_{\mu\nu\alpha}^{(m^2)}(q_1, q_2, q_3) = (q_1 - q_2)_\alpha g_{\mu\nu} + \frac{m^2}{2}\frac{q_{1\mu}q_{2\nu}(q_1 - q_2)_\alpha}{q_1^2 q_2^2} + \text{c.p.}, \qquad (5.28)$$

where c.p. stands for *cyclic permutations* (of momenta and indices). This vertex, or the improper form factor $\widehat{F}$ defined in Eq. (5.30), identically satisfies its Ward identities for any mass m.

If the PT propagator satisfied a Källen–Lehmann representation, we could then proceed to write a spectral integral for the form factor, as in the Abelian gauge theories. It is not surprising that infrared slavery, with its so-called wrong signs, does not permit a positive spectral function, but there still is a dispersion relation of the type of Eq. (5.5). The dispersion relation is just an integral over the massive propagator (we omit writing the η term of Eq. (5.21) because, ultimately, η is set to zero – although this must not be done until the end of any calculation):

$$i\widehat{\Delta}_{\alpha\beta}(q) = \int d\sigma\, \rho(\sigma)Q_{\alpha\beta}\frac{1}{q^2 - \sigma}. \qquad (5.29)$$

Spectral form of the gauge technique   The one-dressed-loop version of this self-energy has a term involving (schematically) an integral over $\Gamma_0\widehat{\Delta}\widehat{\Delta}\widehat{\Gamma}$ plus a seagull term. Provided that the propagators are transverse and $\widehat{\Gamma}$ satisfies its Ward identity, the output self-energy is transverse (and gauge invariant because the inputs to it will all be gauge invariant). The integrand of the one-dressed-loop self-energy integral involves a partly improper form factor, which we define as

$$\widehat{F}_{\mu\nu\alpha}(q_1, q_2, q_3) = \widehat{\Delta}^\rho_\mu(q_2)\widehat{\Delta}^\lambda_\nu(q_3)\widehat{\Gamma}_{\rho\lambda\alpha}(q_2, q_3), \qquad (5.30)$$

with

$$q_1^\mu \widehat{F}_{\mu\nu\alpha}(q_1, q_2, q_3) = \widehat{\Delta}_{\nu\sigma}(q_3) - \widehat{\Delta}_{\mu\rho}(q_2). \qquad (5.31)$$

The expression [9]

$$\widehat{F}^{\text{GT}}_{\mu\nu\alpha}(q_1, q_2, q_3) = \int d\sigma \, \rho(\sigma)\frac{Q^\rho_\mu(q_2)}{q_2^2 - \sigma} \widehat{\Gamma}^{(\sigma)}_{\rho\lambda\alpha}(q_1, q_2, q_3) \frac{Q^\lambda_\nu(q_3)}{q_3^2 - \sigma} \qquad (5.32)$$

satisfies the Ward identity of Eq. (5.27) for any spectral function $\rho(\sigma)$, that is, for any PT propagator.

Algebraic form of the gauge technique   The algebraic form [10] of the gauge technique vertex for NAGTs is considerably more general than the spectral form and can be used not only for QCD-like theories but also for theories with symmetry breaking, at finite temperature, and, in fact, for any physically reasonable circumstances for NAGTs in three or four dimensions. (The three-dimensional version is very useful for the functional Schrödinger equation, and we deal with it in Chapter 6.) We give it here only for the simple circumstance that the gluon PT proper self-energy is diagonal in group space (with group indices assigned in accordance with the momentum argument) and only in $d = 4$; see [10] for generalizations and references to other circumstances. The gauge technique radiative correction to the free vertex (see Eq. (5.25)) is

$$\widehat{\Lambda}_{\mu\nu\alpha}(q_1, q_2, q_3) = -\frac{q_{1\mu}q_{2\nu}}{2q_1^2 q_2^2}(q_1 - q_2)^\rho \widehat{\Pi}_{\rho\alpha}(q_3)$$

$$- [P^\rho_\mu(q_1)\widehat{\Pi}_{\rho\nu}(q_2) - P^\rho_\nu(q_2)\widehat{\Pi}_{\rho\mu}(q_1)]\frac{q_{3\alpha}}{q_3^2} + \text{c.p.}, \qquad (5.33)$$

and this satisfies the Ward identity of Eq. (5.25). (The replacement of $\widehat{\Pi}_{\mu\nu}(q)$ by $m^2 P_{\mu\nu}$ gives – after some algebra – the previous result of Eq. (5.28).) Note the presence of massless longitudinal excitations in this vertex; these decouple if the self-energies vanish at zero momentum.

Both the spectral form and the algebraic form have been used in studies of dynamical gluon mass generation in the light-cone Schwinger–Dyson equations of the pinch technique [9, 10], and a simplified algebraic form has been used in the covariant pinch technique Schwinger–Dyson equations [12]. These equations are detailed in Chapter 6.

References

[1] A. Salam, Renormalizable electrodynamics of vector mesons, *Phys. Rev.* **130** (1963) 1287.

[2] A. Salam and R. Delbourgo, Renormalizable electrodynamics of scalar and vector mesons, II, *Phys. Rev.* **135** (1964) 1398.

[3] J. Papavassiliou, Gauge-invariant four gluon vertex and its Ward identity, *Phys. Rev.* **D47** (1993) 4728.

[4] J. E. King, Transverse vertex and gauge technique in quantum electrodynamics, *Phys. Rev.* **D27** (1983) 1821.

[5] B. J. Haeri, Ultraviolet-improved gauge technique and the effective quark propagator in QCD, *Phys. Rev.* **D38** (1988) 3799.

[6] R. Delbourgo, A critique of the gauge technique, *Austral. J. Phys.* **52** (1999) 681.

[7] J. M. Cornwall and R. E. Norton, Spontaneous symmetry breaking without scalar mesons, *Phys. Rev.* **D8** (1973) 3338.

[8] R. Jackiw and K. Johnson, Dynamical model of spontaneously-broken gauge symmetries, *Phys. Rev.* **D8** (1973) 2386.

[9] J. M. Cornwall, Dynamical mass generation in continuum quantum chromodynamics, *Phys. Rev.* **D26** (1982) 1453.

[10] J. M. Cornwall and W. S. Hou, Extension of the gauge technique to broken symmetry and finite temperature, *Phys. Rev.* **D34** (1986) 585.

[11] J. M. Cornwall, Spontaneous symmetry breaking without scalar mesons, II, *Phys. Rev.* **D10** (1974) 500.

[12] A. C. Aguilar, D. Binosi, and J. Papavassiliou, Gluon and ghost propagators in the Landau gauge: Deriving lattice results from Schwinger-Dyson equations, *Phys. Rev.* **D78** (2008) 025010.

6

Schwinger–Dyson equations in the pinch technique framework

In this chapter, we provide a detailed demonstration of how the application of the PT algorithm at the level of the conventional Schwinger–Dyson series leads to a new, modified Schwinger–Dyson equation (SDE) for the gluon propagator endowed with very special truncation properties. In particular, because of the QED-like Ward identities from the fully dressed Green's functions entering into the gluon SDE, the transversality of the gluon self-energy is guaranteed at each level of the dressed-loop expansion. This result constitutes one of the main objectives of the PT program, namely, the device of a gauge-invariant truncation scheme for the equations governing the nonperturbative dynamics of non-Abelian Green's functions.

Of course, like any propagator SDE, this equation depends on the full three- and four-point gluon vertices, and these in turn depend on infinitely many other Green's functions. We have suggested, in the last chapter, how to truncate the SDE for the PT propagator by using the gauge technique to approximate the three- and four-point PT Green's functions as functionals only of the PT proper self-energy, while maintaining the exact Ward identities demanded by the PT. This approximation can only be useful in the infrared; the gauge technique PT Green's functions clearly fail to be exact at large momenta (although this failure is quantitative, not qualitative, so it should not change the fundamental findings from PT SDEs, except in the numerics).

For the purposes of this book, it would be too much to study thoroughly all the ramifications of combining the gauge technique, which, in its most general form, is quite complicated, with the pinch technique in the all-order SDEs. In fact, this has not yet been done. The emphasis in this chapter will be on the SDEs themselves, not on how to make approximations to them. However, we will present some applications in which a relatively simple version of the gauge technique provides the necessary structure for maintaining gauge invariance and obtaining massive (infrared finite) solutions.

As far as the fundamental underlying dynamics are concerned, there is no qualitative difference between the all-order SDEs presented here and the simple one-loop equations of Chapters 1 and 2: infrared slavery demands the generation of a dynamical gluon mass of about half a GeV. Before going into detail, we really should ask whether it is worth it – is there any solid evidence that the gluon does have a mass? Of course, this cannot be a direct experimental finding because gluons are screened out of physical existence as isolated particles. But for QCD theorists, lattice simulations are the ideal laboratory, and so we begin with summarizing the lattice evidence for the gluon mass. These simulations are done in Euclidean space, which leads us to the following notational changes. We will derive the SDEs in Minkowski space, as we have used so far. But beginning with Section 6.5 and through the rest of the chapter, we will work in Euclidean space, with metric $\delta_{\mu\nu}$, for ease of comparison with lattice data.

6.1 Lattice studies of gluon mass generation

The experimental laboratory for validation of PT results is the computer, and the experiments are simulations. In this case, we ask what simulations of gluon propagators say about the gluon mass. A large number of such simulations have been done by a number of different groups [1, 2, 3, 4, 5, 6, 7, 8, 9, 10, 11, 12, 13, 14, 15, 16, 17, 18, 19, 20, 21]. All these simulations, for technical reasons, are gauge fixed to a Landau gauge[1] rather than using the background-field Feynman gauge, which would then yield the PT propagator. The Landau gauge propagator is not gauge invariant, but it has one crucial feature: if the gluon has a mass, the propagator will be finite and nonvanishing at zero momentum – because it would have a zero-momentum singularity if the gluon were massless. This evidence for a massive gluon is found in all lattice simulations of the Landau-gauge gluon propagator cited earlier.

If it were possible to extend the Landau-gauge propagator to the region of timelike momenta, the position of the pole found in the propagator would be gauge invariant and yield the gluon mass. Such an extrapolation is risky, and neither we nor the simulations' workers have done this systematically and compared all the various results. However, to our eyes, all the propagator results are quite similar and presumably yield gluon masses in a fairly narrow range, and when the authors do give approximate mass values, they all cluster around 500–700 MeV. One group [1, 2] has attempted an extrapolation, making use of the functional form of the older PT propagator found in the first attempt [23] at calculating the PT propagator, which yields a very good fit for a gluon mass of around 600 MeV. It is not clear why a PT propagator should fit the simulated Landau-gauge propagator so closely. This issue needs lattice simulations in a background-field Feynman gauge, which do not yet exist. Figure 6.1 shows recent, and typical, lattice results for the gluon

---

[1] A Landau gauge has Gribov ambiguities [22]; various simulation groups take various approaches to resolving this issue, as detailed in specific papers. The upshot is that all groups claim that their results are more or less free of Gribov ambiguity problems.

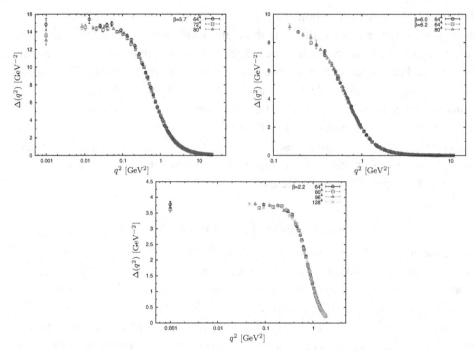

Figure 6.1. Lattice results for the gluon propagator from three different lattice groups. (*Upper left*) Bare lattice $SU(3)$ gluon propagator (data from Bogolubsky et al. [15]). (*Upper right*) $SU(3)$ gluon propagator renormalized at 3 GeV (data from Oliveira and Silva [11]). (*Lower left*) Bare $SU(2)$ propagator (data from Cucchieri and Mendes [19]).

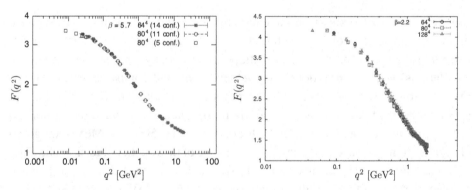

Figure 6.2. (*Left*) Bare lattice $SU(3)$ ghost dressing function. Reprinted with permission from I. L. Bogolubsky, E. M. Ilgenfritz, M. Muller-Preussker, and A. Sternbeck, *Phys. Lett.* **B676** (2009) 69; © 2009 by Elsevier. (*Right*) Bare $SU(2)$ ghost dressing function (data from Cucchieri and Mendes [19]).

propagator $\Delta(q^2)$ obtained by three different lattice groups [15, 19, 11]; Figure 6.2 shows the ghost propagator dressing function $F(q^2)$ (defined as the product of q^2 times the ghost propagator).

Note that in this figure, positive q^2 means a Euclidean (spacelike) momentum. The dressed gluon propagator is *finite* at zero momentum, indicating a gluon mass; if the mass were zero, the zero-momentum value would be infinite. Similarly, the ghost dressing function approaches a finite value at zero momentum. Both these properties will be found in the solutions to the PT SDEs of this chapter. We now go on to the meat of this chapter: the gauge-invariant PT SDEs and gluon mass generation.

6.2 The need for a gauge-invariant truncation scheme for the Schwinger–Dyson equations of NAGTs

The SDEs provide a formal framework for tackling physics problems requiring a nonperturbative treatment. Given that the SDEs constitute an infinite system of coupled nonlinear integral equations for all Green's functions of the theory, their practical usefulness hinges crucially on one's ability to devise a self-consistent truncation scheme that would select a tractable subset of these equations without compromising the physics one hopes to describe.

In Abelian gauge theories, the Green's functions satisfy naive Ward identities: the all-order generalization of a tree-level Ward identity is obtained simply by replacing the Green's functions appearing in it by their all-order expressions. In general, as we have seen, this is not true in non-Abelian gauge theories, where the Ward identities are modified nontrivially beyond tree level and are replaced by more complicated expressions known as Slavnov–Taylor identities: in addition to the original Green's functions appearing at tree level, they involve various composite *ghost operators*.

To appreciate why the Ward identities are instrumental for the consistent truncation of the SDEs, whereas the Slavnov–Taylor identities complicate it, let us first consider how nicely things work in an Abelian case, namely, *scalar* QED (photon), and then turn to the complications encountered in QCD (gluon).

Local gauge invariance (BRST in the case of the gluon) forces $\Pi_{\alpha\nu}(q)$ (photon and gluon alike) to satisfy the fundamental transversality relation

$$q^\alpha \Pi_{\alpha\beta}(q) = 0, \tag{6.1}$$

both perturbatively (to all orders) and nonperturbatively. The SDE governing $\Pi_{\alpha\beta}(q)$ in scalar QED is shown in Figure 6.3. The main question we want to address is the following: can one truncate the right-hand side (rhs) of Figure 6.3, i.e., eliminate some of the graphs, without compromising the transversality of $\Pi_{\alpha\beta}(q)$? The answer is shown already in Figure 6.3: the two blocks of graphs

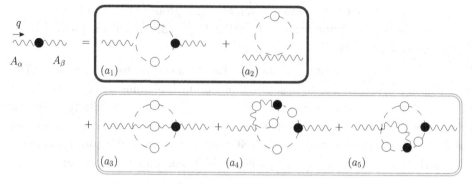

Figure 6.3. The SDE for the photon self-energy in scalar QED. The two boxes enclose a gauge-invariant subset of diagrams.

$[(a_1) + (a_2)]$ and $[(a_3) + (a_4) + (a_5)]$ are *individually transverse*, i.e.,

$$q^\alpha \sum_{i=1}^{2} (a_i)_{\alpha\beta} = 0 \qquad q^\alpha \sum_{i=3}^{5} (a_i)_{\alpha\beta} = 0. \tag{6.2}$$

The reason for this special property is precisely the Ward identities satisfied by the full vertices appearing on the rhs of the SDE; for example, the first block is transverse simply because the *full* photon-scalar vertex Γ_μ (black blob in (a_1)) satisfies the Ward identity:

$$q^\alpha \Gamma_\alpha = e[\mathcal{D}^{-1}(k+q) - \mathcal{D}^{-1}(k)], \tag{6.3}$$

where $\mathcal{D}(q)$ is the *full* propagator of the charged scalar. A similar Ward identity relating the four-vertex with a linear combination of Γ_α forces the transversality of the second block in Figure 6.3. Thus, owing to the simple Ward identities satisfied by the vertices appearing on the SDE for $\Pi_{\alpha\beta}(q)$, one may omit the second block of graphs and still maintain the transversality of the answer intact, i.e., the approximate $\Pi_{\alpha\beta}(q)$ obtained after this truncation satisfies Eq. (6.1).

Let us now turn to the *conventional* SDE for the gluon self-energy, in the R_ξ gauges, given in Figure 6.4. Clearly, by virtue of Eq. (6.1), we must have

$$q^\alpha \sum_{i=1}^{5} (a_i)_{\alpha\beta} = 0. \tag{6.4}$$

However, unlike the Abelian example, the diagrammatic verification of Eq. (6.4), i.e., through contraction of the individual graphs by q^α, is very difficult, essentially because of the complicated Slavnov–Taylor identities satisfied by the vertices involved. For example, the full three-gluon vertex $\Gamma_{\alpha\mu\nu}(q, k_1, k_2)$ satisfies the

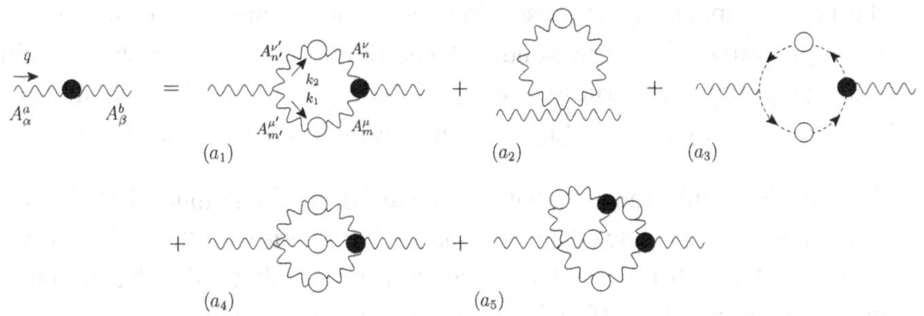

Figure 6.4. Schwinger-Dyson equation satisfied by the gluon self-energy $-\Gamma_{AA}$. The symmetry factors of the diagrams are $s(a_1, a_2, a_5) = 1/2$, $s(a_3, a_6) = -1$, and $s(a_4) = 1/6$.

Slavnov–Taylor identity of Eq. (1.89). In addition, some of the pertinent Slavnov–Taylor identities are either too complicated, such as that of the conventional four-gluon vertex, or cannot be cast in a particularly convenient form, such as that of the conventional gluon-ghost vertex. The main practical consequence of this is that one cannot truncate the rhs of Figure 6.4 in any obvious way without violating the transversality of the resulting $\Pi_{\alpha\beta}(q)$. For example, keeping only graphs (a_1) and (a_2) is not correct even perturbatively because the ghost loop is crucial for the transversality of $\Pi_{\alpha\beta}$ already at one loop; adding (a_3) is still not sufficient for an SDE analysis because (beyond one loop) $q^\alpha[(a_1) + (a_2) + (a_3)]_{\alpha\beta} \neq 0$.

As we will see in what follows, the application of the pinch technique to the conventional SDEs of the NAGTs gives rise to new equations endowed with special properties. The building blocks of the new SDEs are modified Green's functions obeying Abelian all-order Ward identities instead of the Slavnov–Taylor identities satisfied by their conventional counterparts. As a result, and contrary to the standard case explained earlier, the new equation for the gluon self-energy can be truncated gauge invariantly at any order in the dressed-loop expansion.

6.3 The pinch technique algorithm for Schwinger–Dyson equations

At the end of Chapter 4, we saw, in a one-loop context, how the PT algorithm can be translated in Batalin–Vilkovisky language. More generally, the one-loop procedure described there carries over practically unaltered to the corresponding SDEs. This is so because of the following observations:

1. The pinching momenta will be always determined from the tree-level decomposition of Eqs. (1.41), (1.42), and (1.43).
2. Their action is completely fixed by the structure of the Slavnov–Taylor identities they trigger (Eq. (4.27) for the vertex at hand).

3. The kernels appearing in these Slavnov–Taylor identities are the same as those appearing in the corresponding background-quantum identities, making it always possible to write the result of the action of pinching momenta in terms of auxiliary Green's functions appearing in the latter identities.

The only operational difference is that, in the case of the Schwinger–Dyson equations for the three-gluon vertex (and the quark-gluon vertex, in the case in which quarks are included), all three external legs will be off shell. This is, of course, unavoidable, given that these (fully dressed) vertices are nested in the SDE of the off-shell gluon self-energy (see, e.g., diagram (a_1) of Figure 6.4), and their legs inside the diagrams are irrigated by the virtual off-shell momenta. As a result, the equations of motion usually employed to kill some of the resulting pinching terms should not be used in this case; therefore, the corresponding terms, proportional to inverse self-energies, do not drop out but rather, form part of the resulting background-quantum identity.

The PT rules for the construction of SDEs may be summarized as follows [24]:

1. For the SDEs of vertices, with all three external legs off shell, the pinching momenta, coming from the (only) external three-gluon vertex undergoing the decomposition (1.41), generate at most four types of terms: one is a genuine vertexlike contribution that must be included in the final PT answer for the vertex under construction; the remaining three terms will form part of the emerging background-quantum identities (and thus would be discarded from the PT vertex). These latter terms have a very characteristic structure that facilitates their identification in the calculation. Specifically, one of them is always proportional to the auxiliary function $\Gamma_{\Omega A^*}$, whereas the other two are proportional to the inverse propagators of the fields entering into the two legs that did not undergo the PT decomposition.
2. In the case of the new SDE for the gluon propagator, the pinching momenta will only generate pieces proportional to $\Gamma_{\Omega A^*}$, which should be discarded from the PT answer for the gluon two-point function (because they are exactly what cancels against the contribution coming from the corresponding vertices).

6.4 Pinch technique Green's functions from Schwinger–Dyson equations

We are now ready to describe in detail the application of the PT program to the (nonperturbative) case of SDEs. We will concentrate on the SDE of the gluon self-energy,[2] which requires carrying out the PT decomposition of Eq. (1.41)

---

[2] The construction of the (fully off-shell) quark-gluon and three-gluon vertices can be found in [24].

on both sides of the self-energy diagram because both external gluons must be converted into background gluons. This will be achieved through the following three steps [24, 25].

6.4.1 First step: $\Gamma_{AA} \to \Gamma_{\widehat{A}A}$

The first step of the construction is the standard PT step: starting from diagram (a_1) of Figure 6.4, we decompose the tree-level three-gluon vertex according to the usual PT splitting of Eq. (1.41) and concentrate on the Γ^P part. We then get

$$(a_1)^P = ig f^{amn} \int_{k_1} \frac{1}{k_1^2} \Delta_\alpha^\nu(k_2) k_1^\mu \Gamma_{A_\mu^m A_\nu^n A_\beta^b}(k_2, -q). \tag{6.5}$$

At this point, the application of the Slavnov–Taylor identity (4.24) together with Eq. (4.25) and the Faddeev–Popov equation (4.26) results in the following terms:

$$(a_1)^P = ig f^{amn} \int_{k_1} D(k_1) \Delta_\alpha^\nu(k_2) \Gamma_{c^m A_\nu^n A_b^{*\gamma}}(k_2, -q) \Gamma_{A_\gamma A_\beta}(q)$$

$$+ g f^{amn} \int_{k_1} D(k_1) \left[\Gamma_{c^m A_\beta^b A_\alpha^{*n}}(-q, k_2) - i\frac{k_{2\alpha}}{k_2^2} \Gamma_{c^m A_\beta^b \bar{c}^n}(-q, k_2) \right]$$

$$= (r_1)^P + (r_2)^P. \tag{6.6}$$

Clearly, using the SDE of the auxiliary function $\Gamma_{\Omega A^*}$, shown in Eq. (4.44), one has immediately that

$$(r_1)^P = -i\Gamma_{\Omega_\alpha^a A_d^{*\gamma}}(q) \Gamma_{A_\gamma^d A_\beta^b}(q). \tag{6.7}$$

This would be half of the pinching contribution coming from the vertex in the S-matrix pinch technique.

As far as the term $(r_2)^P$ is concerned, its general structure suggests that the first term in the square brackets should give rise to the ghost quadrilinear vertex, whereas the second term in that same bracket, when added to diagram (a_3), should symmetrize the trilinear ghost-gluon coupling. It turns out that this expectation is essentially correct, but its realization is not immediate because we see, for example, that $(r_2)^P$ contains a tree-level $[(k_2^2)^{-1}]$ instead of a full $[D(k_2)]$ ghost propagator and that it is unclear how one would generate, e.g., diagrams (b_7) or (b_9) of Figure 6.6. The solution to this apparent mismatch is rather subtle: one must employ the SDE satisfied by the ghost propagator, as shown in Figure 6.5. This SDE is common to both the R_ξ-gauge and the BFM, given that there are no background ghosts, and reads

$$iD^{dn'}(k_2) = i\frac{\delta^{dn'}}{k_2^2} + i\frac{\delta^{dg}}{k_2^2} \left[-\Gamma'_{c^g \bar{c}^{g'}}(k_2) \right] iD^{g'n'}(k_2), \tag{6.8}$$

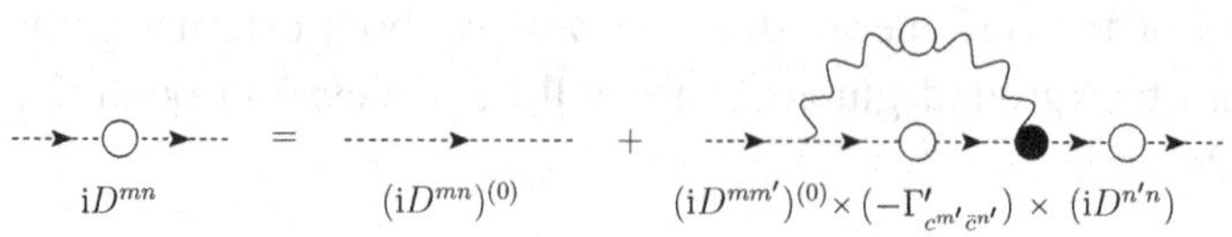

Figure 6.5. The Schwinger–Dyson equation (6.8) satisfied by the ghost propagator.

with $\Gamma'_{c\bar{c}}$ being given by $\Gamma_{c\bar{c}}$ minus its tree-level part; notice that multiplying the preceding equation by k_2^2, using the Faddeev–Popov equation (4.21), and dropping the color factor $\delta^{dn'}$, we can rewrite the ghost SDE as

$$k_{2\alpha} D(k_2) = \frac{k_{2\alpha}}{k_2^2} - \Gamma'_{cA^*_\alpha}(k_2)D(k_2). \tag{6.9}$$

Taking advantage of this last equation, we can rewrite $(r_2)^{\mathrm{P}}$ as the sum of the following two terms:

$$(s_1)^{\mathrm{P}} = -\mathrm{i}g f^{amn} \int_{k_1} k_{2\alpha} D(k_1)D(k_2)\Gamma_{c^m A^b_\beta \bar{c}^n}(-q, k_2) \tag{6.10}$$

$$(s_2)^{\mathrm{P}} = -g f^{amn} \int_{k_1} \mathrm{i}D(k_1) \left[\mathrm{i}\Gamma_{c^m A^b_\beta A^{*n}_\alpha}(-q, k_2) \right.$$

$$\left. + \Gamma'_{cA^*_\alpha}(k_2)D(k_2)\Gamma_{c^m A^b_\beta \bar{c}^n}(-q, k_2) \right]. \tag{6.11}$$

The term $(s_1)^{\mathrm{P}}$ symmetrizes the trilinear ghost-gluon coupling, and one has

$$(s_1)^{\mathrm{P}} + (a_3) = (b_3) \tag{6.12}$$

(with (b_3) shown in Figure 6.6). The term $(s_2)^{\mathrm{P}}$ will instead be responsible for generating all the missing diagrams needed to convert Γ_{AA} into $\Gamma_{\widehat{A}A}$. To see how this happens, we denote by $(s_{2a})^{\mathrm{P}}$ and $(s_{2b})^{\mathrm{P}}$ the two terms appearing in the square brackets of $(s_2)^{\mathrm{P}}$ and concentrate on the first one. Making use of the SDE (4.43) satisfied by the auxiliary function Γ_{cAA^*}, we get

$$(s_{2a})^{\mathrm{P}} = g^2 C_A \delta^{ab} g_{\alpha\beta} \int_{k_1} D(k_1)$$

$$+ g^2 f^{amd} f^{dnr} \int_{k_1} \int_{k_3} D(k_1)\Delta^\rho_\alpha(k_3)D(k_4)\mathcal{K}_{c^m A^b_\beta A^r_\rho \bar{c}^n}(-q, k_3, k_4)$$

$$= (b_4) + (b_7) + (b_8) + (b_{10}). \tag{6.13}$$

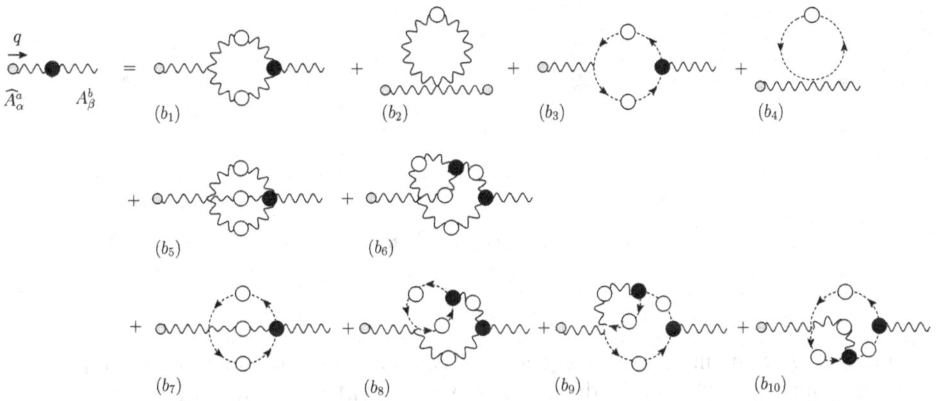

Figure 6.6. Schwinger–Dyson equation satisfied by the gluon self-energy $-\Gamma_{\widehat{A}A}$. The symmetry factors of the diagrams are $s(b_1, b_2, b_6) = 1/2$ and $s(b_5) = 1/6$, and all the remaining diagrams have $s = -1$.

Using the SDE (4.42) satisfied by Γ_{cA^*}, we then obtain

$$(s_{2b})^{\mathrm{P}} = \mathrm{i}g^2 f^{amd} f^{drs} \int_{k_1} \int_{k_3} D(k_1) \Gamma_{c^n A_\rho^r \bar{c}^s}(k_3, k_4) \Delta_\alpha^\rho(k_3) D(k_4) D(k_2)$$

$$\times \Gamma_{c^m A_\beta^b \bar{c}^n}(-q, k_2)$$

$$= (b_9). \tag{6.14}$$

In addition, because diagrams (a_2), (a_4), (a_5), and (a_6) carry over to the corresponding BFM diagrams (b_2), (b_5), (b_6), and (b_{11}), and $(a_1)^{\mathrm{F}} = (b_1)$, we finally find the identity

$$(r_2)^{\mathrm{P}} + \left[(a_1)^{\mathrm{F}} + \sum_{i=2}^{6} (a_i) \right] = \sum_{i=1}^{11} (b_i), \tag{6.15}$$

and therefore

$$-\Gamma_{A_\alpha^a A_\beta^b}(q) = -\mathrm{i}\Gamma_{\Omega_\alpha^a A_d^{*\gamma}}(q) \Gamma_{A_\gamma^d A_\beta^b}(q) - \Gamma_{\widehat{A}_\alpha^a A_\beta^b}(q), \tag{6.16}$$

which coincides with the background-quantum identity (4.31).

6.4.2 Second step: $\Gamma_{\widehat{A}A} \to \Gamma_{A\widehat{A}}$

The second step in the propagator construction is to employ the obvious relation

$$\Gamma_{\widehat{A}_\alpha^a A_\beta^b}(q) = \Gamma_{A_\alpha^a \widehat{A}_\beta^b}(q), \tag{6.17}$$

that is, to interchange the background and quantum legs (the SDE for the self-energy $-\Gamma_{A\widehat{A}}$ is shown in Figure 6.7). This apparently trivial operation introduces

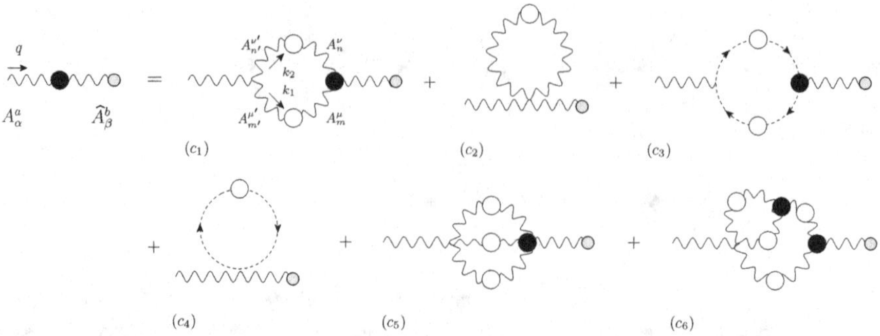

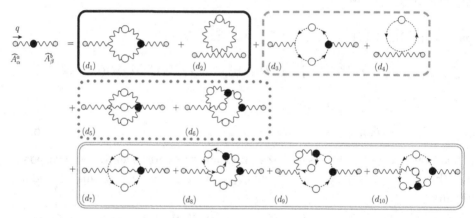

Figure 6.7. Schwinger–Dyson equation satisfied by the gluon self-energy $-\Gamma_{A\widehat{A}}$. The symmetry factors of the diagrams are $s(c_1, c_2, c_6) = 1/2$, $s(c_3, c_4, c_7) = -1$, $s(c_5) = 1/6$.

Figure 6.8. Schwinger–Dyson equation satisfied by the gluon self-energy $-\Gamma_{\widehat{A}\widehat{A}}$. The symmetry factors are the same as the one described in Figure 6.5. The different boxes show gauge-invariant subgroups: one-loop dressed gluon (solid line) and ghost (dashed line) contributions and two-loop dressed gluon (dotted line) and ghost (double line) contributions (see the discussion in Section 6.4.4).

a considerable simplification. First, it allows for the identification of the pinching momenta from the usual PT decomposition of the (tree level) Γ appearing in diagram (c_1) of Figure 6.7 (something not directly possible from diagram (b_1)); thus, from the operational point of view, we remain on familiar ground. In addition, it avoids the need to employ the (formidably complicated) BQI for the four-gluon vertex; indeed, the equality between diagrams (c_5), (c_6), and (c_7) of Figure 6.7 and diagrams (d_5), (d_6), and (d_{11}) of Figure 6.8 is now immediate (as it was before, between diagrams (a_4), (a_5), and (a_6) and (b_5), (b_6), and (b_{11}) of Figures 6.6 and 6.7, respectively).

6.4.3 Third step: $\Gamma_{A\widehat{A}} \to \Gamma_{\widehat{A}\widehat{A}}$

We now turn to diagram (c_1) and concentrate on its pinching part given by

$$(c_1)^P = ig f^{amn} \int_{k_1} \frac{1}{k_1^2} \Delta_\alpha^\nu(k_2) k_1^\mu \Gamma_{A_\mu^m A_\nu^n \widehat{A}_\beta^b}(k_2, -q). \tag{6.18}$$

Notice the appearance of the full BFM vertex $\Gamma_{AA\widehat{A}}$ instead of the standard Γ_{AAA} (in the R_ξ). The Slavnov–Taylor identity satisfied by this vertex can be derived by means of the methods introduced in the previous chapter; the result is[3]

$$k_1^\mu \Gamma_{\widehat{A}_\alpha^a A_\mu^m A_\nu^n}(k_1, k_2) = [k_1^2 D^{mm'}(k_1)] \left\{ \Gamma_{c^{m'} A_\nu^n A_\epsilon^{*e}}(k_2, q) \Gamma_{\widehat{A}_\alpha^a A_\epsilon^e}(q) \right.$$

$$\left. + \Gamma_{c^{m'} \widehat{A}_\alpha^a A_\epsilon^{*e}}(q, k_2) \Gamma_{A_\epsilon^e A_\nu^n}(k_2) \right\}$$

$$- ig f^{amn}(k_1^2 g_{\alpha\nu} - k_{1\alpha} k_{2\nu}), \tag{6.19}$$

where the tree-level term accounts for working with the reduced Slavnov–Taylor functional. This identity can be further manipulated by using Eq. (4.25) and the Faddeev–Popov equation satisfied by $\Gamma_{c\widehat{A}A^*}$ to get

$$k_1^\mu \Gamma_{\widehat{A}_\alpha^a A_\mu^m A_\nu^s}(k_1, k_2) = [k_1^2 D^{mm'}(k_1)] \left\{ \Gamma_{c^{m'} A_\nu^n A_\epsilon^{*e}}(k_2, q) \Gamma_{\widehat{A}_\alpha^a A_\epsilon^e}(q) \right.$$

$$\left. + \Gamma_{c^{m'} \widehat{A}_\alpha^a A_\epsilon^{*e}}(q, k_2)(\Delta^{-1})_{\epsilon\nu}^{en}(k_2) + k_{2\nu} \Gamma_{c^{m'} \widehat{A}_\alpha^a \bar{c}^n}(q, k_2) \right\}$$

$$- ig f^{amn} k_1^2 g_{\alpha\nu}. \tag{6.20}$$

Consider first the terms that appear in braces. When inserted back into Eq. (6.18) (we indicate the resulting expression as $(r_1)^P$), we see that one gets precisely the same contributions found in Eq. (6.6), the only difference being that the A_β^b gluon field appearing there is now the background field $\widehat{A}_\beta^b$. Therefore, following step by step the same procedure used in that case, we find that (see Figure 6.8 for the diagrams corresponding to each (d_i))

$$(r_1)^P + (c_3) = -i\Gamma_{\Omega_\alpha^a A_\epsilon^{*e}}(q) \Gamma_{A_\epsilon^e \widehat{A}_\beta^b}(q)$$

$$+ (d_3) + (d_4) + (d_7) + (d_8) + (d_9) + (d_{10}). \tag{6.21}$$

Finally, adding diagram (c_2) to the remaining tree-level term of Eq. (6.20), one gets

$$(r_2)^P + (c_2) = g^2 C_A \delta^{ab} \int_{k_1} \Delta_{\alpha\beta}(k_1) + (c_2) = (d_2). \tag{6.22}$$

---

[3] Given a Green's function involving background as well as quantum fields, it is clear that if we contract it with the momentum corresponding to a background leg, we will obtain a linear Ward identity (see, e.g., Eqs. (6.28)–(6.31)), whereas if we contract it with the momentum corresponding to a quantum leg, we will obtain a nonlinear Slavnov–Taylor identity.

Putting everything together, and using the standard PT identity $(c_1)^F = (d_1)$, we get

$$(r_1)^P + (r_2)^P + \left[(c_1)^F + \sum_{i=2}^{7}(c_i)\right] = -i\Gamma_{\Omega_\alpha^a A_\epsilon^{*e}}(q)\Gamma_{A_\epsilon^e \widehat{A}_\beta^b}(q) + \sum_{i=1}^{11}(d_i), \quad (6.23)$$

and therefore

$$-\Gamma_{A_\alpha^a \widehat{A}_\beta^b}(q) = -i\Gamma_{\Omega_\alpha^a A_\epsilon^{*e}}(q)\Gamma_{A_\epsilon^e \widehat{A}_\beta^b}(q) - \Gamma_{\widehat{A}_\alpha^a \widehat{A}_\beta^b}(q), \quad (6.24)$$

which is the background-quantum identity (4.32).

6.4.4 The final rearrangement and the new Schwinger–Dyson equation

Pulling together all the intermediate results, we see that the PT procedure has given rise to a new Schwinger–Dyson series:

$$i\Gamma_{A_\alpha A_\beta} + 2\Gamma_{\Omega_\alpha A^{*\gamma}}\Gamma_{A_\gamma A_\beta} - i\Gamma_{\Omega_\alpha A^{*\gamma}}\Gamma_{A_\gamma A_\epsilon}\Gamma_{\Omega_\beta A^{*\epsilon}} = i\sum_{1=1}^{11}(d_i)_{\alpha\beta}. \quad (6.25)$$

By Lorentz-decomposing the auxiliary function $\Gamma_{\Omega A^*}(q)$ according to Eq. (4.34) and trading the gluon two-point function for the gluon propagator, we arrive at the equation (viz. Eq. (4.35))

$$\Delta^{-1}(q^2)[1 + G(q^2)]^2 P_{\alpha\beta}(q) = q^2 P_{\alpha\beta}(q) + i\sum_{1=1}^{11}(d_i)_{\alpha\beta}. \quad (6.26)$$

Equivalently, the new SDE (6.26) can be cast into a more conventional form by isolating on the left-hand side (lhs) the inverse of the unknown quantity, thus writing

$$\Delta^{-1}(q^2)P_{\alpha\beta}(q) = \frac{q^2 P_{\alpha\beta}(q) + i\sum_{1=1}^{11}(d_i)_{\alpha\beta}}{[1 + G(q^2)]^2}. \quad (6.27)$$

As a consequence of the all-order Ward identities satisfied by the full vertices appearing in the diagrams defining the PT-background Feynman gauge self-energy (Figure 6.8), the new SDE (6.27) has a special transversality property: in fact, gluonic and ghost contributions are separately transverse, and, in addition, no mixing between the one- and two-loop dressed diagrams will take place.

To prove this, one needs to know the Ward identity corresponding to the four fully dressed vertices appearing in $\widehat{\Pi}$: $\Gamma_{\widehat{A}AA}$, $\Gamma_{c\widehat{A}\bar{c}}$, $\Gamma_{\widehat{A}AAA}$, and, finally, $\Gamma_{c\widehat{A}\bar{c}A}$. One way to derive them is to differentiate the Ward identity functional (4.14) with respect to the corresponding field combination in which the background field has been replaced by the corresponding gauge parameter ϑ. On the other hand, one can also write the corresponding tree-level Ward identities and then use linearity to generalize them to all orders. In either case, the following results are obtained:

$$q^{\alpha}\Gamma_{\widehat{A}^a_{\alpha}A^m_{\mu}A^n_{\nu}}(k_1, k_2) = gf^{amn}\left[\Delta^{-1}_{\mu\nu}(k_1) - \Delta^{-1}_{\mu\nu}(k_2)\right] \tag{6.28}$$

$$q^{\alpha}\Gamma_{c^n\widehat{A}^a_{\alpha}\bar{c}^m}(q, -k_1) = igf^{amn}\left[D^{-1}(k_2) - D^{-1}(k_1)\right] \tag{6.29}$$

$$q^{\alpha}\Gamma_{\widehat{A}^a_{\alpha}A^b_{\beta}A^m_{\mu}A^n_{\nu}}(k_1, k_2, k_3) = gf^{adb}\Gamma_{A^d_{\beta}A^m_{\mu}A^n_{\nu}}(k_2, k_3)$$

$$+ gf^{adm}\Gamma_{A^d_{\mu}A^b_{\beta}A^n_{\nu}}(k_1, k_3)$$

$$+ gf^{adn}\Gamma_{A^d_{\nu}A^b_{\beta}A^m_{\mu}}(k_1, k_2) \tag{6.30}$$

$$q^{\alpha}\Gamma_{c^n\widehat{A}^a_{\alpha}A^b_{\beta}\bar{c}^m}(q, k_3, -k_1) = gf^{adb}\Gamma_{c^nA^d_{\beta}\bar{c}^m}(q + k_3, -k_1)$$

$$+ gf^{adm}\Gamma_{c^nA^b_{\beta}\bar{c}^d}(k_3, q - k_1)$$

$$+ gf^{adn}\Gamma_{c^dA^b_{\beta}\bar{c}^m}(k_3, -k_1). \tag{6.31}$$

We shall consider then the one-loop dressed gluonic contributions given by the combination $(d_1) + (d_2)$ of Figure 6.8. Using the Ward identity (6.28), we get

$$q^{\beta}(d_1)^{ab}_{\alpha\beta} = -g^2 C_A \delta^{ab} q_{\alpha} \int_k \Delta^{\mu}_{\mu}(k), \tag{6.32}$$

whereas by simply computing the divergence of the tree-level vertex $\Gamma_{\widehat{A}\widehat{A}AA}$, we get

$$q^{\beta}(d_2)^{ab}_{\alpha\beta} = g^2 C_A \delta^{ab} q_{\alpha} \int_k \Delta^{\mu}_{\mu}(k). \tag{6.33}$$

Thus, clearly,

$$q^{\beta}[(d_1) + (d_2)]^{ab}_{\alpha\beta} = 0. \tag{6.34}$$

Exactly the same procedure shows that for the one-loop-dressed ghost contribution,

$$q^{\beta}(d_3)^{ab}_{\alpha\beta} = -2g^2 C_A \delta^{ab} q_{\alpha} \int_k D(k) = -q^{\beta}(d_4)^{ab}_{\alpha\beta}, \tag{6.35}$$

and therefore that

$$q^{\beta}[(d_3) + (d_4)]^{ab}_{\alpha\beta} = 0. \tag{6.36}$$

For the two-loop dressed contributions, the proof is to a certain extent more involved. Begin with the gluonic contributions. Using the Ward identity (6.30), we see that diagram (d_5) would give rise, in principle, to three different terms.

However, it is straightforward to prove that these terms are all equal modulo rela-beling of momenta and Lorentz or color indices. Thus, recalling that this diagram carries a symmetry factor of $1/6$, we get

$$q^\beta (d_5)^{ab}_{\alpha\beta} = \frac{i}{2} g f^{bmn} \Gamma^{(0)}_{\widehat{A}^a_\alpha A^m_{\mu'} A^g_{\gamma'} A^e_{\epsilon'}} \int_k \int_\ell \Delta^{\epsilon'\epsilon}(k) \Delta^{\gamma'\gamma}(\ell+k) \Gamma_{A^g_\gamma A^e_\epsilon A^n_\mu}(k, \ell)$$

$$\times \Delta^{\mu'\mu}(\ell+q), \tag{6.37}$$

while the remaining graph gives (after making use of the full Bose symmetry of the three-gluon vertex)

$$q^\beta (d_6)^{ab}_{\alpha\beta} = \frac{i}{2} g f^{bmn} \Gamma^{(0)}_{\widehat{A}^a_\alpha A^m_{\mu'} A^g_{\gamma'} A^e_{\epsilon'}} \int_k \int_\ell \Delta^{\epsilon'\epsilon}(k) \Delta^{\gamma'\gamma}(\ell+k) \Gamma_{A^g_\gamma A^e_\epsilon A^n_{\nu'}}(k, \ell)$$

$$\times \left[\Delta^{\nu'\nu}(\ell) g^{\mu'}_\nu - \Delta^{\mu'\mu}(\ell+q) g^{\nu'}_\mu \right]. \tag{6.38}$$

Now, on the one hand, the first term in the square brackets must integrate to zero because the integral is independent of q, and therefore the free Lorentz index α cannot be saturated. On the other hand, the second term is exactly equal but opposite in sign to the one appearing in Eq. (6.37) so that we obtain

$$q^\beta \left[(d_5) + (d_6) \right]^{ab}_{\alpha\beta} = 0. \tag{6.39}$$

Finally, we turn to the two-loop dressed ghost contributions. Using the Ward identity (6.31), we see that the divergence of diagram (d_7) gives us three terms, namely,

$$q^\beta (d_7)^{ab}_{\alpha\beta} = -i \Gamma^{(0)}_{c^{m'} \widehat{A}^a_\alpha A^{r'}_{\rho'} \bar{c}^{n'}} \int_k \int_\ell D^{m'm}(\ell+k) D^{n'n}(\ell+q) \Delta^{\rho'\rho}_{r'r}(k)$$

$$\times \left[g f^{ber} \Gamma_{c^n A^e_\rho \bar{c}^e}(k-q, -\ell-k) + g f^{ben} \Gamma_{c^e A^r_\rho \bar{c}^m}(k, -\ell-k) \right.$$

$$\left. + g f^{ben} \Gamma_{c^n A^r_\rho \bar{c}^n}(k, -q-\ell-k) \right]. \tag{6.40}$$

Each one of these three terms can be easily shown to cancel exactly against the individual divergences of the remaining three graphs. To see this in detail, let us consider, for example, diagram (d_{10}) and use the Ward identity (6.29) to obtain

$$q^\beta (d_{10})^{ab}_{\alpha\beta} = -i g f^{bmn} \Gamma^{(0)}_{c^m \widehat{A}^a_\alpha A^r_\rho \bar{c}^d} \int_k \int_\ell \Delta^{\rho'\rho}(k) \Gamma_{c^d A^r_\rho \bar{c}^n}(k, -q-\ell-k)$$

$$\times D(\ell+q) \left[D(\ell+k) - D(\ell+k+q) \right]. \tag{6.41}$$

We see that the second term in the square brackets will integrate to zero, whereas the first term will cancel exactly the third term appearing in the square brackets of Eq. (6.40). It is not difficult to realize that the same pattern will be encountered when calculating the divergence of diagrams (d_8) and (d_9) so that one has the identity

$$q^\beta \left[(d_7) + (d_8) + (d_9) + (d_{10}) \right]_{\alpha\beta}^{ab} = 0, \tag{6.42}$$

which then concludes our proof of the special transversality properties of the new Schwinger–Dyson series.

This special property has far-reaching practical consequences for the treatment of the Schwinger–Dyson series [24, 25]. Specifically, it furnishes a systematic truncation scheme that preserves the transversality of the answer. For example, keeping only the diagrams in the first group, we obtain the truncated SDE

$$\Delta^{-1}(q^2) P_{\alpha\beta}(q) = \frac{q^2 P_{\alpha\beta}(q) + i[(d_1) + (d_2)]_{\alpha\beta}}{[1 + G(q^2)]^2}, \tag{6.43}$$

and from Eq. (6.34), we know that $[(d_1) + (d_2)]_{\alpha\beta}$ is transverse, i.e.,

$$[(d_1) + (d_2)]_{\alpha\beta} = (d-1)^{-1} [(d_1) + (d_2)]_\mu^\mu P_{\alpha\beta}(q). \tag{6.44}$$

Thus, the transverse projector $P_{\alpha\beta}(q)$ appears *exactly* on both sides of Eq. (6.43); one may susbsequently isolate the scalar cofactors on both sides, obtaining a scalar equation of the form

$$\Delta^{-1}(q^2) = \frac{q^2 + i[(d_1) + (d_2)]_\mu^\mu}{[1 + G(q^2)]^2}. \tag{6.45}$$

A truncated equation similar to Eq. (6.43) may be written for any other of the four groups previously isolated, or for sums of these groups, without compromising the transversality of the answer. The price one has to pay for this advantageous situation is rather modest and consists in considering the additional equation determining the scalar function $G(q^2)$; notice, however, that one can approximate this function via a dressed-loop expansion without jeopardizing the transversality of $\Pi_{\alpha\beta}(q)$, given that $[1 + G(q^2)]^2$ affects only the size of the scalar prefactor.

Let us conclude by noticing that in going from Eq. (6.26) to Eq. (6.27), one essentially chooses to retain the original propagator $\Delta(q)$ as the unknown quantity to be dynamically determined from the Schwinger–Dyson equation. There is, of course, an alternative strategy: one may define a new *variable* from the quantity appearing on the lhs of Eq. (6.26), namely,

$$\widehat{\Delta}(q) \equiv \left[1 + G(q^2) \right]^{-2} \Delta(q), \tag{6.46}$$

which leads to a new form for Eq. (6.26):

$$\widehat{\Delta}^{-1}(q^2)P_{\alpha\beta}(q) = q^2 P_{\alpha\beta}(q) + i \sum_{i=1}^{11} (d_i)_{\alpha\beta}. \tag{6.47}$$

Obviously, the special transversality properties established earlier hold as well for Eq. (6.47); for example, one may truncate it gauge invariantly as

$$\widehat{\Delta}^{-1}(q^2)P_{\alpha\beta}(q) = q^2 P_{\alpha\beta}(q) + i[(d_1) + (d_2)]_{\alpha\beta}. \tag{6.48}$$

Should one opt for treating $\widehat{\Delta}(q)$ as the new unknown quantity, then an additional step must be carried out: one must use Eq. (6.46) to rewrite the entire rhs of Eq. (6.47) in terms of $\widehat{\Delta}$ instead of Δ, i.e., carry out the replacement $\Delta \to [1 + G]^2 \, \widehat{\Delta}$ *inside* every diagram on the rhs of Eq. (6.47) that contains Δs.

Therefore, whereas Eq. (6.43) furnishes a gauge-invariant approximation for the conventional gluon self-energy $\Delta(q)$, Eq. (6.47) is the gauge-invariant approximation for the effective PT self-energy $\widehat{\Delta}(q)$. The crucial point is that one may switch from one to the other by means of Eq. (6.46). For practical purposes, this means, for example, that one may get a gauge-invariant approximation not just for the PT quantity (background Feynman gauge) but also for the conventional self-energy computed in the Feynman gauge. Equation (6.46) plays an instrumental role in this entire construction, allowing one to convert the Schwinger–Dyson series into a dynamical equation for either $\widehat{\Delta}(q)$ or $\Delta(q)$.

6.4.5 Truncation of the pinch technique Schwinger–Dyson equation

The new Schwinger–Dyson series projected by the pinch technique has both theoretical and practical advantages. On the one hand, the main theoretical advantage is that the various fully dressed graphs organize themselves into gauge-invariant subsets, thus allowing for a systematic gauge-invariant truncation. On the other hand, at the practical level, this property reduces the number of coupled SDEs that one has to consider to maintain the gauge (BRST) symmetry of the theory to only two: the one for the gluon self-energy, given by, e.g., the first gauge-invariant subset only (i.e., $[(d_1) + (d_2)]_{\alpha\beta}$ in Figure 6.8), and the Schwinger–Dyson equation satisfied by the full three-gluon vertex. This is to be contrasted to what happens within the conventional formulation where the equations for all vertices must be considered, or else the transversality of the gluon propagator is violated (which is what usually happens).

However, even the PT analysis does not furnish a simple diagrammatic truncation, analogous to that of the gluon self-energy, for the Schwinger–Dyson equation satisfied by the three-gluon vertex $\Gamma_{\widehat{A}_\alpha A_\mu A_\nu}(k_1, k_2)$. Thus, if one were to truncate the equation for the three-gluon vertex by discarding some of the graphs appearing in it, the validity of the all-order Ward identity (6.28) would be violated; this, in turn, would lead immediately to the violation of the transversality property of the subset kept, thus making the entire truncation scheme collapse.

One should adopt the following strategy instead: given that the proposed truncation scheme hinges crucially on the validity of Eq. (6.28), one should start with an approximation that manifestly preserves it. The way to enforce this, therefore, is through the gauge technique, as described in Chapter 5.

6.5 Solutions of the pinch technique Schwinger–Dyson equations and comparison with lattice data

We will next focus on a particular application of the powerful machinery offered by the truncation scheme introduced in the previous section. In particular, we will study gauge invariantly the coupled SDEs of the gluon and ghost propagators and compare the results obtained to recent lattice data. To this end, we will do the following:

1. We will truncate (gauge invariantly) the gluon propagator SDE by keeping only the one-loop dressed contributions; that is, we only consider the first two blocks appearing in Figure 6.8. Recall that when evaluating these diagrams, one must employ the BFM Feynman rules. Note also the following crucial point: the (background) three-gluon vertex $\Gamma_{\widehat{A}AA}$ and the ghost vertex $\Gamma_{c\widehat{A}\bar{c}}$ appearing in the corresponding diagrams must be fully dressed and satisfy the correct Ward identities, namely, Eqs. (6.28) and (6.29), respectively, to enforce the transversality of the resulting gluon self-energy.

2. For the ghost SDE, we use the equation shown in Figure 6.5. Notice that the ghost vertex appearing in the SDE is the conventional one, $\Gamma_{cA\bar{c}}$; therefore, one may employ a different approximation than the one used for the background ghost vertex. In particular, according to lattice studies [26, 27], the vertex $\Gamma_{cA\bar{c}}$ may be accurately approximated simply by its tree-level expression. The ability to employ a different treatment for the two ghost vertices, without compromising gauge invariance, is indicative of the versatility of the new truncation scheme introduced earlier.[4]

3. For the SDE (4.44) governing the dynamics of the auxiliary function $\Gamma_{\Omega A^*}$, the vertex Γ_{AcA^*} will be approximated by its tree-level value.[5]

---

[4] Notice also that, whereas the background ghost vertex satisfies a linear Ward identity, which can be solved through a gauge technique ansatz, the conventional vertex satisfies a Slavnov–Taylor identity of rather limited usefulness.

[5] This vertex has not been studied on the lattice; the only constraints are the ones imposed by the Faddeev–Popov equation (4.26), relating it with the conventional ghost vertex, that are indeed satisfied by our tree-level choice for both vertices.

We then have the following equations:

$$\Delta^{-1}(q^2)P_{\alpha\beta}(q) = \frac{q^2 P_{\alpha\beta}(q) + i\sum_{1=1}^{4}(d_i)_{\alpha\beta}}{[1 + G(q^2)]^2},$$

$$F^{-1}(q^2) = 1 + g^2 C_A \int_k \left[1 - \frac{(k \cdot q)^2}{k^2 q^2}\right] \Delta(k)D(k+q),$$

$$G(q^2) = \frac{g^2 C_A}{d-1} \int_k \left[(d-2) + \frac{(k \cdot q)^2}{k^2 q^2}\right] \Delta(k)D(k+q),$$

$$L(q^2) = \frac{g^2 C_A}{d-1} \int_k \left[1 - d\frac{(k \cdot q)^2}{k^2 q^2}\right] \Delta(k)D(k+q). \qquad (6.49)$$

The concrete dynamics that will give rise to an infrared finite gluon propagator, signaling the dynamical generation of a gluon mass, are inserted into the first of the preceding equations through an appropriate gauge technique ansatz for the fully dressed vertices $\Gamma_{\widehat{A}_\alpha A_\mu A_\nu}(k_1, k_2)$ and $\Gamma_{c\widehat{A}_\alpha \bar{c}}(q, k_1)$, appearing in the diagrams (d_1) and (d_3) of Figure 6.8. Specifically, one expresses $\Gamma_{\widehat{A}_\alpha A_\mu A_\nu}(k_1, k_2)$ and $\Gamma_{c\widehat{A}_\alpha \bar{c}}(q, k_1)$ as a function of the gluon and ghost self-energy, respectively, such that they automatically satisfy the two Ward identities (6.28) and (6.29). In addition, as already explained in the previous chapter, the vertices must contain longitudinally coupled massless (bound state) poles, as is required for triggering the Schwinger mechanism. We will use the following simplified gauge technique ansatze:

$$\Gamma_{\widehat{A}_\alpha A_\mu A_\nu}(k_1, k_2) = \Gamma^{(0)}_{\widehat{A}_\alpha A_\mu A_\nu}(k_1, k_2) + i\frac{q_\alpha}{q^2}\left[\Pi_{\mu\nu}(k_2) - \Pi_{\mu\nu}(k_1)\right]$$

$$\Gamma_{\widehat{A}_\alpha c\bar{c}}(k_1, k_2) = \Gamma^{(0)}_{\widehat{A}_\alpha c\bar{c}}(k_1, k_2) - i\frac{q_\alpha}{q^2}\left[L(k_2^2) - L(k_1^2)\right], \qquad (6.50)$$

where L denotes the ghost self-energy, $D^{-1}(p^2) = p^2 - iL(p^2)$. It is elementary to verify that the preceding two vertices indeed satisfy the correct Ward identities (6.28) and (6.29), respectively.

The resulting expression in the Landau gauge[6] for the gluon SDE is too long to be reported here [28]; rather, we show in Figure 6.9 the solution of the full system (6.49) when $G(q^2)$ is approximated by its one-loop expression ($L(q^2)$ is not needed in this case), comparing it with the corresponding lattice data.

Note that, whereas there is good qualitative agreement with the lattice, there is a significant discrepancy (a factor of 2) in the intermediate region of momenta. This,

---

[6] The projection to the Landau gauge is a subtle exercise because one cannot directly set $\xi = 0$ in the integrals given the presence of terms proportional to $1/\xi$ in the background three-gluon vertex. Instead, one has to use the expressions for general ξ, carry out explicitly the set of cancellations produced when the terms proportional to ξ generated by the identity $k^\mu \Delta_{\mu\nu}(k) = -i\xi k_\nu/k^2$ are used to cancel $1/\xi$ terms, and set $\xi = 0$ only at the very end.

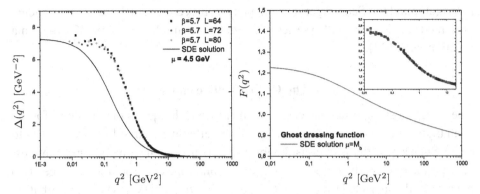

Figure 6.9. (*Left*) The numerical solution for the gluon propagator from the PT modified SDE (solid line) compared to the lattice data of [12]. (*Right*) The ghost-dressing function $p^2 D(p^2)$ obtained from the SDE. In the inset are the lattice data for the same quantity; notice the absence of any enhancement in both cases.

of course, may not come as a surprise, given that (1) the two-loop dressed part of the SDE for the gluon propagator has been omitted (the last two blocks in Figure 6.8) and (2) the auxiliary function $G(q^2)$ has been evaluated at the one-loop level only. Even though this omission has not introduced artifacts (because it was done gauge invariantly), the terms left out are expected to modify precisely the intermediate region, given that both the infrared and ultraviolet limits of the solutions are already captured by the one-loop dressing terms considered.

The finiteness of the ghost-dressing function F in the infrared has an important theoretical consequence related with the so-called Kugo–Ojima confinement criterion, which is already mentioned at the beginning of Chapter 4. In the Kugo–Ojima confinement picture (in covariant gauges), the absence of colored asymptotic states from the physical spectrum of the theory is due to the so-called quartet mechanism [29]. A sufficient condition for the realization of this mechanism (and the meaningful definition of a conserved BRST charge) is that the correlation function $u(q^2)$ defined in Eq. (4.51) satisfies the condition $u(0) = -1$. In addition, as first noted by Kugo [30], in Landau gauge, $u(0)$ is related to the infrared behavior of the ghost-dressing function $F(q^2)$ through the identity $F^{-1}(0) = 1 + u(0)$. This is none other than the second identity of Eq. (4.50), under the additional assumption that $L(0) = 0$. In fact, from the last of Eqs. (6.49), it is straightforward to establish that if both F and Δ are infrared finite, this condition is indeed fulfilled [31, 32]. Therefore, the Kugo–Ojima confinement scenario predicts a divergent ghost-dressing function and vice versa. Interestingly, the same prediction about $F^{-1}(0)$ is obtained when implementing the Gribov–Zwanziger horizon condition [22, 33]: in the infrared region, the ghost propagator diverges more rapidly than at tree level. Evidently, these affirmations are at odds not only with the recent lattice results of the ghost propagator and the Schwinger–Dyson analysis based on the PT truncation scheme presented here but also with a series of direct lattice simulations of the Kugo–Ojima function itself [34].

6.6 The QCD effective charge

One of the most important successes of the pinch technique is that it allows for the unambiguous extension of the concept of the effective charge [35] from QED to QCD. Such a quantity is of considerable theoretical and phenomenological interest because once correctly defined, it provides a continuous interpolation between two physically distinct QCD regimes: the deep ultraviolet (UV), where perturbation theory works well, and the deep infrared (IR), where nonperturbative techniques must be employed. In fact, the effective charge is intimately connected with two phenomena that are of central importance to QCD: asymptotic freedom in the UV and dynamical gluon mass generation in the IR.

6.6.1 The prototype: The QED effective charge

The quantity that serves as the field-theoretic prototype for guiding our analysis is the effective charge of QED. Consider the bare dressed photon propagator between conserved external currents:

$$\Delta_0^{\mu\nu}(q^2) = \frac{g^{\mu\nu}}{q^2[1 + \Pi_0(q^2)]}, \tag{6.51}$$

where the dimensionless quantity $\Pi(q^2)$ is the vacuum polarization, which is independent of the gauge-fixing parameter to all orders. The expression $\Delta_0^{\mu\nu}(q^2)$ is renormalized multiplicatively according to $\Delta_0^{\mu\nu}(q^2) = Z_A \Delta^{\mu\nu}(q^2)$, where Z_A is the wave function renormalization of the photon ($A_0 = Z_A^{1/2} A$). Imposing the on-shell renormalization condition for the photon, we obtain $\Pi(q^2) = Z_A[1 + \Pi_0(q^2)]$, where $Z_A = 1 - \Pi_0(0)$ and $\Pi(q^2) = \Pi_0(q^2) - \Pi_0(0)$; clearly $\Pi(0) = 0$.

The renormalization procedure introduces, in addition, the standard relations between renormalized and unrenormalized electric charge, $e = Z_e^{-1} e^0 = Z_f Z_A^{1/2} Z_1^{-1} e_0$, where Z_e is the charge renormalization constant, Z_f the wave function renormalization constant of the fermion, and Z_1 the vertex renormalization. From the famous QED relation $Z_1 = Z_f$, a direct consequence of the Ward identity $q^\mu \Gamma_\mu^0(p, p + q) = S_0^{-1}(p + q) - S_0^{-1}(p)$, it follows immediately that $Z_e = Z_A^{-1/2}$.

Given these relations between the renormalization constants, we can now form the following combination:

$$e_0^2 \Delta_0^{\mu\nu}(q^2) = e^2 \Delta^{\mu\nu}(q^2), \tag{6.52}$$

which is invariant under the renormalization group (RG); that is, it maintains the same form before and after renormalization. After pulling out the kinematic factor $(1/q^2)$, one may define the QED effective charge $\alpha_{\text{eff}}(q^2)$, namely,

$$\alpha_{\text{eff}}(q^2) = \frac{\alpha}{[1 + \Pi(q^2)]}, \tag{6.53}$$

where α is the fine-structure constant.

The QED effective charge of Eq. (6.53) is independent of the gauge-fixing parameter and invariant under the renormalization group to all orders in perturbation theory. Furthermore, given that $\Pi(0) = 0$, at low energies, the effective charge matches on to the fine-structure constant: $\alpha_{\text{eff}}(0) = \alpha = 1/137.036 \cdots$.

In addition, for asymptotically large values of q^2, i.e., for $q^2 \gg m_f^2$, where m_f denotes the masses of the fermions contributing to the vacuum polarization loop ($f = e, \mu, \tau, \ldots$), $\alpha(q^2)$ matches on to the running coupling $\bar{\alpha}(q^2)$ defined from the RG. At the one-loop level,

$$\alpha_{\text{eff}}(q^2) \xrightarrow{q^2 \gg m_f^2} \bar{\alpha}(q^2) = \frac{\alpha}{1 - (\alpha\beta_1/2\pi)\log(q^2/m_f^2)}, \tag{6.54}$$

where $\beta_1 = 2/3 n_f$ is the coefficient of the QED beta function for n_f fermion types.

6.6.2 The QCD effective charge

The main difficulty in the generalization of the QED effective charge to QCD (or any other NAGT) is that, unlike the QED vacuum polarization, the conventional QCD gluon self-energy depends explicitly on the gauge-fixing parameter. In addition, with the exception of some special gauges, $Z_1 \neq Z_f$, and therefore the gluon self-energy does not capture, in general, the leading renormalization group logarithms.

The pinch technique solves the preceding difficulties at once: the new gluon self-energy is independent of the gauge-fixing parameter, whereas the Abelian ward identities satisfied by the PT Green's functions (e.g., quark-gluon vertex) restore the crucial equalities $\widehat{Z}_1 = \widehat{Z}_f$ and $Z_g = \widehat{Z}_A^{-1/2}$. In addition, the PT self-energy is process independent [36] (see also Chapter 1) and can be Dyson resummed to all orders [37, 38, 39, 40, 41] (see Chapter 11). Therefore, the construction of the universal and RG-invariant combination analogous to that of Eq. (6.52) is immediate because the quantity

$$\widehat{d}_0(q^2) = g_0^2 \widehat{\Delta}_0(q^2) = g^2(\mu^2)\widehat{\Delta}(q^2, \mu^2) = \widehat{d}(q^2), \tag{6.55}$$

is manifestly RG invariant (viz., μ independent). Then, by virtue of the identity (4.35), we can write equivalently

$$\widehat{d}(q^2) = g^2(\mu^2)\frac{\Delta(q^2, \mu^2)}{\left[1 + G(q^2, \mu^2)\right]^2}. \tag{6.56}$$

It is important to obtain a quantitative confirmation of the μ independence of $\widehat{d}(q^2)$ using as ingredients the individually μ-dependent solutions for $\Delta(q^2, \mu^2)$ and $G(q^2, \mu^2)$ obtained from the system of SDEs given in Eq. (6.49). At this point, one notices a shift in our philosophy, which is, however, dictated only by practical considerations. Specifically, as repeatedly stated, the genuine PT propagator $\widehat{\Delta}(q^2, \mu^2)$ is the one obtained in the Feynman gauge of the BFM, whereas the system of Eq. (6.49) that we consider is formulated in the Landau gauge mainly because the relevant lattice simulations are almost exclusively carried out in this latter gauge. Therefore, we study for the rest of this section the generalized PT effective charge in the BFM Landau gauge (see Section 2.3.4). As is well known, and contrary to what happens in the conventional R_ξ gauges, at one loop in perturbation theory, the coefficient multiplying the renormalization group logarithms does not depend on the specific value of ξ_Q (see Eq. (2.71)). As a result, the asymptotic (ultraviolet) behavior of the two charges (Feynman vs. Landau gauges) will be identical. Note, however, that this is not necessarily so in the infrared; in fact, the freezing value obtained for the Landau gauge effective charge is significantly elevated compared to the value of about 0.5 that one finds for the genuine PT effective charge.

Returning to Eq. (6.56), one may clearly see in Figure 6.10 that all quantities obtained from the aforementioned SDEs, and in particular $\Delta(q^2, \mu^2)$ and $G(q^2, \mu^2)$, display a sizable dependence on the choice of the renormalization point μ. However, when $\Delta(q^2, \mu^2)$ and $G(q^2, \mu^2)$ are used as inputs to form the special combination (6.56), their net μ dependence cancels almost completely against that of $g^2(\mu)$; the latter dependence is obtained from the four-loop beta function corresponding to the minimal-subtraction scheme that we employ for renormalizing the SDEs [42] (see Figure 6.11). Thus one ends up with a nearly μ-independent quantity, as is clearly shown in Figure 6.12.

The next step is to extract out of the dimensionful $\widehat{d}(q^2)$ a dimensionless quantity that would correspond to the QCD effective charge. In the perturbative regime, when momenta are asymptotically large, it is clear that the mass scale is saturated simply by q^2, the bare gluon propagator, and the effective charge is defined by pulling a factor $1/q^2$ out of the corresponding RG-invariant quantity exactly as

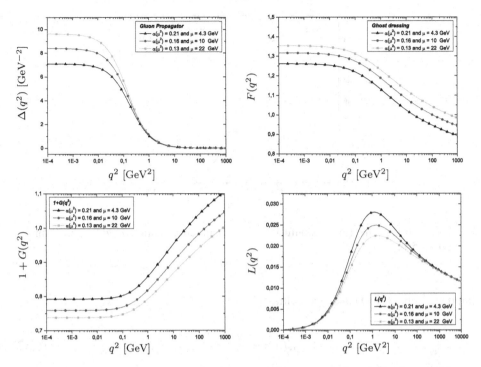

Figure 6.10. The μ dependence of the quantities (*top left*) $\Delta(q^2)$, (*top right*) $F(q^2)$, (*bottom left*) $G(q^2)$, and (*bottom right*) $L(q^2)$ as obtained from solving the system of SDEs (6.49).

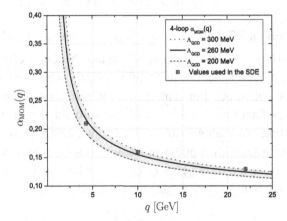

Figure 6.11. The perturbative running coupling in the MOM scheme up to four loops, $\alpha_{\mathrm{MOM}}(q^2)$, for different values of Λ_{QCD}. Black squares represent the values used for $\alpha(\mu^2)$.

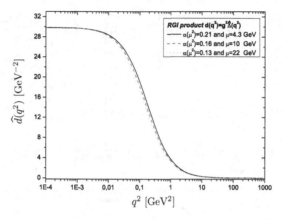

Figure 6.12. The (dimensionful) RG-invariant combination $\widehat{d}(q^2)$ defined in Eq. (6.56).

happens in the QED case. We define

$$\widehat{d}(q^2) = \frac{\overline{g}^2(q^2)}{q^2},$$

(6.57)

with $\overline{g}^2(q^2)$ being the RG-invariant effective charge of QCD; then, at one loop,

$$\overline{g}^2(q^2) = \frac{g^2}{1 + bg^2 \ln \left(q^2/\mu^2\right)} = \frac{1}{b \ln \left(q^2/\Lambda_{\text{QCD}}^2\right)},$$

(6.58)

where Λ_{QCD} denotes an RG-invariant mass scale of a few hundred MeV.

On the other hand, given that the gluon propagator becomes effectively massive in the IR, particular care is needed in deciding exactly what combination of mass scales ought to be pulled out. For example, if one insists on defining the effective charge by trivially factoring out $1/q^2$, the result obtained will be a completely unphysical coupling, vanishing in the deep IR, where QCD is supposed to be strongly coupled. Instead, the correct procedure in such a case is to factor out a massive propagator of the form $[q^2 + m^2(q^2)]^{-1}$, where $m^2(q^2)$ is the dynamical momentum-dependent mass; that is, one must set [23, 31]

$$\widehat{d}(q^2) = \frac{\overline{g}^2(q^2)}{q^2 + m^2(q^2)}.$$

(6.59)

Clearly, for $q^2 \gg m^2(q^2)$, the expression on the rhs of Eq. (6.59) goes over to that of Eq. (6.57).

Even though the aforementioned procedure of factoring out a massive propagator was spelled out long ago in the original work on dynamical gluon mass generation [23], owing to a variety of recent developments, this issue deserves some further clarification. To that end, it is instructive to compare the situation with the more familiar, and conceptually more straightforward, case of the electroweak sector of the SM, where the corresponding gauge bosons (W and Z) are also massive, albeit through an entirely different mass-generation mechanism; indeed, despite the difference in their origins, the masses act in a very similar fashion at the level of the RG-invariant quantity associated with the corresponding gauge boson.

In the case of the W-boson, the quantity corresponding to that of Eq. (6.59) would read (Euclidean momenta)

$$\widehat{d}_w(q^2) = \frac{\overline{g}_w^2(q^2)}{q^2 + M_w^2},\tag{6.60}$$

with

$$\overline{g}_w^2(q^2) = g_w^2(\mu)\left[1 + b_w g_w^2(\mu)\int_0^1 dx \ln\left(\frac{q^2 x(1-x) + M_w^2}{\mu^2}\right) - \cdots\right]^{-1},\tag{6.61}$$

where $b_w = 11/24\pi^2$ and the ellipses denote the contributions of the fermion families. Clearly, $\widehat{d}_w(0) = \overline{g}_w^2(0)/M_w^2$, with

$$\overline{g}_w^2(0) = g_w^2(\mu)\left[1 + b_w g_w^2(\mu)\ln\left(\frac{M_w^2}{\mu^2}\right)\right]^{-1}.\tag{6.62}$$

Evidently, in the deep IR, the coupling freezes at a constant value; Fermi's constant is in fact determined as[7] $4\sqrt{2}G_F = \overline{g}_w^2(0)/M_w^2$.

This property of the freezing of the coupling can be reformulated in terms of what in the language of the effective field theories is referred to as *decoupling* [43]. At energies that are small compared to their masses, the particles appearing in the loops (in this case, the gauge bosons) cease to contribute to the running of the coupling. Possibly large logarithmic constants, e.g., $\ln(M_w^2/\mu^2)$, may be reabsorbed in the renormalized value of the coupling. Of course, the decoupling, as described earlier should not be misinterpreted to mean that the running coupling vanishes; instead, as already mentioned, it freezes at a constant, nonzero value. In other words, the decoupling does not imply that the theory becomes free (noninteracting) in the IR.

It would certainly be incorrect in such cases to insist on the perturbative prescription and simply factor out a $1/q^2$. Even though one is merely redistributing a given

---

[7] Note that in the case of QCD, the corresponding combination, $\overline{g}^2(0)/m^2(0)$, would be similar to a Nambu–Jona–Lasinio type of coupling [1]: at energies below the gluon mass m, the tree-level amplitude of four quarks starts looking a lot like that of a four-Fermi interaction.

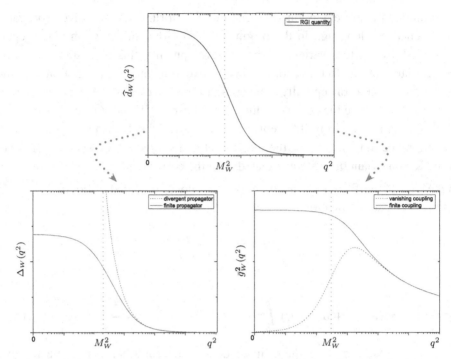

Figure 6.13. A sketch plot that shows how the same RG-invariant quantity in the presence of the mass scale M_w^2 can be decomposed in two different ways: one giving a divergent propagator and a vanishing coupling and the other, giving a finite propagator and a finite coupling.

function, namely, $\widehat{d}_w(q^2)$, into two pieces, factoring out $1/q^2$ deprives both of any physical meaning. Indeed, the effective coupling so defined would be given by the expression $\widetilde{g}_w^2(q^2) = q^2 \widehat{d}_w(q^2)$, and so $\widetilde{g}_w^2(0) = 0$; evidently, one would be attempting to describe weak interactions in terms of a massless, IR-divergent, gauge-boson propagator and a vanishing effective coupling (see the dotted lines in Figure 6.13). Correspondingly, given that the gluon propagator is finite in the IR, if this latter (wrong) procedure were to be applied to QCD, it would furnish a completely unphysical coupling, namely, one that vanishes in the deep infrared, where QCD is expected to be (and is) strongly coupled.

After this long detour, let us return again to the basic equation (6.59). To actually determine the effective charge from $\widehat{d}(q^2)$, some additional information about the concrete running of $m^2(q^2)$ must be provided. The actual running of the mass may be determined dynamically through elaborate considerations that we will not present here. Instead, we will assume that $m^2(q^2)$ displays a power-law running, as shown by the early work of Lavelle [44] (see Chapter 2) and as has been independently confirmed within an entirely different formalism [45]. In Figure 6.14, we show the

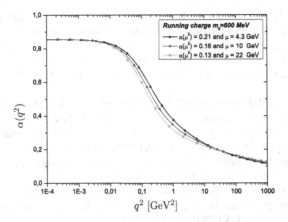

Figure 6.14. The effective charge $\alpha(q^2)$ obtained from the RG-invariant $\widehat{d}(q^2)$ using Eqs. (6.59) and (6.63) for a value of the effective gluon mass $m_0 = 600$ MeV.

effective charge, $\alpha = \overline{g}^2(q^2)/4\pi$, obtained from the $\widehat{d}(q^2)$ shown in Figure 6.14 (*left*) after using Eq. (6.59) with a running gluon mass of the form

$$m^2(q^2) = \frac{m_0^4}{q^2 + m_0^2}. \tag{6.63}$$

One observes the *freezing* of the coupling at a finite nonvanishing value,[8] which is a direct consequence of the appearance of the dynamical mass in the RG logarithm.

References

[1] C. Alexandrou, P. de Forcrand, and E. Follana, The gluon propagator without lattice Gribov copies, *Phys. Rev.* **D63** (2001) 094504.

[2] C. Alexandrou, P. de Forcrand, and E. Follana, The gluon propagator without lattice Gribov copies on a finer lattice, *Phys. Rev.* **D65** (2002) 114508.

[3] P. Boucaud et al., The infrared behaviour of the pure Yang-Mills Green functions, arXiv hep-ph/0507104.

[4] P. Boucaud et al., Short comment about the lattice gluon propagator at vanishing momentum, arXiv hep-lat/0602006.

[5] P. Boucaud, J. P. Leroy, A. Le Yaouanc, J. Micheli, O. Pene, and J. Rodriguez-Quintero, On the IR behaviour of the Landau-gauge ghost propagator, *JHEP* **0806** (2008) 099.

[6] P. O. Bowman, U. M. Heller, D. B. Leinweber, M. B. Parappilly, and A. G. Williams, Unquenched gluon propagator in Landau gauge, *Phys. Rev.* **D70** (2004) 034509.

---

[8] This means, in particular, that when quark masses are ignored, QCD becomes conformally invariant in the deep IR. The existence of such a conformal window has been advocated for certain applications of the AdS/CFT correspondence in QCD [46].

[7] W. Kamleh, P. O. Bowman, D. B. Leinweber, A. G. Williams, and J. Zhang, Unquenching effects in the quark and gluon propagator, *Phys. Rev.* **D76** (2007) 094501.

[8] A. Cucchieri, T. Mendes, O. Oliveira, and P. J. Silva, Just how different are $SU(2)$ and $SU(3)$ Landau-gauge propagators in the IR regime?, *Phys. Rev.* **D76** (2007) 114507.

[9] O. Oliveira and P. J. Silva, Infrared gluon and ghost propagators from lattice QCD: Results from large asymmetric lattices, *Eur. Phys. J.* **A31** (2007) 790.

[10] O. Oliveira, P. J. Silva, E. M. Ilgenfritz, and A. Sternbeck, The gluon propagator from large asymmetric lattices, *PoS* **LAT2007** (2007) 323.

[11] O. Oliveira and P. J. Silva, The lattice infrared Landau gauge gluon propagator: The infinite volume limit, arXiv 0910.2897 [hep-lat].

[12] I. L. Bogolubsky, E. M. Ilgenfritz, M. Muller-Preussker, and A. Sternbeck, The Landau gauge gluon and ghost propagators in 4D $SU(3)$ gluodynamics in large lattice volumes, *PoS* **LAT2007** (2007) 290.

[13] E. M. Ilgenfritz, M. Muller-Preussker, A. Sternbeck, A. Schiller, and I. L. Bogolubsky, Landau gauge gluon and ghost propagators from lattice QCD, *Braz. J. Phys.* **37** (2007) 193.

[14] I. L. Bogolubsky, V. G. Bornyakov, G. Burgio, E. M. Ilgenfritz, M. Müller-Preussker, P. Schemel, and V. K. Mitrjushkin, The Landau gauge gluon propagator: Gribov problem and finite-size effects, *PoS* **LAT2007** (2007) 318.

[15] I. L. Bogolubsky, E. M. Ilgenfritz, M. Müller-Preussker, and A. Sternbeck, Lattice gluodynamics computation of Landau gauge Green's functions in the deep infrared, *Phys. Lett.* **B676** (2009) 69.

[16] A. Cucchieri and T. Mendes, What's up with IR gluon and ghost propagators in Landau gauge? A puzzling answer from huge lattices, *PoS* **LAT2007** (2007) 297.

[17] A. Cucchieri and T. Mendes, Constraints on the IR behavior of the gluon propagator in Yang-Mills theories, *Phys. Rev. Lett.* **100** (2008) 241601.

[18] A. Cucchieri and T. Mendes, Landau-gauge propagators in Yang-Mills theories at beta = 0: Massive solution versus conformal scaling, *Phys. Rev.* **D81** (2010) 016005.

[19] A. Cucchieri and T. Mendes, Numerical test of the Gribov-Zwanziger scenario in Landau gauge, *PoS* **QCD-TNT09** (2010) 026.

[20] A. Sternbeck, L. von Smekal, D. B. Leinweber, and A. G. Williams, Comparing $SU(2)$ to $SU(3)$ gluodynamics on large lattices, *PoS* **LAT2007** (2007) 340.

[21] Y. B. Zhang, J. L. Ping, X. F. Lu, and H. S. Zong, Unquenched effects and quark mass dependence of lattice gluon propagator in infrared region, *Commun. Theor. Phys.* **50** (2008) 125.

[22] V. N. Gribov, Quantization of non-Abelian gauge theories, *Nucl. Phys.* **B139** (1978) 1.

[23] J. M. Cornwall, Dynamical mass generation in continuum QCD, *Phys. Rev.* **D26** (1982) 1453.

[24] D. Binosi and J. Papavassiliou, New Schwinger-Dyson equations for non-Abelian gauge theories, *JHEP* **0811** (2008) 063.

[25] D. Binosi and J. Papavassiliou, Gauge-invariant truncation scheme for the Schwinger-Dyson equations of QCD, *Phys. Rev.* **D77**(R) (2008) 061702.

[26] A. Cucchieri, T. Mendes, and A. Mihara, Numerical study of the ghost-gluon vertex in Landau gauge, *JHEP* **0412** (2004) 012.

[27] E. M. Ilgenfritz, M. Müller-Preussker, A. Sternbeck, and A. Schiller, Gauge-variant propagators and the running coupling from lattice QCD, arXiv hep-lat/0601027.

[28] A. C. Aguilar, D. Binosi, and J. Papavassiliou, Gluon and ghost propagators in the Landau gauge: Deriving lattice results from Schwinger-Dyson equations, *Phys. Rev.* **D78** (2008) 025010.

[29] T. Kugo and I. Ojima, Local covariant operator formalism of non-Abelian gauge theories and quark confinement problem, *Prog. Theor. Phys. Suppl.* **66** (1979) 1.

[30] T. Kugo, The universal renormalization factors Z(1)/Z(3) and color confinement condition in non-Abelian gauge theory, talk given at the International Symposium on BRS Symmetry on the Occasion of Its 20th Anniversary, Kyoto, Japan (1995).

[31] A. C. Aguilar, D. Binosi, J. Papavassiliou, and J. Rodriguez-Quintero, Non-perturbative comparison of QCD effective charges, *Phys. Rev.* **D80** (2009) 085018.

[32] A. C. Aguilar, D. Binosi, and J. Papavassiliou, Indirect determination of the Kugo-Ojima function from lattice data, *JHEP* **0911** (2009) 066.

[33] D. Zwanziger, Fundamental modular region, Boltzmann factor and area law in lattice gauge theory, *Nucl. Phys.* **B412** (1994) 657.

[34] A. Sternbeck, The infrared behavior of lattice QCD Green's functions, arXiv hep-lat/0609016, and references therein.

[35] M. Gell-Mann and F. E. Low, Quantum electrodynamics at small distances, *Phys. Rev.* **95** (1954) 1300.

[36] N. J. Watson, Universality of the pinch technique gauge boson self-energies, *Phys. Lett.* **B349** (1995) 155.

[37] J. Papavassiliou and A. Pilaftsis, Gauge invariance and unstable particles, *Phys. Rev. Lett.* **75** (1995) 3060.

[38] J. Papavassiliou and A. Pilaftsis, A gauge independent approach to resonant transition amplitudes, *Phys. Rev.* **D53** (1996) 2128.

[39] J. Papavassiliou and A. Pilaftsis, Gauge-invariant resummation formalism for two point correlation functions, *Phys. Rev.* **D54** (1996) 5315.

[40] N. J. Watson, The gauge-independent QCD effective charge, *Nucl. Phys.* **B494** (1997) 388.

[41] D. Binosi and J. Papavassiliou, The QCD effective charge to all orders, *Nucl. Phys. Proc. Suppl.* **121** (2003) 281.

[42] P. Boucaud et al., Artifacts and $<A^2>$ power corrections: Re-examining Z_ψ^2 and Z_v in the momentum-subtraction scheme, *Phys. Rev. D* **74** (2006) 034505.

[43] T. Appelquist and J. Carazzone, Infrared singularities and massive fields, *Phys. Rev.* **D11** (1975) 2856.

[44] M. Lavelle, Gauge invariant effective gluon mass from the operator product expansion, *Phys. Rev.* **D44** (1991) 26.

[45] D. Dudal, J. A. Gracey, S. P. Sorella, N. Vandersickel, and H. Verschelde, A refinement of the Gribov-Zwanziger approach in the Landau gauge: Infrared propagators in harmony with the lattice results, *Phys. Rev.* **D78** (2008) 065047.

[46] S. J. Brodsky and G. F. de Teramond, Light-front hadron dynamics and AdS/CFT correspondence, *Phys. Lett.* **B582** (2004) 211.

7

Nonperturbative gluon mass and quantum solitons

7.1 Notation

In this chapter and Chapters 8 and 9, we work in Euclidean space with metric $\delta_{\mu\nu}$ so that a (real) Euclidean four-vector has $q^2 > 0$; timelike vectors in Minkowski space have negative squared length in this metric. To avoid a profusion of is and coupling constants g, we introduce the notation

$$\mathcal{A}_\mu = \frac{g}{i} A_\mu = \frac{g}{i} A_\mu^a t^a, \tag{7.1}$$

where (see Chapter 1) the t^a are the elements of the Lie algebra in the fundamental representation. Other changes include the following:

$$\frac{1}{q^2 - m^2 + i\epsilon} \to \frac{1}{q^2 + m^2}, \tag{7.2}$$

$$-g_{\mu\nu} + \frac{q_\mu q_\nu}{q^2} \to \delta_{\mu\nu} - \frac{q_\mu q_\nu}{q^2},$$

$$\mathcal{D}_\mu \to \partial_\mu + \mathcal{A}_\mu,$$

where the left-hand sides refer to Minkowski space and the right-hand sides to Euclidean space. The field strength tensor is

$$\mathcal{G}_{\mu\nu} = [\mathcal{D}_\mu, \mathcal{D}_\nu] \tag{7.3}$$

so that the action is

$$S = -\frac{1}{2g^2} \int d^4x \, \mathrm{Tr} \, \mathcal{G}_{\mu\nu}^2. \tag{7.4}$$

7.2 Introduction

We have seen in previous chapters that infrared instability of NAGTs leads to generation of a dynamical gluon mass that removes the infrared singularities of

144

perturbation theory. But mere removal of these singularities, important as it is, is far from the major role of gluon mass generation. In this and the next chapter, we survey the important implications of gluon mass generation for QCD-like theories. The most direct connection is that an infrared-effective action with dynamical gluon mass predicts a number of quantum solitons not present in the massless classical gauge theory. Condensates (i.e., a finite vacuum density) of these solitons have successfully explained[1] (if only qualitatively) many nonperturbative effects, as seen on the lattice, including confinement; generation of nonintegral topological charge; and chiral symmetry breaking (although we will only briefly consider this last topic).

A quantum soliton is a localized finite-energy configuration of gauge potentials arising from an effective action that summarizes quantum effects not present in the classical action; in our case, the effect is a gauge-invariant dynamical gluon mass. Like all effective actions, the one we use is not intended to substitute for the standard NAGT action, nor can it necessarily be quantized itself; it is treated classically. An effective action is merely a summary statement of a particular set of quantum effects; the underlying action of our NAGTs is always the conventional one as used, for example, in the Schwinger–Dyson equations to find the gluon mass. Perhaps the premier example in other fields of a quantum effective action is the Ginzburg–Landau scalar field action for superconductivity – which, in fact, has gluon mass generation (the Meissner effect) and solitons (the Abrikosov vortex) closely related to some of the ones we use. But the Ginzburg–Landau effective action is far removed from the original action, involving electrons and ions, that describes a superconductor.

7.2.1 Condensates and solitons

There is only one kind of soliton in classical QCD: the instanton (and its relatives, calorons and sphalerons). It is certainly possible, in principle, that there could be a finite density of instantons in the QCD vacuum. In that case, we speak of a *condensate* of instantons. Some of the properties of the instanton condensate would be captured in VEVs such as $-\langle \text{Tr} \, \mathcal{G}_{\mu\nu}^2 \rangle$ or other combinations of gauge-invariant field strengths that could possess a VEV, that are also referred to as condensates.[2] At a simple and incomplete level of understanding, one could find a formula relating the instanton density to the value of this VEV. But even if this VEV were to vanish,

---

[1] Instantons and the like in NAGTs have not proven suitable for explaining nonperturbative phenomena because it is far from clear how to tame the infrared instabilities of these solitons, which are allowed in arbitrarily large sizes classically.

[2] The phenomenological existence of such condensates was proposed long ago; see Novikov et al. [1], for example, who make reference to earlier works.

it would not in any essential way affect the possibility of the existence of one or a few instantons in the QCD vacuum.

Things are very different with the *quantum* solitons we discuss in this and the next chapter. Unless there is a condensate, as measured by a finite VEV, the solitons themselves will not exist. And once these solitons can exist, it is natural to think that these solitons themselves supply the condensate VEV. So the solitons and the condensates must exist together in a self-consistent way: it is impossible to have, for example, just one center vortex as the only soliton in the vacuum; there must be a condensate of solitons to have a condensate VEV. Although we study quantum solitons, such as center vortices, as if they were isolated, they exist only by virtue of a finite gluon mass, which summarizes the effects of a condensate whose properties we do not investigate further here. In any case, to be an important contributor to nonperturbative effects, any soliton must have a finite density in the vacuum (i.e., be part of a condensate).

One side of the self-consistent relation between quantum solitons and condensates is already clear from Lavelle's equation for the large-momentum behavior of the running gluon mass, given in Chapter 2 for the Euclidean PT propagator:

$$\partial^{-1}(q) \rightarrow q^2 + c_d \frac{\langle -\text{Tr} \, \mathcal{G}_{\mu\nu}^2 \rangle}{q^2}, \tag{7.5}$$

where both the constant c_d and the VEV are positive in $d = 3, 4$. If there is no condensate, there is no quantum soliton because gluon mass is essential for the soliton, and the condensate is essential for the gluon mass. The other side of the quantum soliton-condensate relation amounts to the statement that the entropy of a soliton of given action must exceed the action so that solitons can condense. There is a good thermodynamic analog. The canonical partition function $\mathcal{Z}$ of, say, a crystalline solid is

$$\mathcal{Z} = \exp[-\beta(U - TS)], \tag{7.6}$$

where β is the inverse temperature $1/T$, U the free energy, and S the entropy (so $U - TS$ is the Helmholtz free energy, minimized in equilibrium for an isolated system at constant temperature). In the QCD analog, S is still the entropy (of gauge configurations of given action), βU is the action, and g^2 is analogous to T. A strongly coupled gauge theory is like a condensed-matter system at high temperature. Clearly, a crystalline solid will melt at high-enough temperature, essentially because the entropy becomes so great as new configurations open up for the melted material that TS exceeds U. This melted phase does support a condensate of solitons (dislocations). Similarly, in QCD, we assume that the coupling is so large that the entropy of, say, a center vortex (per unit length in $d = 3$ or area in $d = 4$) exceeds the action per unit length or area. In this case, the free energy can decrease further by formation of more center vortices, and a condensate forms (its formation arrested only by interactions among the center vortices).

Various estimates, not given here, suggest that this dominance of entropy over action does take place for QCD in $d = 4$, and we can actually prove the existence of a condensate in $d = 3$, as shown in the next chapter. So $d = 3$ QCD-like theories confine just as do $d = 4$ theories.

We now take up the connection between a condensate of center vortices and confinement. Then, in the next chapter, we discuss the role of nexuses, monopole-like objects whose world lines lie on center vortex surfaces (and disorient them), in generation of nonintegral topological charge as well as the QCD sphaleron, a saddlepoint soliton associated with tunneling just as the usual sphaleron describes a tunneling barrier in electroweak theory.

7.2.2 What does confinement really mean?

Any proposed mechanism of confinement must be tested in detail; it is not enough to say vaguely that the mechanism keeps quarks and gluons from actually materializing. We are fortunate that computers allow us to look (at least in principle) at detailed tests not readily available in the real world, where Wilson loops are not much larger than the QCD scale length, quark loops can be broken by pair production, and the area law forces (linearly rising potential) may in certain cases be far from the most important part of the QCD binding forces.

To conduct these tests, we imagine an idealized world of NAGT gauge potentials but no matter fields at all. Confinement will be probed with large Wilson loops whose relevant length scales are large compared to the NAGT length scale in various group representations. Following is a list of some of the important tests of confinement. To keep this book to a manageable length, we will discuss just the first four criteria in this chapter (and finite-temperature issues in Chapters 9 and 10). For the other four, it is not always the case that both a detailed center-vortex picture and lattice simulations exist; for example, although there are lattice simulations of k-string tensions, there is yet no good theoretical understanding of the details of such string tensions in any confinement picture, and there are no lattice simulations on multiple Wilson loops, as called for in item 7. Where a center-vortex picture and lattice simulations both exist for a given criterion, there is good agreement. It would take us far too much space to detail this agreement. For a comprehensive review of this and other results up to 2003, see Greensite[2].

1. The leading term in the VEV of a fundamental-representation Wilson loop, all of whose scale lengths are large compared to the QCD length scale, is of the form $\exp[-K_F A]$, where A is the area of a minimal surface spanning the loop and K_F is the usual string tension.

2. For $SU(3)$ baryons made of static quarks, the baryonic area law is a Y-law, with the three-quark Wilson loop of the form $\exp\{-K_F[\sum A_i]\}$, where K_F is the fundamental mesonic string tension and the A_i, $i = 1, 2, 3$, are minimal areas of three Wilson loops having a common line. This is a nontrivial statement because the string tensions could have been a sum of pairwise forces. (The analogs for $SU(N)$, $N > 3$, depend on certain as yet uncalculated properties of k-string tensions; see criterion 5.)

3. The Wilson loop for the adjoint representation (or other representations with N-ality[3] zero) shows no area law extending to arbitrarily large distance; instead it shows *screening*. This means that the leading term is a perimeter term of the form $\exp[-ML]$ for some mass $M \sim m$ and loop perimeter L. However, there is an area law and a so-called breakable string for intermediate distances. At a distance roughly corresponding to storing enough energy in the string to allow formation of a gluon pair, the string breaks as this pair is formed by tunneling. Because it takes an energy of about $2m$ for gluon-pair formation, this breaking occurs at some finite distance, and the breakable string plays a role up to this distance.

4. It should explain the general characteristics of hybrids, which differ from qqq or $\bar{q}q$ states by having one or more extra valence gluons. The center-vortex picture is in good agreement with lattice simulations.

5. There should be plausible explanations for finite-temperature phenomena, including deconfinement, the mechanism for the thermal phase transition, chiral symmetry restoration, and generation of a magnetic mass in $d = 3$.

6. For $SU(N)$ with $N > 3$, there are k-string tensions K_k, subject to $K_k = K_{N-k}$, depending only on the N-ality k. The real test is to calculate the K_k, which has not yet been done accurately for any picture of confinement. (For center vortices, there are different areal densities $\rho_k = \rho_{N-k}$ that have yet to be calculated; their knowledge is tantamount to knowing the K_k.)

7. The exceptional Lie groups G_2, F_4, and E_8, whose centers are trivial, have no confined representations, only screened ones. In the center vortex picture,

---

[3] The N-ality is an integer equal to the number $n - \bar{n}$ mod N, where n is the number of fundamental representations and $\bar{n}$ is the number of antifundamental representations of $SU(N)$ whose product forms the given representation.

these groups each have a vortex solution that corresponds to the trivial element (the identity) of the gauge group; these behave just as explained in item 3 for N-ality = 0 representations in the gauge group $SU(N)$.[4]

8. A confinement mechanism should explain certain special cases such as for two disjoint, parallel, fundamental-representation Wilson loops as a function of separation between them: does the compound Wilson loop behave as if it were rings with soap films stretched on them?

There is not space here to detail all these criteria, and we concentrate on the first four. As for the rest; some but not all finite-temperature effects will be discussed in Chapters 9 and 10. It is worth pointing out here the basic reason for a finite-temperature transition to a deconfined phase in the center vortex picture [3]. As the length β of the Euclidean time direction shrinks with increasing temperature, the vortices of size m^{-1} are squeezed in this direction. When (up to a factor we will not estimate) $m \sim T$, vortices tend to point in the time direction because the action of spacelike vortices increases, and their entropy decreases because of the squeezing. Vortices pointing in the time direction cannot link to a Polyakov loop, so there is a transition to deconfinement. These vortices are not entropically constrained in the three spacelike directions and can condense, providing for mass generation in this $d = 3$ space.[5] Essentially, the same explanation was given later and confirmed by lattice simulations: confining vortices fail to condense (percolate) above a critical temperature [4, 5]. For interesting but not definitive work on k-string tensions, see Greensite and Olejnik [6] and [7], who show that center vortices are consistent but not yet predictive with what is known (from the lattice) about such tensions. The exceptional gauge group G_2 is discussed by Holland et al. [8, 9] and Pepe [10]. The last criterion – essentially about the tension in surfaces spanning Wilson loops – is particularly challenging for center vortices because the basic topological confinement mechanism of linkages of loops and center vortices is independent of any surface spanning a Wilson loop. Some progress has been made, using center vortices, on this problem (see also [11] and [12]), but there are no lattice results for comparison.

---

[4] Aside from the triviality of the center, there is a physical explanation why there is screening but not confinement in such theories. Consider G_2; all of its representations can be decomposed into sums of $SU(3)$ representations. The gluons lie in $8 \oplus 3 \oplus \bar{3}$, and so a string in any G_2 representation can break by production of $3, \bar{3}$ gluons. It takes three gluons to shield a G_2 "quark."

[5] For bosons, dynamical mass generation as discussed here is not damped by finite-temperature effects, whereas for fermions mass generation is damped, so the superconducting mass gap for superconductors vanishes above a certain temperature.

7.3 The quantum solitons

Three fundamental types of solitons are extrema of the NAGT effective action. The
first is the center vortex of the present chapter.[6] In the next chapter, we present
its close relative, the nexus [15, 16, 17, 18, 19, 20, 21], and modified versions of
the instanton [14] and sphaleron [22]. Of these, the most important are the center
vortex and the nexus, a monopole-like object whose world line is in fact embedded
in a center vortex and therefore does not exist independently of center vortices.

All the solitons have idealized three-dimensional (static) realizations displayed by
dropping the time integration in the $d = 4$ effective action S_{eff} and extremalizing
the resulting $d = 3$ massive effective Hamiltonian[7]:

$$H_{\text{eff}} = -\frac{1}{2g^2} \int d^3x \, \text{Tr} \, \mathcal{G}_{ij}^2 - \frac{m^2}{g^2} \int d^3x \, \text{Tr} \left[U^{-1} \mathcal{D}_i U \right]^2 . \tag{7.7}$$

The coupling g^2 is evaluated at some low mass scale such as m^2 itself. Rescale the
potential as done in Derrick's theorem: $\mathcal{A}_i(\vec{x}) \to (1/\rho)\mathcal{A}_i(\rho\vec{x})$. Then the first term
of H_{eff} scales like $1/\rho$ and the mass term scales like $m^2\rho$. Using ρ as a variational
parameter shows that S_{eff} must have solitons of size $\rho \sim m^{-1}$ and that H_{eff} itself
scales like m/g^2. Note that the value of H_{eff} scales with the length of the z-axis, or
more generally with the length of the closed string defining the vortex. Similarly, in
$d = 4$, the vortex effective action scales with the area of the closed vortex surface
(because of the reinstated integral over the time direction). All the solitons have
$d = 4$ counterparts, whose form we will discuss as needed. However, it is in their
$d = 3$ realization that we see most clearly their topological structure.

The gauge potential and the U-matrix are coupled at large distances. So that the
energy of a soliton can be infrared finite, the integrand of both the $\mathcal{G}^2$ term and
the mass term in S_{eff} must vanish at infinity. These requirements imply that $\mathcal{A}_i$
approaches a pure gauge at infinite distance, and this gauge is precisely U itself:

$$x_i \to \infty : \qquad \mathcal{A}_i \to U \partial_i U^{-1}, \tag{7.8}$$

as one sees from Eq. (7.11). Generally, the subleading terms vanish exponentially
rapidly near infinity. The topological properties of a soliton are characterized by the
behavior of U at large distance. These topological properties are most succinctly

---

[6] The center vortex was introduced as a mathematical construct by 't Hooft [13]. The first dynamical model based
on gluon mass is given in [14]. For a review of the lattice properties of center vortices, see Greensite [2].
[7] This has the same form as the effective action $S_{\text{eff}(3)}$ of $d = 3$ gauge theory, except that the coupling g_3^2 has
dimensions of mass; thus, $S_{\text{eff}(3)}$ is dimensionless.

expressed as certain homotopies, whose mathematical truth we take for granted in this book.[8] For us, the interesting question is the physical realization of these homotopies. So for each soliton, we begin by postulating a form for U that not only carries the homotopy but also provides instructions for the kinematic structure of the soliton in spatial and group coordinates.

All we need to know about homotopies at this point is that the homotopy group Π_1, which applies to the center vortex, describes the ways in which a closed one-dimensional string can be mapped on to a topological space. In physical terms, this is the homotopy that tells us the consequences of the topological linkage of the closed string of a center vortex (at a given time) with a Wilson loop. We find these consequences by studying the usual Wilson loop for a closed curve Γ but without taking its trace:

$$\mathcal{W}_\Gamma = P \exp \left[\oint_\Gamma dx_\mu \, \mathcal{A}^\mu(x) \right]. \tag{7.9}$$

To preserve the single-valuedness of the gauge potential $\mathcal{A}_\mu(x)$ in the presence of vortices, we must – as shown later in this chapter – quantize the magnetic flux carried by center vortices, which causes the group matrix $\mathcal{W}_\Gamma$ to lie in the center of the group (the necessarily Abelian subgroup that commutes with all group elements). If $\mathcal{W}_\Gamma$ is in the fundamental representation, its elements are of the form of powers of $\exp[2\pi i/N]$ times the $N \times N$ identity matrix for $SU(N)$. But in representations with N-ality zero, such as the adjoint representation, the elements of $\mathcal{W}_\Gamma$ are only the identity matrix itself.

Because the center group is Abelian, center vortices themselves appear at first glance to be simply Abelian objects and so are described by group matrices in the Cartan subalgebra of $SU(N)$. To the extent that center vortices are Abelian, they very much resemble earlier vortex constructions in physics, notably the Abrikosov vortex of type II superconductors and its relativistic generalization, the Nielsen–Olesen vortex of the Abelian Higgs model. In fact, as one might expect for a non-Abelian group, center vortices are much more complicated than these earlier vortices and actually must be thought of as participating fully in non-Abelian processes. But we need not study this now; it is the subject of the next chapter.

---

[8] Roughly speaking, homotopy groups $\Pi_n(X)$ classify the different ways an n-sphere can be mapped to a topological space X. The group Π_1 – the fundamental group – tells about the ways of mapping closed $d = 1$ loops into the space. For us, such a closed loop is a Wilson loop. As we will see in Chapter 8, Π_2 is relevant for nexus-vortex intersections (a source of topological charge) and Π_3 for various carriers of topological charge. For a review of homotopies and other topological subjects aimed at physicists, see Mermin [23].

7.4 The center vortex soliton

The field equations for S_{eff} are

$$[\mathcal{D}_j, \mathcal{G}_{ji}] - m^2 \tilde{\mathcal{A}}_i = 0 \tag{7.10}$$

$$[\mathcal{D}_i, \tilde{\mathcal{A}}_i] = 0,$$

where the first equation comes from varying $\mathcal{A}_i$, the second equation from varying U, and $\tilde{\mathcal{A}}_i$ is defined by

$$\tilde{\mathcal{A}}_i = \mathcal{A}_i + (\partial_i U)U^{-1}. \tag{7.11}$$

It is important to note that the U equation of motion is not really independent; it is required by the identity

$$[\mathcal{D}_i, [\mathcal{D}_j, \mathcal{G}_{ij}] \equiv 0. \tag{7.12}$$

The equation for U has an interesting interpretation. After some algebra, one finds that it can be written

$$\partial_i(U\mathcal{A}_i U^{-1} + U\partial_i U^{-1}) = 0. \tag{7.13}$$

(The analogous equation also holds in $d = 4$.) This says that if the original gauge potential is in the Landau gauge (as, e.g., happens for center vortices), there is a gauge transformation U that preserves this gauge – in other words, a Gribov ambiguity. So if there is a condensate of center vortices, there is automatically a condensate of Gribov ambiguities.

There are solutions to the U equation of motion in perturbation theory, but they are of no interest for the solitons. They do, however, exhibit the long-range longitudinally coupled scalar excitations that are essential if a gauge-invariant gluon mass is to be generated. Because the gauge potential incorporates a power of the coupling g, this is an expansion in powers of $\mathcal{A}_i$ itself. Write $U = \exp \omega$, where ω is an anti-Hermitean function on the Lie algebra, and find [24]:

$$\omega = -\frac{1}{\nabla^2}\partial \cdot \mathcal{A} + \frac{1}{\nabla^2}\left\{\left[\mathcal{A}_i, \partial_i \frac{1}{\nabla^2}\partial \cdot \mathcal{A}\right] + \frac{1}{2}\left[\partial \cdot \mathcal{A}, \frac{1}{\nabla^2}\partial \cdot \mathcal{A}\right] + \cdots\right\}. \tag{7.14}$$

Clearly, this is a very complicated object because of the non-Abelian nature of the group. But for the standard Abelian center vortex that we now take up, U is a diagonal matrix that commutes with all its derivatives.

7.4.1 The standard Abelian center vortex

The simplest case [14] for the center vortex is an Abelian thick string extending along the entire z-axis. (We consider that the z-axis is a closed string, identifying the points at $\pm\infty$.) In this case, the center vortex is (with some important limitations) the Nielsen–Olesen vortex of the Abelian Higgs model in the limit of infinite Higgs-field mass. Only the massless Goldstone fields survive, to be eaten by the gauge field. The appropriate ansatz for U (equivalent to the angular, or Goldstone, part of the Higgs–Goldstone field) is

$$U = \exp[iQ\phi] \qquad U\partial_i U^{-1} = \frac{Q}{i\rho}\widehat{\phi}_i, \qquad (7.15)$$

where ϕ, ρ are cylindrical coordinates and $\widehat{\phi}_i$ is a unit vector in the ϕ direction; Q is diagonal and one of a set of matrices in the Cartan subalgebra whose detailed properties we will soon give. Because Q is the only group matrix appearing in the standard center vortex, this vortex has no obvious non-Abelian properties; Q itself is equivalent to an Abelian charge and U to a simple phase factor.

To make the connection to the Abelian Higgs model, note that in that model, the Higgs–Goldstone field ψ is of the form $\psi = \chi \exp[i\phi]$. The real (Hermitean) field $\chi(\rho)$ is the Higgs field; it has a vacuum expectation value v. The Goldstone field is ϕ and is massless. The coupling to the canonical Abelian gauge potential A_j is

$$\frac{1}{2}|\partial_j \psi - iGA_j \psi|^2, \qquad (7.16)$$

with G standing for the Abelian charge of the Higgs field; G is replaced by the diagonal matrix Q in the NAGT version. Suppose we freeze the field χ by taking the limit of infinite Higgs self-coupling λ at fixed v (or equivalently, at infinite Higgs mass). In the scalar coupling term $\lambda(|\psi|^2 - v^2)^2$, the limit tells us that $\chi = v$ everywhere. This is a singular limit because for finite λ, the equations of motion say that the field χ vanishes at the center of the vortex (along the z-axis, in our case). But in this limit, the Higgs-gauge potential coupling simply reduces to the Abelian version of the mass term in Eq. (7.7), with gauge particle mass[9] Gv. In the case of NAGTs with no Higgs fields, the effective vanishing of the Higgs field along the vortex axis is replaced by the requirement, coming from solving the Schwinger–Dyson equations, that the dynamical gluon mass vanish at large momentum or short distance. For the sake of brevity, we will not actually impose this condition on the solitons of this book; failure to do so leaves some logarithmic divergences in the gauged nonlinear sigma model part of the action that will be resolved when the short-distance vanishing of the mass is accounted for.

As is the case with all gauge solitons, the form of the gauge potential $\mathcal{A}_i$ follows from the form of the pure-gauge part. Equation (7.15) suggests the following simple

---

[9] In the non-Abelian case, it is also possible to interpret the gauged nonlinear sigma model mass term as coming from N frozen Higgs fields in the fundamental representation of $SU(N)$ [14].

kinematics for $\mathcal{A}_i$:

$$\mathcal{A}_i = \frac{Q}{i}\widehat{\phi}_i F(\rho). \qquad (7.17)$$

With this ansatz, $\partial_i \mathcal{A}_i = 0$, so the potential is in the Landau gauge; of course, any other gauge can be reached by a regular gauge transformation.

The corresponding pure-gauge potential $U\partial_i U^{-1}$ has no gauge fields almost everywhere, but there is a Dirac string (in this case, the z-axis) where there is a singular field strength.[10] We will see that the short-range parts of $\mathcal{A}_i$, the only parts depending on the mass, precisely cancel this string so that there is no such string in the full gauge potential $\mathcal{A}_i$. But the would-be Dirac string is essential for describing the topology and homotopy.

The U equation is

$$\nabla^2 \phi = 0 \qquad (7.18)$$

and is satisfied by the usual polar angle, except for the z-axis.[11] However, because ϕ is discontinuous along the z-axis, there is a singular magnetic field along this axis, or in other words, a Dirac string:

$$B_i = \epsilon_{ijk}\left[\nabla_j \times \frac{Q}{i\rho}\widehat{\phi}_k\right] = \left(\frac{Q}{i}\right) 2\pi \widehat{z}_i \delta(x)\delta(y). \qquad (7.19)$$

There must be such a singularity to this solution of Laplace's equation somewhere in three-space. We now show that the equations of motion automatically cancel this singularity with an equal and opposite contribution from the short-range parts of $\mathcal{A}_i$. These equations are as follows:

$$\nabla^2 \mathcal{A}_i = m^2 (\mathcal{A}_i - \partial_i \phi), \qquad (7.20)$$

and they have the solution

$$\mathcal{A}_i = \frac{Q}{i}\widehat{\phi}_i \left[-mK_1(m\rho) + \frac{1}{\rho}\right], \qquad (7.21)$$

where K_1 is the Hankel function of the first kind with an imaginary argument. As advertised, near the Dirac string ($\rho \simeq 0$), the singularity of the Hankel function exactly cancels the $1/\rho$ term; moreover, at large ρ, the K_1 term vanishes exponentially, and $\mathcal{A}_i$ approaches the required pure gauge.

---

[10] In other dimensions d, there is a Dirac hypersurface of dimension $d - 2$, that is, of codimension 2.

[11] There are no perturbative contributions from the series in Eq. (7.14).

7.4.2 The general center vortex

To describe a center vortex generally, specify a closed line (surface) in $d = 3(4)$[12] and a matrix in the Cartan subalgebra that gives rise to a center element on parallel transport of the vortex's gauge function around a closed loop. We will only describe $k = 1$ vortices explicitly here because higher-k vortices are in some sense composed of $k = 1$ vortices.

In $d = 3$, the generalized center vortex is

$$A_i(x; j) = \frac{2\pi Q_j}{i} \epsilon_{iab} \oint_\Gamma dz_a \, \partial_b \left[\Delta_m(x - z) - \Delta_0(x - z)\right], \quad (7.22)$$

where $\Delta_{m(0)}$ is the free $d = 3$ propagator for mass $m(0)$ and Γ is a closed curve; the coordinate z traces out the curve. It may not be obvious, but the massless propagator term is the generalization of the pure gauge U of Eq. (7.18) but with a Dirac string along the closed curve Γ.[13] With the identification of the Δ_0 term with the U term, it is easy to check that the field equations are satisfied by the vortex of Eq. (7.22). A simple calculation gives the field strength:

$$\mathcal{G}_{ab} = \frac{2\pi Q_j}{i} \epsilon_{abc} \oint_\Gamma dz_c \, m^2 \Delta_m(x - z). \quad (7.23)$$

This has an integrable logarithmic singularity on the Dirac string, which would be removed if we were to account for the short-distance vanishing of the mass. There is no singularity in the classical $\mathcal{G}^2$ action term, but the mass term has a logarithmic short-distance divergence that would be cured by the short-distance vanishing of the mass.

The generalization to $d = 4$ is equally simple:

$$A_\mu(x; j) = \frac{2\pi Q_j}{i} \frac{1}{2} \epsilon_{\mu\nu\alpha\beta} \partial_\nu \oint_\Gamma d\sigma_{\alpha\beta}(z) \left[\Delta_m(x - z) - \Delta_0(x - z)\right], \quad (7.24)$$

where now the integral is over a closed two-surface Γ of surface element $d\sigma_{\alpha\beta}(z)$ and the propagators are four-dimensional. The field strengths are

$$\mathcal{G}_{\mu\nu}(x; j) = \frac{2\pi Q_j}{i} \oint_\Gamma d\tilde{\sigma}_{\mu\nu}(z) m^2 \Delta_m(x - z), \quad (7.25)$$

where $d\tilde{\sigma}_{\mu\nu}$ is the dual surface element.

The massive propagators, in either dimension, approach the massless ones in the limit $m|x - z| \ll 1$, so the propagator singularities cancel on the would-be Dirac string or surface. (This shows, by the way, that for finiteness, one must choose

---

[12] For general dimensions, the Dirac singularity of the pure-gauge part of a center vortex has codimension 2 or ordinary dimension $d - 2$. So in $d = 2$, vortices are point particles.

[13] A function F_i can be both a gradient and a curl, provided it satisfies Laplace's equation.

the same string or surface for the massive and massless terms in the vortex.) This cancellation gets rid of the Dirac string-surface in the full soliton. However, it will prove very useful for us later to have expressions for these Dirac string-surface field strengths because they give the cleanest view of the topology of these vortices and their interactions. From the field strengths in Eqs. (7.23) and (7.25) and from the propagators expressed as Fourier integrals, the Dirac would-be singularities emerge in the limit $m \to \infty$ – or otherwise said, the limit of zero vortex thickness – and give

$$d = 3: \qquad \mathcal{G}_{ab} = \frac{2\pi Q_j}{i} \epsilon_{abc} \oint_\Gamma dz_c \, \delta_3(x - z) \qquad (7.26)$$

$$d = 4: \qquad \mathcal{G}_{\mu\nu}(x; j) = \frac{2\pi Q_j}{i} \oint_\Gamma d\tilde{\sigma}_{\mu\nu}(z) \, \delta_4(x - z),$$

so they are just integrals of delta functions over the string-surface.

So far we have not explained why the string-surface must be closed. Suppose otherwise and consider the integral in $d = 3$:

$$\mathcal{A}_i(x) = \epsilon_{ij3} \partial_j \int_{-\infty}^0 dz \, \Delta_0(x - z), \qquad (7.27)$$

which represents part of the center vortex but with an open string. Do the integral to find precisely the gauge potential of a Dirac monopole with a magnetic field $\sim 1/r^2$ and a Dirac string along the negative z-axis. We know there are no such long-range monopoles in QCD, so to exclude them, we use a closed string. (The same argument works for $d = 4$, using a surface with boundary rather than a closed surface. The boundary is a closed loop, the world line of the monopole.) This argument is certainly correct for Abelian vortices but not for non-Abelian vortices, as we describe in the next chapter for vortex junctions.

7.4.3 The Q-matrices and the center-vortex homotopy

Relation of confinement to homotopy   Now we can see how to associate a homotopy with the group carried by Wilson loops themselves. A Wilson loop depends on a closed curve and a map of this curve to group elements:

$$\mathcal{W}_\Gamma^R = P \exp \left[\oint_\Gamma dx_i \, \mathcal{A}_i \right], \qquad (7.28)$$

where Γ is a closed curve and R labels the group representation for the loop. The argument in the exponent, $\mathcal{A}_i[x(t)]$, is a map of the closed curve described by $x_j(t)$, as t runs from 0 to 1, say, to an element of the Lie algebra of the gauge group. Exponentiation turns this into a map of the loop to the gauge group, and so the homotopy is the group of (Wilson) loops to the gauge group G, or $\Pi_1(G)$. Of course, Π_1 tells us to look for a wrapping of closed strings around this gauge group. One must be careful to appreciate that the gauge group G for gluons is not $SU(N)$ but $SU(N)/Z_N$, where Z_N, the group of integers, is the center of the gauge group. As far as gluons go, any group element of $SU(N)$ transforming it has exactly the same effect as if any element of the center group in this element were discarded. Equivalently, for the adjoint representation, every element of the center group is represented by the identity. So the homotopy associated with the center vortex is

$$\Pi_1(SU(N)/Z_N) \simeq Z_N, \tag{7.29}$$

which is not at all the same as $\Pi_1(SU(N)) \simeq \mathbb{I}$. These equations are not entirely trivial, but we will not pause to prove them here by the usual mathematical techniques. Instead, we give a physical demonstration.

Let the curve Γ be any closed path in the presence of a center vortex with Dirac string along the z-axis, as in the standard form of Eq. (7.21), and let all parts of the loop be far from the z-axis. At the Wilson loop, the only surviving part of the gauge potential is its pure-gauge part, as in Eq. (7.8), in which case, $\mathcal{W}_\Gamma^R$ is the product $U(i)U^{-1}(f)$, where i, f are the (identical) initial and final points of the contour Γ, respectively. These are the same points, but different curves between them gives different results. If the curve does not enclose the z-axis, then $\mathcal{W}_\Gamma^R = \mathbb{I}$, where $\mathbb{I}$ is the identity matrix. But if the curve does enclose it (is linked to it), the result is

$$\mathcal{W}_\Gamma^R = \exp[2\pi i Q], \tag{7.30}$$

where Q is the representative of the group generator in representation R.

Here is the physics of the homotopy group. One of the most basic requirements we can place on the gauge potentials of an NAGT is that they be single valued. Imagine transporting a localized gluon wave function once around a curve Γ linked to the z-axis (and thus to the center vortex of Eq. (7.21)) and back to the starting position. This has to yield the same wave function. When Γ is far from the z-axis, the only contribution comes from the pure-gauge function U. As the gluon wave function is transported around the closed loop, it suffers a gauge transformation:

$$\mathcal{A}_i \to V\mathcal{A}_i V^{-1} + V\partial_i V^{-1}, \tag{7.31}$$

where V is the group element

$$V = P \exp\left[\oint dx_i \, U\partial_i U^{-1}\right]. \tag{7.32}$$

With $U = \exp[iQ\phi]$, this phase factor is just $V = U(\phi = 2\pi)U^{-1}(0) = \exp[2\pi i Q]$. The only elements of $SU(N)$ that commute with all other elements are elements of the center, and we conclude that $\exp(2\pi i Q)$ is in the center of the group.

The Q-matrices   What matrices Q realize the center of the group in the fundamental representation? For $SU(N)$, the center is generated by the matrix[14] $Z = \exp(2\pi i/N)$ and contains the elements $Z(k) \equiv Z^k, k = 1, 2, \ldots N$. There are center vortices for every value of k, which we call k-vortices. The magnetic flux carried by a k-vortex is just k (in units of $2\pi/N$); because multiples of 2π are irrelevant, we need only define flux *mod N*. There are many traceless diagonal matrices Q for any value of k, and we need indices to distinguish them. For $k = 1$, the generator Z is the exponential of any of the matrices $Q_j, j = 1 \ldots N$:

$$Q_j = \text{diag}\left(\frac{1}{N}, \frac{1}{N}, \ldots \frac{1}{N}, -1+\frac{1}{N}, \frac{1}{N}, \ldots\right) \quad j = 1, 2, \ldots N, \quad (7.33)$$

with the -1 in the jth position.[15] Each of these matrices obeys

$$\exp[2\pi i Q_j] = \exp[2\pi i/N] \quad (7.34)$$

and so has unit flux. Because each Q_j can be transformed into any other by an element of the permutation group, one may think of the index j as a label for group collective coordinates of the vortex. These matrices are linearly dependent; the sum of all N of them vanishes. They obey the trace formula

$$\text{Tr}(Q_i Q_j) = \delta_{ij} - \frac{1}{N}. \quad (7.35)$$

It might be thought that for a k-vortex, one replaces Q_j by kQ_j; however, the corresponding vortex has higher energy than one with the same space-time configuration but using the matrix $Q_j(k)$ defined as the sum of any k distinct Q_j. The energy of a $k = 1$ vortex (or action, in $d = 4$) scales with the factor

$$\frac{1}{g^2}\text{Tr}\,Q_j^2 = \frac{N-1}{Ng^2}. \quad (7.36)$$

A k-vortex described by $Q_j(k)$ has flux k and energy proportional to

$$\frac{1}{g^2}\text{Tr}\,Q_j(k)^2 = \frac{k(N-k)}{Ng^2}. \quad (7.37)$$

Note that this is symmetric under the exchange $k \leftrightarrow N - k$, which is equivalent to replacing a k-vortex by its antivortex, or conjugate vortex with flux $-k$ mod N.

---

[14] The identity matrix $\mathbb{I}$ is understood.
[15] Various useful properties of these matrices are given in [7].

This antivortex is just the $N - k$-vortex. Note also that this action factor is less than the action factor of k widely separated unit vortices, so in some sense, k-vortices are bound composites of unit vortices.

In $SU(N)$ for odd N, there are $(N - 1)/2$ vortices plus their antivortices or conjugate vortices identical in all dynamical properties (except for their fluxes, which are opposite in sign) to those of their vortex partners. For even N, there is in addition a self-conjugate vortex. This property has special implications for $SU(2)$ and $SU(3)$, which are the only unitary groups with exactly one dynamically distinct vortex. For $N > 3$, there is always more than one type with generically different space-time density and other dynamical properties. Little is known about center vortices for $N > 3$.

One might think that center vortices are irrelevant at large N because their action S_C increases with N in the 't Hooft limit of Ng^2 fixed, and so $\exp(-S_C)$ vanishes strongly. However, for the leading N behavior, the collective-coordinate integral over the gauge group grows with N at exactly the rate needed to compensate for the action [25] (this also happens for other solitons with action growing like a power of N such as instantons [26]). What happens at nonleading orders remains to be settled, but there can well be the necessary cancellations that cause center vortices to persist at large N. We will, for other reasons, only discuss $N = 2, 3$ explicitly, and so the large-N behavior is a secondary issue.

7.4.4 Confinement

Confinement is a *topological* property of center vortices, with the homotopy of Eq. (7.29) directly realized as a linking of closed center-vortex Dirac strings with a Wilson loop.

If center vortices are to be responsible for confinement, there must be a condensate of them – that is, a finite density of vortices.[16] Such a condensate forms when the configurational (and other) entropy of vortices per unit length (in $d = 3$)[17] exceeds their action per unit length, in which case, a finite fraction of the vortices will have infinite length. A little thought shows that the vortex density is an *areal* density with the same codimension as the vortices. So in $d = 2$, k-vortices, which are point particles there, have a density ρ_k per unit area. By imagining that $d = 2$ is projected from higher dimensions, one sees that, in all dimensions, the vortex

---

[16] Recall that as long as the entropy exceeds the action, the vortices grow bigger and bigger; at some point, they fill enough of the space so that the entropy of an additional vortex segment diminishes because of the unavailability of unoccupied space. If this new segment were to try to occupy space that already had a vortex, the action would increase. The result is that growth terminates at a finite vortex density where the increase in action just balances the decrease in entropy.

[17] In other dimensions, substitute a closed hypersurface of codimension 2.

density is an area density. Although it is true that the density ρ_k of vortices and the density $\rho_{-k} \equiv \rho_{N-k}$ of conjugate vortices are always equal, there is no reason for the density to be independent of k; however, there are in fact good reasons for ρ_k to depend nontrivially on k. In $SU(2)$ and $SU(3)$, there is only one density, so these two groups give the easiest description of confinement (and fortunately, both are relevant to the real world).

The simplest semirealistic picture of center-vortex confinement begins with a flat Wilson loop, whose boundary is the flat curve Γ, *in the fundamental representation* of $SU(N)$; we drop the superscript R indicating this representation. Take this loop to be a sum of touching squares, each of side λ, and Γ is similarly approximated by a polygon of sides λ. The length λ is the correlation length of vortices such that there is only one vortex per λ-square. Any such square is pierced by a single vortex with probability p, and clearly

$$p = \rho \lambda^2. \tag{7.38}$$

Now assume that piercings in different λ-squares are statistically independent, which is not really true. However, correcting for this effect gives a quantitative but not a qualitative change in the picture of confinement. We wish to calculate the VEV $\langle W_\Gamma \rangle$ of this flat Wilson loop, bounded by the (also flat) curve Γ. Our interest is only in the area law part of this VEV, which comes solely from the long-range pure-gauge parts of the vortex (the massless propagator terms in Eqs. (7.22) and (7.24)). The massive propagator terms at best contribute perimeter terms to the VEV.

Begin by calculating the fundamental-representation Wilson loop (not yet its VEV) in the presence of a single k-vortex, described by the matrix $Q_j(k)$ and closed contour C. We need the trace of the path-ordered exponential:

$$W_\Gamma = \frac{1}{N} \mathrm{Tr}_F \, \mathcal{W}_\Gamma = \frac{1}{N} \mathrm{Tr}_F \, P \exp \left[\oint_\Gamma dx_i \, \mathcal{A}_i \right], \tag{7.39}$$

and with the help of Eq. (7.22), we find

$$W_\Gamma = e^{2\pi i Q_j(k) Lk}, \tag{7.40}$$

where

$$Lk = \oint_\Gamma dx_i \, \epsilon_{ijk} \oint_C dy_j \, \partial_k \Delta_0(x - y). \tag{7.41}$$

The expression for Lk happens to be the canonical expression for the Gauss linking number of two closed curves in $d = 3$ (which is why we call it Lk).[18] It is an integer counting the signed number of times that the curve C is linked to the curve Γ (see Kaufmann [27]).

The linking number has another interesting interpretation as an Abelian Chern–Simons term (see Chapter 9). The result, then, for the Wilson loop is

$$W_\Gamma = Z(k)^{Lk}. \tag{7.42}$$

Only the link number mod N has any significance for the Wilson loop, which, for a specific vortex, is a certain element of the center group. This is the dynamical realization of the homotopy in Eq. (7.29) that we sought.

For an ensemble of vortices, under the assumption of vortex independence, the Wilson loop VEV is an average of center-group elements, each of which is a product over the vortices in a particular member of the ensemble of the form $\prod_i Z_i^{Lk_i}$, where Z_i is an element of the center of the gauge group, as specified by the properties of the ith vortex (and the group representation of the loop itself).

The rest of this section, deals explicitly only with $SU(2, 3)$ for reasons given earlier. For the fundamental Wilson loop in $SU(2)$, the only nontrivial element of the center has $Z_i = -1$. In the preceding product, Lk_i is the Gauss linking number of this vortex with the Wilson loop. The Gauss linking number, a topological invariant, can be written through an integration by parts as an intersection number of the vortex and any surface spanning the Wilson loop; its (integral) value is independent of the choice of surface.[19] In the $SU(2)$ case, the necessary average is

$$\left\langle \exp\left[i\pi \sum_i Lk_i \right] \right\rangle. \tag{7.43}$$

We now make explicit the assumption that p is the probability that a vortex is actually linked once to a flat Wilson loop. When an odd number of vortices is linked once, the Wilson loop has value -1, and when an even number is linked, the value is $+1$. If the assumption is true, the area law follows from multiplying the probabilities $\bar{p} - p$ for all the λ-squares of the spanning surface so that

$$\langle W_\Gamma \rangle = (\bar{p} - p)^{A_\Gamma/\lambda^2} = \exp\left[-|\ln(1 - 2p)| \frac{A_\Gamma}{\lambda^2} \right], \tag{7.44}$$

where A_Γ/λ^2 is the number N_Γ of λ-squares in the Wilson loop. Another useful way of expressing this area law is to write out the combinatorics for vortex occupancy

---

[18] In other dimensions, the Wilson loop for a center vortex yields the Gauss linking number for linking of a closed 1-curve with a $d - 2$-dimensional closed hypersurface.

[19] Which raises the interesting question of why a *specific* surface occurs in the VEV; for a flat Wilson loop, the obvious and correct choice is the flat surface spanning the loop.

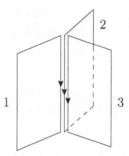

Figure 7.1. A baryonic Wilson loop in $SU(3)$ is composed of three simple Wilson loops sharing a common central line (expanded in the figure). The central line is invisible to $SU(3)$ center vortices.

of N_Γ sites of a surface spanning a Wilson loop Γ,

$$\langle W_\Gamma \rangle = \bar{p}^{N_\Gamma} - N_\Gamma \bar{p}^{N_\Gamma-1} p + \frac{N_\Gamma(N_\Gamma-1)}{2} \bar{p}^{N_\Gamma-2} p^2 + \cdots$$

$$= (\bar{p} - p)^{N_\Gamma} = (1 - 2p)^{A_\Gamma/\lambda^2}, \tag{7.45}$$

as before. Here each term represents the number of ways of arranging empty and once-filled λ-squares.

The result for $SU(3)$ is also an area law but with a different string tension because of different center-group phase factors. It is easy to see that $\bar{p} - p$ should be replaced by $\bar{p} + \cos(2\pi/3)p = 1 - (3/2)p$, where $\cos(2\pi/3)$ is the average of the vortex and conjugate vortex center elements.

We have assumed no correlations other than mutual repulsion between vortices, but in fact, other correlations do exist. In particular, a vortex has a finite chance of reentering a Wilson-loop spanning surface a few steps after piercing it the first time; this dilutes the effective density of actually linked vortices below the density ρ of vortices piercing the flat spanning surface, as shown in [11]. This dilution has been observed in lattice simulations with center vortices [28].

Another issue for $SU(3)$ is the form of the area law for baryons formed from infinitely heavy static quarks; the corresponding Wilson loop is shown in Figure 7.1.

There are a priori two interesting possibilities: a sum of areas between each of the three quark pairs (e.g., surface 4, spanning quark lines 1 and 3, with folds at the two dotted lines) or a three-bladed minimal area joining each quark line to a central Steiner line, often termed the *Y-area law*. In Figure 7.1, the Steiner line is shown as three coincident lines from each of three elementary Wilson loops. A straightforward extension of the linking arguments already given for the elementary Wilson loop shows that the second possibility is correct [29]. Note that in $SU(3)$,

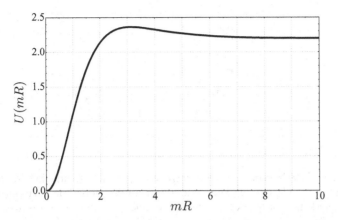

Figure 7.2. Adjoint potential $U(mR)$ in a $d = 2$ center-vortex model.

these three coincident lines always give the identity element of the center group. The Y-area law has been confirmed on the lattice [30, 31, 32, 33].

7.4.5 Screening

Physically, screening means that gluons or other N-ality zero fields have a string between them, as do quarks, but the string breaks when enough energy has been stored in it to materialize another gluon pair. Of course, this is the same thing that happens for quarks when they are included as dynamical fields. A model of this adjoint string breaking has been worked out in [25], with results for the potential $V(R) = (K_F/m)U(mR)$ shown in Figure 7.2. The model is in $d = 2$, but the NAGT action used is not the conventional $d = 2$ action; instead, it has a mass term added (and so is the same as Eq. (7.7), except for the integration, which is over two-space, and the coupling, which has dimensions of mass). The center vortex is a point particle described by the wave function

$$A_i(x) = \left(\frac{2\pi Q}{ig} \right) \epsilon_{ik} \partial_k [\Delta_M(x - x_0) - \Delta_0(x - x_0)], \qquad (7.46)$$

where x_0 is the center of the vortex. A condensate of these is simply a condensate of point particles in the plane. A simple closed Wilson loop is linked to a vortex if the vortex is inside the loop and is unlinked otherwise. However, if the Wilson loop is in the adjoint representation, linkages contribute trivially to the loop VEV, and we can drop the massless pure-gauge part of Eq. (7.46). The short-range massive parts contribute only if they are within a distance $\sim 1/m$ of the loop, whether inside it or outside. Consider now the contribution of these vortices to an adjoint Wilson rectangle with spatial width R. After integrating over all vortex positions x_0 and

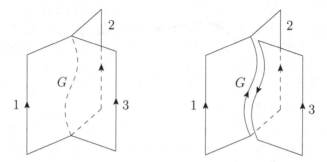

Figure 7.3. (*Left*) Wilson loop for a $qqqG$ configuration (quarks labeled 1, 2, and 3; gluon labeled G). (*Right*) The same loop with the gluon line decomposed into another quark and antiquark line, with quark line 3 singled out, as discussed in the text.

other collective coordinates [34, 25], the adjoint potential of Figure 7.2 emerges. It is the same for $SU(2)$ and $SU(3)$. Note that the center-vortex potential is always positive, is roughly linear in R up to distances $\sim 1/m$, and then breaks (becomes constant) at a distance of about 1 fm. The breaking height of about $2.4K_F/m$ should be of order $2m$ to materialize a gluon pair at breaking, suggesting that $m \simeq K_F^{1/2}$ – a value that is hardly quantitative but in the right ballpark.

Studying the adjoint potential on the lattice has been difficult, apparently because of a poor overlap between the physical gluonic state and the corresponding adjoint Wilson loop and because of the large distance to string breaking. But it is claimed that breaking has been observed, at least in $d = 3$ $SU(2)$ gauge theory (see, e.g., Kratochvila and de Forcrand [35]).

7.4.6 Hybrids

Finally, the role of a dynamical gluon mass (along with center vortices) is apparent in lattice simulations [36] of a heavy-quark baryonic hybrid, a bound state of three infinitely heavy quarks and a single valence gluon. The corresponding Wilson loop is shown in Figure 7.3.

In Figure 7.3 (*left*), the gluon occupies what would be the Steiner line for an ordinary baryon, but of course, the gluon world line, not being infinitely massive, fluctuates. This gluon line can be decomposed into a $q\bar{q}$ pair, as shown on the right, indicating mixing between the hybrid and a baryon plus meson. In the lattice simulations [36], the authors claim a clear signal of a mass of some 600 MeV for the valence gluon; estimates [37] based on a modified form of the Y-law for normal baryons, plus shorter-distance corrections, are in good agreement with the lattice data over a range of separations of the three quarks.

References

[1] V. A. Novikov, M. A. Shifman, A. I. Vainshtein, and V. I. Zakharov, Are all hadrons alike?, *Nucl. Phys.* **B191** (1981) 301.

[2] J. Greensite, The confinement problem in lattice gauge theory, *Prog. Part. Nucl. Phys.* **51** (2003) 1.

[3] J. M. Cornwall and W. S. Hou, Extension of the gauge technique to broken symmetry and finite temperature, *Phys. Rev.* **D34** (1986) 585.

[4] M. Engelhardt, K. Langfeld, H. Reinhardt, and O. Tennert, Deconfinement in $SU(2)$ Yang-Mills theory as a center vortex percolation transition, *Phys. Rev.* **D61** (2000) 054504.

[5] R. Bertle, M. Faber, J. Greensite, and S. Olejnik, The structure of projected center vortices in lattice gauge theory, *JHEP* **9903** (1999) 019.

[6] J. Greensite and S. Olejnik, K-string tensions and center vortices at large N, *JHEP* **0209** (2002) 039.

[7] J. M. Cornwall, A three-dimensional scalar field theory model of center vortices and its relation to k-string tensions, *Phys. Rev.* **D70** (2004) 065005.

[8] K. Holland, P. Minkowski, M. Pepe, and U. J. Wiese, Exceptional confinement in $G(2)$ gauge theory, *Nucl. Phys.* **B668** (2003) 207.

[9] K. Holland, P. Minkowski, M. Pepe, and U. J. Wiese, Confinement without a center: The exceptional group $G(2)$, *Nucl. Phys. Proc. Suppl.* **119** (2003) 652.

[10] M. Pepe, Confinement and the center of the gauge group, *PoS* **LAT2005** (2006) 017.

[11] J. M. Cornwall, Probing the center-vortex area law in $d = 3$: The role of inert vortices, *Phys. Rev.* **D73** (2006) 065004.

[12] J. M. Cornwall, Relativistic center-vortex dynamics of a confining area law, *Phys. Rev.* **D69** (2004) 065019.

[13] G. 't Hooft, On the phase transition towards permanent quark confinement, *Nucl. Phys.* **B138** (1978) 1.

[14] J. M. Cornwall, Quark confinement and vortices in massive gauge-invariant QCD, *Nucl. Phys.* **B157** (1979) 392.

[15] J. M. Cornwall, Nexus solitons in the center vortex picture of QCD, *Phys. Rev.* **D58** (1998) 105028.

[16] M. N. Chernodub, M. I. Polikarpov, A. I. Veselov, and M. A. Zubkov, Aharonov-Bohm effect, center monopoles and center vortices in $SU(2)$ lattice gluodynamics, *Nucl. Phys. B Proc. Suppl.* **73** (1999) 575.

[17] B. L. G. Bakker, A. I. Veselov, and M. A. Zubkov, The simple center projection of $SU(2)$ gauge theory, *Phys. Lett.* **B497** (2001) 159.

[18] M. Engelhardt, Center vortex model for the infrared sector of Yang-Mills theory – topological susceptibility, *Nucl. Phys.* **B585** (2000) 614.

[19] M. Engelhardt and H. Reinhardt, Center projection vortices in continuum Yang-Mills theory, *Nucl. Phys.* **B567** (2000) 249.

[20] R. Bertle, M. Engelhardt, and M. Faber, Topological susceptibility of Yang-Mills center projection vortices, *Phys. Rev.* **D64** (2001) 074504.

[21] H. Reinhardt, Topology of center vortices, *Nucl. Phys.* **B628** (2002) 133.

[22] J. M. Cornwall, Semiclassical physics and confinement, in *Deeper Pathways in High-Energy Physics: Proceedings of the Conference, Coral Gables, Florida, 1977*, edited by B. Kursonoglu et al. (Plenum, New York, 1977), 683.

[23] N. D. Mermin, The topological theory of defects in ordered media, *Rev. Mod. Phys.* **51** (1979) 591.

[24] J. M. Cornwall, Spontaneous symmetry breaking without scalar mesons II, *Phys. Rev.* **D10** (1974) 500.

[25] J. M. Cornwall, Center vortices and confinement vs. screening, *Phys. Rev.* **D57** (1998) 7589.

[26] D. J. Gross and A. Matytsin, Instanton induced large N phase transitions in two- and four-dimensional QCD, *Nucl. Phys.* **B429** (1994) 50.

[27] L. Kaufmann, *On Knots* (Princeton University Press, Princeton, 1987).

[28] K. Langfeld, O. Tennert, M. Engelhardt, and H. Reinhardt, Center vortices of Yang-Mills theory at finite temperatures, *Phys. Lett.* **B452** (1999) 301.

[29] J. M. Cornwall, Center-vortex baryonic area law, *Phys. Rev.* **D69** (2004) 065013.

[30] T. T. Takahashi, H. Matsufuru, Y. Nemoto, and H. Suganuma, Three-quark potential in $SU(3)$ lattice QCD, *Phys. Rev. Lett.* **86** (2001) 18.

[31] T. T. Takahashi, H. Matsufuru, Y. Nemoto, and H. Suganuma, Detailed analysis of the three-quark potential in SU(3) lattice QCD, *Phys. Rev.* **D65** (2002) 114509.

[32] C. Alexandrou, P. de Forcrand, and O. Jahn, The ground state of three quarks, *Nucl. Phys. B Proc. Suppl.* **119** (2003) 667.

[33] H. Ichie, V. Bornyakov, T. Streuer, and G. Schierholz, Flux tubes of two- and three-quark system in full QCD, *Nucl. Phys.* **A721** (2003) C899.

[34] J. M. Cornwall, Finding dynamical mass in continuum QCD, in *Workshop on Non-perturbative Quantum Chromodynamics: Proceedings of the Conference, Stillwater, Oklahoma, 1983*, edited by K. A. Milton and A. Samuel (Birkhäuser, Boston, 1983), 119.

[35] S. Kratochvila and P. de Forcrand, String breaking with Wilson loops?, *Nucl. Phys. B Proc. Suppl.* **119** (2003) 670.

[36] T. T. Takahashi and H. Suganuma, Detailed analysis of the gluonic excitation in the three-quark system in lattice QCD, *Phys. Rev.* **D70** (2004) 074506.

[37] J. M. Cornwall, Baryonic hybrids: Gluons as beads on strings between quarks, *Phys. Rev.* **D71** (2005) 056002.

8

Nexuses, sphalerons, and fractional topological charge

8.1 Introduction to nexuses and junctions

So far, it may appear that center vortices are embedded Abelian objects. But center vortices can be extended to non-Abelian objects in several ways. We describe two: the first we call junctions, representing the merging and branching of vortex lines (or sheets, in $d = 4$) without the necessity of monopole-like objects called nexuses; the second are nexuses themselves [1, 2, 3, 4], which are modifications of 't Hooft–Polyakov monopoles but with their magnetic flux bundled into tubes that are parts of center vortices. The most interesting property of nexuses is that, along with center vortices, they admit the formation of quantum lumps of nonintegral topological charge [5, 6, 7, 8, 9, 10].[1] Nexuses do not change the picture of confinement given in this book in any material way, although this is not completely obvious. But they enter in a crucial way into a reinterpretation of Polyakov's [12] discussion of confinement in the $d = 3$ Georgi–Glashow model, as we indicate at the end of this section [2, 13].

Note that we continue to use the notation of Chapter 7. All sections are in Euclidean space, except for Section 8.3.2, which is in Minkowski space.

8.1.1 Junctions

Junctions are thick points ($d = 3$) or lines ($d = 4$) where a vortex can branch into other vortices [14, 15]. It is easy to draw them in $d = 3$, where they look like vacuum Feynman graphs. Figure 8.1 shows a simple example with four junctions in $SU(3)$, where three lines meet at each junction (up to N lines can meet at an $SU(N)$ junction, each associated with a distinct flux matrix Q_j).

---

[1] In $SU(2)$, sphalerons (see Section 8.3) can carry half-integral topological charge [11] as lines that form when an ordinary instanton is split in half and the halves are pulled apart.

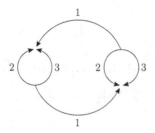

Figure 8.1. A junction and an antijunction in $SU(3)$. The numbers labeling the lines are the values of the index i in the flux matrix Q_i of each vortex line.

Suppose that one of the loops with lines labeled 2 and 3 meets the line labeled 1 at the origin 0. In the neighborhood of the origin, the junction term of the gauge potential is

$$\mathcal{A}_i(x) = \frac{2\pi}{i}\epsilon_{ijk}\partial_j \sum_{a=1}^{3}\int_0 dz(a)_k \, Q_a \{\Delta_M[z(a) - x] - \Delta_0[z(a) - x]\}, \quad (8.1)$$

where the Q_a are the three $k = 1$ matrices of $SU(3)$ discussed earlier.[2] Because $\sum^3 Q_j = 0$, the objection to open vortex lines, raised in Section 7.4.2, no longer applies; the would-be monopole charge, which is the sum of the Q_i, vanishes. For $N > 3$, there are configurations where some of the lines have higher flux: a k-vortex arises from associating a sum $Q_1 + Q_2 + \cdots$ with a single line integral.[3] Because of this feature, junction lines are not topologically stable, but they can be *entropically* stable because the total configurational entropy of two or more junction lines is greater than the entropy of a fewer number. Whether they actually are entropically stable depends on a comparison of action and entropy, which we do not attempt here.

Every junction has an action above and beyond the action per length of the vortices to which it is attached. Generally, this action depends on the geometry of the junction (e.g., it vanishes if lines 2 and 3 of Figure 8.1 are collapsed into one). The case in which the three junction lines meet at right angles has the $d = 3$ value $2\pi m/g^2$ [1].

8.1.2 Nexuses, magnetic charge, and topological charge

In NAGTs where all gluons have mass (not necessarily equal, so we contemplate Higgs–Kibble effects here, as in the Georgi–Glashow model), the radially

---

[2] Note that the lines are oriented, and they either all go in or all go out of the junction at the origin. We need not specify the upper limits of integration, which are irrelevant to the present discussion.
[3] For $SU(3)$, there is no nontrivial junction with $k > 1$.

Figure 8.2. An $SU(2)$ nexus, showing two tubes of field lines.

symmetric long-range magnetic field of the 't Hooft–Polyakov monopole is squeezed into two or more flux tubes, as shown in Figure 8.2. These flux tubes, which must close either at infinity or on an antinexus, have magnetic flux quantized in the center of the gauge group, just as for center vortices, and for exactly the same reason: gluon wave functions must be single valued on transport around one of the tubes. In fact, these tubes are nothing but pieces of center vortices, divided up by closed nexus and antinexus world lines. So in the simplest case of two flux tubes, each tube has the same Π_1 homotopy exemplified in Eq. (7.40), except that each tube has a different representative of the set of matrices $Q_j(k)$. It will turn out that the total flux of the two tubes together is just that of a 't Hooft–Polyakov monopole for every $SU(N)$.

The field lines from the nexus shown in Figure 8.2 must close, which requires the presence of an antinexus. So a simple $d = 4$ case of a center vortex with nexuses is a torus, with a nexus world line and an antinexus world line (nonintersecting) wrapped around it. These world lines effectively divide the center vortex into regions of different orientation; the center vortex as a whole is nonorientable. This, it turns out, is crucial to the formation of topological charge.

Let us reduce the topological charge to its topological essentials by saving only the long-range pure-gauge parts of vortices and nexuses, which have singular fields on their Dirac surfaces and lines. There are several ways to think of this topological charge:

1. It is formed when a nexus world line links to a center-vortex surface [7, 8, 9, 10, 16].
2. It is measured by the usual $\int \mathcal{G}\tilde{\mathcal{G}}$ topological charge integral, which can be interpreted both as the signed intersection number of nonorientable surfaces and as the vortex-nexus linking number [7, 8, 9, 10, 16]. A special case of this occurs when vortices carry twist or writhe [5, 6, 7, 8, 9, 10, 16].
3. It can be interpreted in terms of a monopole magnetic charge, as defined by a standard integral of the type $\int \vec{B} \cdot d\vec{S}$ over a closed two-sphere [17].

In all cases, the topological charge is divided into nonintegral parts. Generically, two closed surfaces intersect in an even number of points[4]; in this case, the topological

---

[4] The points actually have an extension of size $\sim 1/m$ if we save short-range parts of the gauge fields.

charge associated with each such point is quantized in units of $1/N$, but the total charge is integral if the surfaces are compact. Later we will see that self-intersection effects from twist and writhe [18] can lead to topological charge that is nonintegral but otherwise of any size.

Standard textbooks say that topological charge is manifested through instantons, which are compact lumps of integrally quantized topological charge. Actually, there is no reason for any given compact lump of topological charge to have any particular value, integral or otherwise. It may (and does) happen that the compact lumps carry nonintegral charge, but in such a way that the *global* topological charge, integrated over all Euclidean four-space, is an integer. This integrality result, however, is not automatic but depends on the assumption of *compactification* of the three-space bounding $d = 4$ space at infinity to a three-sphere S^3. The gauge potentials carrying topological charge now involve a map from the three-sphere to these potentials. The gauge group $SU(N)$ is either $SU(2)$ or has $SU(2)$ as a proper subgroup, and $SU(2)$ is topologically equivalent to S^3. So these maps are just maps of S^3 to S^3. Another way of speaking of these maps is through the homotopy

$$\Pi_3(SU(2)) \sim Z. \tag{8.2}$$

This one is easy; it just says that all maps of S^3 onto itself consist of an integral number of wrappings of one sphere onto another.[5] So in a $d = 4$ space whose boundary can be compactified, the *total* topological charge has to be an integer. However, this does not require that compact lumps of topological charge have integral charge, as instantons do, and we have already seen that nonintegral lumps do exist in the forms of nexus-vortex intersections and related objects. Global compactification simply requires that the sum of all the charges, integral or nonintegral, be an integer.

8.2 Nexuses in $SU(N)$

8.2.1 The SU(2) nexus

The first step [1, 16] is to find the gauge representative of an $SU(2)$ nexus in $d = 3$. There are infinitely many choices; a simple one is

$$U = \exp\left[\frac{i}{2}\phi\vec{\tau}\cdot\hat{r}\right], \tag{8.3}$$

where the $\vec{\tau}$ are Pauli matrices and other symbols have their usual meaning. Later we will see that the generalization to $SU(N)$ is quite straightforward. From this

---

[5] Because any $SU(N)$ has $SU(2)$ as a subgroup, it turns out that $\Pi_3(SU(N)) \simeq Z$ for all N, so our arguments about integrality of topological charge apply for all gauge groups.

gauge representative, we can find the Dirac-string fields; they are

$$\frac{1}{2}\epsilon_{ijk}\mathcal{G}_{ij} = -\left(\frac{\tau_3}{2i}\right)\widehat{z}_i\epsilon(z)2\pi\,\delta(x)\delta(y). \tag{8.4}$$

These differ crucially from the corresponding Abelian expression by the factor $\epsilon(z)$, showing that the field lines reverse direction at the origin, which is where this Dirac nexus sits.

This Dirac nexus is beginning to show features like those of the nexus in Figure 8.2. To find the appropriate kinematics, form the gauge representative $\mathcal{A}_i \to U\partial_i U^{-1}$, and by inspection, choose for the full potential

$$\mathcal{A}_j = \frac{\epsilon_{jak}}{2i}\tau_a\widehat{r}_k[F - 1 + G\cos\phi] + \frac{1}{2i}(\tau_i - \widehat{r}_i\tau\cdot\widehat{r})G\sin\phi + \widehat{\phi}_j\frac{\tau\cdot\widehat{r}}{2i}B_1, \tag{8.5}$$

with

$$F = F(\rho, z); \qquad G = G(\rho, z); \qquad B_1 = B_1(\rho). \tag{8.6}$$

The function B_1 carries the thick flux tube of the center vortex (but with oppositely directed flux on the two halves of the z-axis), and so this kinematics describes a compound of a thick center-vortex flux tube and the nearly pointlike core of the nexus, just as in Eq. (7.21):

$$B_1 = \frac{1}{\rho} - mK_1(m\rho). \tag{8.7}$$

Choose boundary conditions to make the gauge potential approach the pure gauge based on the gauge representative of Eq. (8.3):

$$\rho, z \to \infty: \qquad F \to 0, \ G \to -1 \tag{8.8}$$

$$\rho, z \to 0: \qquad F \to +1, \ G \to 0.$$

There is no analytic solution to these coupled, nonlinear, partial differential equations, and no one has yet solved them numerically. However, there is a simple and useful variational approximation [1] using trial functions with a single variational parameter λ:

$$F = \frac{\lambda^2}{\lambda^2 + r^2}; \qquad G = -\frac{\rho r}{\lambda^2 + \rho r}. \tag{8.9}$$

These obey the correct boundary conditions. To find the nexus energy, calculate the entire Hamiltonian with these functions and subtract from it the energy of the pure vortex. One finds, after carrying out the usual variational steps, a nexus energy $3.22(4\pi m)/g^2$.

8.2.2 The SU(N) nexus

All this generalizes to $SU(N)$, although (just as with junctions) there are many new geometries. Note that a nexus, as the boundary between two regions of a vortex with differing field strengths, cannot have its tubes of chromomagnetic field separated into two bundles arbitrarily. It is essential that a center vortex decorated with a nexus give rise to precisely the same element of the center group, as found by transporting the gauge representative around a closed curve linking with the vortex for each flux tube. So for any nexus that has exactly two flux tubes, as in Figure 8.2, if one of the tubes carries flux matrix Q_1, e.g., then the other must carry another Q_j (and similarly for higher-flux matrices).

An elementary calculation shows that for any two choices of Q_k, their difference $Q_i - Q_j$ is a Pauli matrix τ_3 for an embedded $SU(2)$. This means that all entries are zero, except for one $+1$ (in the jth position along the diagonal) and one -1 (in the ith position). So we can write, e.g.,

$$Q_1 = -\frac{1}{2}\tau_3 + R(12), \qquad Q_2 = \frac{1}{2}\tau_3 + R(12),$$

$$R(12) = \mathrm{diag}\left(-\frac{1}{2} + \frac{1}{N}, -\frac{1}{2} + \frac{1}{N}, \frac{1}{N}, \ldots\right). \tag{8.10}$$

The matrix R_{12} commutes with the generators of the embedded $SU(2)$. Now it is elementary to find a gauge representative of a two-tube nexus:

$$U = e^{(i\phi\tau\cdot\hat{r}/2)}e^{i\phi R(12)}, \tag{8.11}$$

where, of course, the Pauli matrices are in the embedded $SU(2)$. The magnetic charge of the nexus can be identified with the eigenvalues of the embedded τ_3, which are ± 1, as would be required for a 't Hooft–Polyakov monopole.

8.2.3 Nexus magnetic charge

How do we detect the nexus magnetic charge and relate it to topological charge? Because there is a strong connection between the nexus and the 't Hooft–Polyakov monopole, the procedure [17] somewhat resembles that for the 't Hooft–Polyakov monopole. The main difference is that there is no Higgs–Kibble field for the nexus in a QCD-like theory. In the 't Hooft–Polyakov monopole, the presence of this Higgs–Kibble field in the adjoint of $SU(2)$ breaks the gauge symmetry to $SU(2)/U(1)$, a space homotopic to the two-sphere S^2, and the surviving long-range magnetic field can be identified with, say, the 3 direction in group space. After a suitable projection onto this unbroken $U(1)$ subspace, the magnetic charge of the

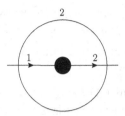

Figure 8.3. The inner black dot represents a nexus that we label A, and the lines represent its associated flux tubes with fluxes described by $Q_{1,2}$. The outer circle represents the plain vortex surface B with flux described by Q_2.

't Hooft–Polyakov monopole is measured through the integral

$$\int_{\Gamma} d\vec{S} \cdot \vec{B} = Q_{\text{mag}}, \tag{8.12}$$

where Γ is an arbitrary closed surface surrounding the monopole; the integral yields an integral magnetic charge Q_{mag}. This corresponds to the homotopy that maps this broken gauge group onto the two-sphere:

$$\Pi_2(SU(2)/U(1)) = \Pi_2(S^2) = Z, \tag{8.13}$$

where Z is the group of integers.

There is a sense in which nexuses also display this homotopy, although this is suspicious because there is no symmetry breaking for the nexus for QCD-like theories, and the homotopy $\Pi_2(G)$ is trivial for every non-Abelian gauge group G. What in fact happens is that nexuses really display *topological charge* and the homotopy above [17] is simply a disguised form of the usual topological charge integral:

$$Q_{\text{topo}} = -\frac{1}{16\pi^2} \int d^4x \, \text{Tr} \, \mathcal{G}_{\mu\nu} \widetilde{\mathcal{G}}_{\mu\nu}. \tag{8.14}$$

We evaluate this integral for the generic intersection of the static nexus already displayed with Dirac fields $\mathcal{G}_{\mu\nu}^{(A)}$ (see Eq. (8.4)) and a plain center vortex that we call (B), as in Figure 8.3. The static nexus is the horizontal line with incoming flux matrix Q_1 on the left and $-Q_2$ on the right. The center vortex surface (B) is a closed surface with the topology of S^2 characterized by the matrix Q_2. The vortex and the nexus intersect at two points, and these are where the topological charge density is located. The topological charge of the overlap between (A) and (B) is

$$Q_{\text{topo}} = -\frac{1}{8\pi^2} \int d^4x \, \text{Tr} \, \widetilde{\mathcal{G}}_{\mu\nu}^{(B)} \mathcal{G}_{\mu\nu}^{(A)} = \frac{1}{4\pi i} \int d\sigma_{\mu\nu} \, \text{Tr} \, Q_2 \mathcal{G}_{\mu\nu}^{(A)}, \tag{8.15}$$

where in the second equality, we replaced $\tilde{\mathcal{G}}_{\mu\nu}^{(B)}$ by its Dirac-surface form. For $SU(2)$, there are only two Q-matrices: $Q_1 = -Q_2 = \tau_3/2$. Clearly, this second equality is precisely the magnetic charge integral Q_{mag}, which we now see is equal to Q_{topo}; both are equal to unity.

For general $SU(N)$, give nexus (A) the Q-matrices Q_a, Q_c, and give the vortex (B) the Q-matrix Q_b. Now the trace factor is

$$\text{Tr}\, Q_b (Q_a - Q_c) = \delta_{ab} - \delta_{cb}. \tag{8.16}$$

This, of course, has only the integral values $0, \pm 1$, vanishing if b is not equal to either a or c. The topological charge depends very much on the surface surrounding the nexus, unlike the purely artificial surface used to define the magnetic charge of a 't Hooft–Polyakov monopole. As advertised, the total topological charge is integral, with a fractional charge of $\text{Tr}\, Q_b Q_a = \delta_{ab} - (1/N)$ at the crossing of flux line a with the vortex surface. All this can be generalized to more complicated vortices and nexuses, but we will not do that here.

8.2.4 Topological charge as an intersection number for nonorientable vortex surfaces

Here we display the intersection number form of Q_{topo} [7, 8, 9, 10, 16]. Start with the vortex field strength in the Dirac-surface limit:

$$\mathcal{G}_{\mu\nu}^A(x) = \frac{2\pi Q_A}{i} \int d\tilde{\sigma}_{\mu\nu}^A(z)\, \delta(x - z(A)), \tag{8.17}$$

where Q_A is one of the flux matrices Q and $d\tilde{\sigma}_{\mu\nu}^A(z)$ is the dual surface element for the surface A characteristic of the vortex. The standard topological charge of Eq. (8.14) is, in terms of the sum of vortex field strengths,

$$Q_{\text{topo}} = \sum_{A,B} \text{Tr}(Q_A Q_B) I(A, B), \tag{8.18}$$

where $I(A, B)$ is an *intersection number*:

$$I(A, B) = \epsilon_{\mu\nu\alpha\beta} \int \frac{1}{2} d\sigma_{\mu\nu}^A \frac{1}{2} d\sigma_{\alpha\beta}^B \,\delta(z(A) - x(B)). \tag{8.19}$$

The intersection number is ± 1 for every transverse intersection of a point on surface A with a point on surface B (transverse means that the normals to the surfaces at the point of intersection span Euclidean four-space).

We are on the road to getting lumps of fractional topological charge localized at the intersection points because the trace factor $\text{Tr}(Q_A Q_B)$ always has a denominator of N for $SU(N)$. Unfortunately, at this stage of the game, we always get zero

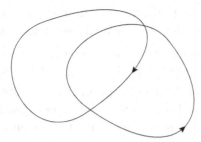

Figure 8.4. Two closed oriented lines in $d = 2$ have a total intersection number of zero because the two intersections have opposite orientation and cancel.

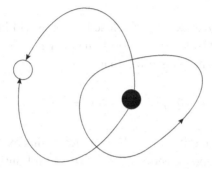

Figure 8.5. Two closed lines in $d = 2$, one with an $SU(2)$ nexus-antinexus pair. They have a total intersection number of 1 because the two intersections have the same orientation.

from Eq. (8.18) [7, 8, 9, 10, 16, 17]. The reason is that when two *closed oriented* surfaces intersect, the total intersection number is zero. One can see this from the corresponding geometry in two dimensions, as shown in Figure 8.4. We will give a formal proof that the total intersection number is zero in $d = 4$ very shortly.

For any pair of ordinary vortices, we can factor out the trace factor in Eq. (8.18), and then the resulting charge is zero. So intersections do not seem very promising for generating topological charge. We can fix the problem by remembering nexuses that in effect change the Q matrices, while preserving the group-center element associated with a vortex as one moves around a given vortex surface. The simplest case to illustrate is $SU(2)$, where a nexus simply changes the orientation of the chromomagnetic flux. Figure 8.5 shows two intersecting, two-dimensional closed curves, but this time, the one on the left has a nexus and an antinexus, reversing the orientation at each. The plain curve, on the right, encloses (is linked to) the nexus, and because the orientation reverses on passing the nexus, the intersection numbers are $+1$ at both intersections. The Q-trace is $1/2$, and now the topological charge is

$1/2 + 1/2 = 1$. The topological charge is still localized at the intersection points and is fractional at these points, but the total topological charge is unity.

The same thing – appearance of lumps of topological charge quantized in units of $1/N$, with integral total charge – happens in $d = 4$ and for any $SU(N)$ [5, 6, 7, 8, 9, 10, 16, 17]. Look at the simplest case, where a single nexus world line on one vortex is linked to a closed vortex with no nexuses. Nexus world lines that are unlinked contribute nothing and are omitted. With this understanding, write the dual field strength of vortex A with nexuses as a sum:

$$\tilde{\mathcal{G}}^A_{\mu\nu}(x) = \frac{2\pi Q_a}{i} \int_{S_a} d\sigma_{\mu\nu}\, \delta(x - z(\sigma)) + \frac{2\pi Q_b}{i} \int_{S_b} d\sigma_{\mu\nu}\, \delta(x - z(\sigma)), \quad (8.20)$$

where S_a is a surface bounded by the closed nexus world line Γ on one side, and S_b is a surface bounded by the same world line on the other side. These boundaries have opposite orientation, in the sense that

$$\partial S_a = \Gamma \qquad \partial S_b = -\Gamma. \quad (8.21)$$

Equation (8.20) is not literally correct because terms that exhibit the corresponding antinexus world line, and possibly other nexuses and antinexuses, are omitted. However, they are irrelevant to the topological charge if they are not linked to the surface of the second vortex. This equation has another flaw: taken literally, it violates the Bianchi identities. This is because there really is no mathematically accurate way of modeling a nexus as an Abelian object; it is essentially non-Abelian. One overcomes this Bianchi identity problem by smoothing the transition from Q_a to Q_b over a region of size $\sim 1/m$, as in the nexus; this smoothing has no effect on topological properties coming from long-range effects.

Let $\mathcal{A}^C_\mu(x)$ be the Dirac-singular part of the gauge potential of a second vortex with no nexuses. Its Dirac gauge potential is an integral over a closed surface S_c:

$$\mathcal{A}^C_\mu(x) = -\frac{2\pi Q_c}{i} \epsilon_{\mu\nu\alpha\beta} \oint_{S_c} \frac{1}{2} d\sigma'_{\alpha\beta}\, \Delta_0(x - y(\sigma')). \quad (8.22)$$

The topological charge is expressible as a linking between the vortex and the nexus world line, completely analogous to the linking of a vortex and a Wilson loop responsible for confinement. The Gauss formula for linking vortex surface S_c and the closed nexus world line Γ is familiar from confinement:

$$Lk = \oint_\Gamma dz_\mu \epsilon_{\mu\nu\alpha\beta} \oint_{S_c} \frac{1}{2} d\sigma'_{\alpha\beta}\, \partial_\nu \Delta_0(z - y(\sigma')). \quad (8.23)$$

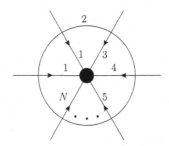

Figure 8.6. An $SU(N)$ nexus split into N lines. The circle labeled 2 schematically represents the plain vortex (B) of the text, and the vortex flux lines emerge from the nexus (A). The intersections of the lines and the circle are points of topological charge density.

Next, we show that the topological charge is essentially this link number by applying Stokes's theorem. Consider the expression

$$Q_{\text{topo}} = \frac{1}{2\pi i} \oint_\Gamma dx_\mu \, \text{Tr} \left[(Q_a - Q_b) A_\mu^C(x) \right] \qquad (8.24)$$

for the topological charge. Using Stokes's theorem on Eq. (8.24) yields an expression that is easily converted into the fundamental topological charge integral of Eq. (8.14), evaluated with the field strengths from Eq. (8.20) and the curl of Eq. (8.22). The minus sign in the trace factor comes from the opposite orientations, as given in Eq. (8.21).

We can also conclude from Eq. (8.24) that the total intersection number of closed oriented surfaces is zero by replacing the individual trace factors $\text{Tr} \, Q_a Q_c$ and $\text{Tr} \, Q_b Q_c$ by unity.

We have earlier seen how to divide topological charge into parts $1 - (1/N)$ and $1/N$. Is it possible to divide this topological charge further into N constituents, each of charge $1/N$? The answer is yes. In the center vortex of Figure 8.3, decompose the flux matrix Q_2 on the right-hand side (rhs) of the nexus as

$$Q_2 = -Q_1 - Q_3 \cdots - Q_N, \qquad (8.25)$$

and associate a flux tube with each of the terms in this equation, as shown in Figure 8.6.

Each intersection of a line coming from the nexus (A) with the circle (B) is a point of topological charge $1/N$, with the sum being unity, as before. Is it probable that an elementary nexus would split into N lines, as shown? Not if action alone were the only consideration because the N lines each have an action per unit area the same as that of the two lines of an elementary nexus. But as we have learned, entropy is equally important, so this N-line splitting should not be less probable than a 2-line nexus.

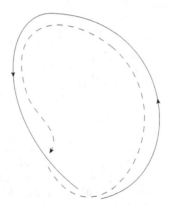

Figure 8.7. A curve turned into a ribbon by adding the curve of the dotted line.

How might one see the effect of fractional topological charge, especially because topological charge is integrally quantized globally? Perhaps the most important way is to study the topological susceptibility χ, which is quadratic in Q_{topo}. This is defined as $\chi = \langle Q_{\text{topo}}^2 \rangle / V_4$, where V_4 is the volume of space-time. Witten [19] and Veneziano [20] have given a large-N formula relating the η' mass to χ, which suggests that the vacuum energy as a function of the θ-angle depends not on θ but on θ/N, as topological charge fractionation would give. This is discussed further in [16].

But topological charge fractionation into units of $1/N$ is, unfortunately, not the whole story. Center vortices need not intersect at points to generate nonintegral topological charge; they can do so by twist and writhe [5, 6, 7, 8, 9, 10, 16, 17].

Twist and writhe   An ordinary two-dimensional ribbon can link to itself (in $d = 3$) by twist and writhe, which means by deformations such that the two edges of the ribbon would be linked (knotted) if the rest of the ribbon were missing.[6] Twist and writhe contribute to the Chern–Simons number in somewhat the same way that intersections contribute to the topological charge [5, 6, 7, 8, 9, 10]; that is, for a vortex with twist, writhe, or both, the integral for N_{cs} in Eq. (8.31) is nonvanishing, and this integral need not be integral or a multiple of $1/N$.

Intuitively, twist comes from forming this ribbon from a long, open paper strip then twisting one end a certain number of times before closing the strip by joining one end to the other. (A half twist leads to a nonorientable Möbius strip not considered here.) For a mathematical curve, twist and writhe need further definition, which can be done by supplying the curve with an infinitesimally close partner, as shown in Figure 8.7. The combination forms a ribbon whose twist and writhe are well defined but not unique (they depend on the partner curve).

---

[6] See Kaufmann [18] for general properties of knots and related subjects.

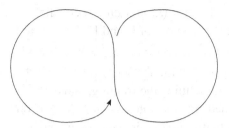

Figure 8.8. A $d = 3$ projection of a center vortex with writhe.

Writhe seems intuitively to be different from twist, but it is not. Figure 8.8 shows a closed curve with writhe. Playing with actual paper ribbons will show that twist and writhe are interconvertible without tearing the paper.

Topological charge can be generated from twist and writhe only if there is a difference in the Chern–Simons number at two boundaries that we can identify as referring to (Euclidean) times of $\pm\infty$ so that if the vortex is to change its Chern–Simons number, it must reconnect by crossing itself. This crossing may be essentially Abelian and easily envisaged by imagining a motion picture of a closed loop crossing itself, or it can be essentially non-Abelian and call for a deeper level of visualization.

Consider, then, the closed Dirac string of a $d = 3$ vortex. There is a famous theorem of such $d = 3$ knotted curves,

$$Lk = Tw + Wr, \tag{8.26}$$

where Lk, a topological number and an integer, is the self-linking number, and the terms on the right, neither of which is an integer or of topological character, are the twist Tw and writhe Wr, respectively. The integral that defines Lk is just the one used earlier (see Eq. (7.41)) for the linkage of two distinct curves but with only one curve in it:

$$Lk = \oint_\Gamma dx_i \oint_\Gamma dx'_j \epsilon_{ijk} \partial_k \Delta_0(x - x'). \tag{8.27}$$

With just one curve Γ, it is inevitable that the points where $x = x'$ are possibly singular. Some form of regulator is needed. The standard one is *ribbon framing*, as in Figure 8.7. The original curve is turned into a ribbon by adding a second curve Γ' infinitesimally separated from Γ and not intersecting it; the self-link number is defined as the Gauss link integral for these two nonintersecting curves. So Γ' is the first curve displaced by an infinitesimal amount:

$$\Gamma': \qquad x'_i(s) = x_i(s) + \epsilon n_i(s), \tag{8.28}$$

where ϵ is infinitesimal and $n_i(s)$ is a unit-vector field. The self-link number is defined as the mutual link number of Γ and Γ'. This is, to be sure, an integer and a topological invariant, but it depends on this new unit-vector field.

This ribbon framing would not make sense for real-world center vortices because we have no good way of defining the framing, and even the topologically invariant self-link is only defined modulo integers that depend on the framing. But real-world center vortices have a finite thickness, as we know. This thickness removes all ambiguity from the limiting process of defining self-linkage. The idea [5] is to write the Chern–Simons number, which is the same as the linking number, for a vortex using both the massive and the massless propagators that occur in the vortex wave function. It then turns out that the Chern–Simons number for a plain unit-flux vortex becomes

$$N_{\text{cs}} = \text{Tr}\, Q_i^2 \oint_\Gamma dx_i \oint_\Gamma dx'_j \, \epsilon_{ijk}\partial_k \Delta_0(x - x')F(M|x_i - x'_i|), \tag{8.29}$$

where

$$F(u) = \frac{1}{2}\int_0^u dv\, v^2 e^{-v}. \tag{8.30}$$

For small u, $F(u) \sim u^3$, and this is more than enough to cancel the singularities at $x_i = x'_i$ in the rest of the integrand in Eq. (8.29). Since $F(\infty) = 1$, vortex segments that are far apart contribute as usual to the Chern–Simons number, and nothing is changed.

The topological charge contained between two time slices is the difference between two Chern–Simons numbers. For a gauge potential on a fixed-time slice, this number is

$$N_{\text{cs}} = -\frac{1}{8\pi^2}\int d^3x\, \epsilon_{ijk}\text{Tr}\left(\mathcal{A}_i\partial_j\mathcal{A}_k + \frac{2}{3}\mathcal{A}_i\mathcal{A}_j\mathcal{A}_k\right). \tag{8.31}$$

This number is not gauge invariant; under the gauge transformation

$$\mathcal{A}_i \to U\mathcal{A}_iU^{-1} + U\partial_iU^{-1}, \tag{8.32}$$

we find

$$N_{\text{cs}} \to N_{\text{cs}} + \frac{1}{8\pi^2}\int d^3x\, \epsilon_{ijk}\text{Tr}\left[\frac{1}{3}U^{-1}\partial_iUU^{-1}\partial_jUU^{-1}\partial_kU - \partial_i(\mathcal{A}_jU^{-1}\partial_kU)\right]. \tag{8.33}$$

If the original gauge potential is zero, so that we are calculating the Chern–Simons number of a pure gauge transformation, the integral in Eq. (8.33) is supposed to be an integer, as prescribed by the homotopy of Eq. (8.2). However, this requires an extra assumption: that the three-space over which one integrates the Chern–Simons density is *compact*, which means that the gauge $U(\vec{x})$ approaches a constant independent of direction as $r \to \infty$. There seems to be no elementary physical reason for assuming compactness, and in Section 8.3, devoted to the sphaleron, we examine this assumption further (the sphaleron naturally has $N_{cs} = 1/2$, seemingly violating compactness).

Nexuses in the Georgi–Glashow model In a famous paper, Polyakov [12] explained confinement in the $d = 3$ Georgi–Glashow model as due to 't Hooft–Polyakov monopoles, with a long-range spherically symmetric magnetic field, thereby exemplifying dual superconductivity as a confining mechanism. (The Georgi–Glashow model is an $SO(3)$ NAGT coupled to a Higgs–Kibble field in the adjoint representation, which gives masses to two charged gauge bosons, leaving the third one, which we call the photon, massless.) In fact [2, 13], confinement in the Georgi–Glashow model is actually an example of center-vortex confinement with asymmetric nexuses, whose world lines lie in center-vortex sheets, as we have already shown for QCD-like gauge theories with no symmetry breaking. Polyakov works in the limit $v \gg g_3$, where v is the Higgs–Kibble VEV and g_3 is the gauge coupling. In this, the semiclassical limit, the 't Hooft–Polyakov monopole has a very large action. Because there must be a monopole condensate, Polyakov points out that a Meissner mass is induced for the photon, just as in ordinary superconductivity. This mass, however, is exponentially small in v/g_3 and is ignored by Polyakov, who then can claim that the semiclassical excitations of the gauge field are indeed 't Hooft–Polyakov monopoles. But as long as v is finite, the 't Hooft–Polyakov monopole becomes a nexus because its magnetic field can no longer be long range. The size of the nexus is exponentially large and, at distances scaled by other parameters of the theory, looks very much like a 't Hooft–Polyakov monopole. Nonetheless, as a matter of principle, for (fundamental representation) Wilson loops that are large compared to the nexus thickness, confinement is by the center vortices in which the nexus is embedded. This becomes clear [13] as the VEV v is reduced; at some point, when $v \leq g_3$, the Higgs–Kibble mass of the charged gauge bosons, proportional to vg_3, is too small to avoid infrared instability, a dynamical gauge-boson mass of $\mathcal{O}(g_3^2)$ is induced, and the nexus (and center vortices) begin to look like the symmetrical ones of a QCD-like theory.

8.3 The QCD sphaleron

There are three gauge-field configurations known as sphalerons. Usually the word *sphaleron*[7] refers to a massive spherically symmetric $d = 3$ electroweak soliton

---

[7] Coined by Klinkhamer and Manton [21].

with a gauge-boson mass driven by a Higgs field [22]. The sphaleron's topological properties were first noted for electroweak theory by Manton [22], where it occurred as a classical saddlepoint on a noncontractible loop in the $d = 3 + 1$ configuration space of gauge potentials, describing the top of the tunneling barrier of minimum energy between vacua with topological charges differing by unity. There is another sphaleron in classical NAGTs corresponding to the saddlepoint at the top of the potential barrier tunneled through by instantons. This classical object is massless but has an arbitrary length scale set by the collective size coordinate of its associated instanton.

There is also a quantum sphaleron in QCD-like NAGTs [23] that differs from both of the preceding sphalerons while retaining the saddlepoint character; we call it the QCD sphaleron. The gauge-boson mass is dynamical, there is no Higgs–Kibble field, and there is no symmetry breaking. The QCD sphaleron has a fixed size determined by the gluon mass, and this size actually corresponds to an upper limit for the size of instantons and sphalerons. This sort of upper size limit is routinely seen in computer simulations in which instantons are identified and built into models [24] of the instanton liquid, where the size scale corresponds to a mass of 600 MeV. The QCD sphaleron may exist transiently as some sort of glue ball, and it also is a mediator between charge-changing events but not of the usual topological charge. Instead, the charge associated with the QCD sphaleron is that of the color-singlet axial current, giving the change in the flavor sum of chiralities.

Both the classical sphaleron and the QCD sphaleron can be embedded in Euclidean four-space ($d = 4$) or in Minkowski space ($d = 3 + 1$), and these embeddings will be the emphasis in the present chapter. (Chapter 9 discusses the sphaleron as a $d = 3$ object more fully.) In $d = 4$, the classical sphaleron is a cross section of an instanton that solves the classical field equations. In $d = 3 + 1$, there is no known embedding in a solution of the field equations, but one can find embeddings that have all the desired properties of an evolution of topological charge in Minkowski time t [25, 26] and define a corresponding Chern–Simons number $N_{\text{cs}}(t)$. By symmetry of the tunneling process from topological charge zero to charge unity, we should assign a Chern–Simons number or topological charge of $1/2$ to the sphaleron and describe the tunneling, in Minkowski time, as a smooth evolution of $N_{\text{cs}}(t)$ from 0 to 1, passing through $1/2$ at the top of the barrier. In the course of this smooth evolution, N_{cs} is clearly nonintegral because it gets contributions from gauge potentials that are not pure gauge with nonvanishing field strengths. Like the other solitons we discuss, the QCD sphaleron is fundamentally a $d = 3$ object but is unstable in isolation. After discussing this basic QCD sphaleron, we defer to Chapter 9 for further discussion of the QCD sphaleron as a pure $d = 3$ object and will show [27] that, considered purely as a $d = 3$ object, the $N_{\text{cs}} = 1/2$ sphaleron is closely connected to the properties of knots or closed $d = 1$ strings, embedded in three dimensions, that are linked.

There are many potential physical applications of sphalerons, both in electroweak theory and in QCD. Some arise through the connection, via the anomaly, of topological charge and the divergence of a current. In electroweak theory, this current is the sum of the baryonic (B) and leptonic (L) current and leads to B + L violation. In QCD, the current with an anomalous divergence is the $U_A(1)$ current, and helicity conservation is violated. In both cases, the violations have a tunneling interpretation. Sphaleronic configurations are also important in estimating the (lack of) overlap at high energy between few-particle states and many-particle states; see the references in Chapter 9. It would take almost another book to detail such applications.

8.3.1 The QCD sphaleron as a $d = 3$ object

The ansatz for the gauge function U of the sphaleron is the well-known one for spherically symmetric solitons,

$$U = \exp\left[\frac{\mathrm{i}}{2}\beta(r)\vec{\tau}\cdot\hat{r}\right], \tag{8.34}$$

differing from Eq. (8.3) of the nexus only in the choice of a rotation angle, which for the sphaleron is radially symmetric. Forming the pure-gauge representative $\mathcal{A}_i \to U\partial_i U^{-1}$, we infer the standard kinematics for spherically symmetric solitons:

$$\mathcal{A}_j = \frac{\epsilon_{jak}}{2\mathrm{i}}\tau_a\hat{r}_k\left[\frac{\phi_1(r)-1}{r}\right] - \frac{1}{2\mathrm{i}}(\tau_j - \hat{r}_j\tau\cdot\hat{r})\left[\frac{\phi_2(r)}{r}\right] + \hat{r}_j\frac{\tau\cdot\hat{r}}{2\mathrm{i}}H_1(r). \tag{8.35}$$

The boundary conditions are as follows:

$$\phi_1(\infty) = \cos\beta(\infty); \qquad \phi_2(\infty) = -\sin\beta(\infty); \qquad H_1 \to \left.\frac{\mathrm{d}\beta}{\mathrm{d}r}\right|_{r=\infty}. \tag{8.36}$$

For future reference (see Chapter 9), we note that the Chern-Simons number of U is

$$N_{\mathrm{cs}}\{U\} = \frac{1}{2\pi}[\beta(\infty) - \beta(0)]. \tag{8.37}$$

The equations of motion have a finite-energy solution [23] for the special choice $\beta(r) = \pi$. There is no analytic solution, but there is an analytic approximation [11, 23] based on a variational approach that gives excellent agreement with numerical calculations. Use the trial functions

$$\phi_1(r) = \frac{a^2 - r^2}{a^2 + r^2}; \qquad \beta = \pi; \qquad \phi_2 = H_1 = 0, \tag{8.38}$$

where the length a is a variational parameter. Of course, the true $\phi_1 + 1$ vanishes exponentially as $r \to \infty$, but our trial wave function vanishes only like $1/r^2$. The variational mass turns out to be $5.44(4\pi m/g^2)$, which is within half a percent of the numerical answer, in which 5.44 is replaced by 5.41.[8]

One might think that if β is a half-integral multiple of π, the CS number is also a half-integral. But Eq. (8.37) shows that N_{cs} vanishes. Only when we embed this sphaleron in a $d = 3 + 1$ context will we find a Chern–Simons number of $1/2$. In any event, we can change the Chern–Simons number arbitrarily by making a spherical gauge transformation, although at the price of foregoing compactness.

As a solution of the spherical field equations, this $\beta = \pi$ sphaleron is an extremum, but it is a saddlepoint and therefore has a maximum for some parameters in a space orthogonal to a space in which the minimum lies (in our case, this space is just the space of the trial parameter a). For example, Ref. [11] exhibits trial functions yielding finite energy and having $\beta = \beta_0$ for any fixed angle β_0. Let ϕ_{1c} be the exact solution for the $\beta = \pi$ sphaleron, and define

$$\phi \equiv \phi_1 + i\phi_2 = \frac{1}{2}(1 + \phi_{1c}) + \frac{1}{2}e^{i\beta_0}(1 - \phi_{1c}). \tag{8.39}$$

Also, take $H_1 = 0$. The new ϕ obeys the boundary conditions of Eq. (8.36) with $\beta = \beta_0$, and the associated gauge potential, constructed from $\phi - 1$, smoothly changes toward zero as $\beta_0 \to 0 \bmod 2\pi$. The trial mass function is

$$M_s(\beta_0) = \frac{1}{2}(1 - \cos \beta_0)M_s(\beta_0 = \pi), \tag{8.40}$$

where $M_s(\beta_0 = \pi)$ is the sphaleron mass. So there is a maximum at $\beta_0 = \pi$, and smoothly reducing β_0 to zero reduces the soliton to nothing.

8.3.2 Sphalerons in four-dimensional Minkowski space

There is a very simple but apparently accurate description [25, 26] of this minimum-height barrier that has only one dynamical degree of freedom, a simple scalar function of time called $\lambda(t)$. It nicely extends the trial function of Eq. (8.38) to time-dependent configurations. Bitar and Chang [25] suggested that the standard expressions for a classical instanton could be used in Minkowski space with the simple replacement of t by $\lambda(t)$ and the insertion of $\dot{\lambda}(t)$ at a particular place. These expressions, in terms of the spherically symmetric space components of Eq. (8.35), are

$$\phi_1 = \frac{\lambda^2 + a^2 - r^2}{\lambda^2 + a^2 + r^2}; \qquad \phi_2 = \frac{-2\lambda r}{\lambda^2 + a^2 + r^2}; \qquad H_1 = \frac{2\lambda}{\lambda^2 + a^2 + r^2}. \tag{8.41}$$

---

[8] Warning: in Cornwall [26], an incorrect value was used in place of 5.44. This paper also has several typos.

To this list, we add a time component of the gauge potential,

$$\mathcal{A}_0 \equiv \frac{1}{2i}\vec{\tau} \cdot \hat{x} H_2; \qquad H_2 = \frac{-2\dot{\lambda}}{\lambda^2 + a^2 + r^2}, \qquad (8.42)$$

and take $\beta = 2\arctan(r/\lambda)$. This choice for β is equivalent to making a spherical gauge transformation of the $d = 3$ spherical decomposition by an angle $\alpha = -\pi + 2\arctan(r/\lambda)$ that carries λ as a parameter with no particular dynamical significance in $d = 3$.

If λ is replaced by t, these expressions are exactly those for an instanton in $d = 4$ of size a, which is arbitrary for the classical instanton. However, for the QCD sphaleron and its embedding, a has a different interpretation and is determined by the gluon mass m. As in Bitar and Chang, these embedding functions for the QCD sphaleron are used in Minkowski space, not Euclidean space. Of course, in Minkowski space, they are neither solutions of the equations of motion nor self-dual, but they are still useful because they represent the tunneling barrier itself quite accurately. Because λ in some sense is a replacement for time t, we require that λ be an odd function of t and monotone increasing in t, and we impose the conditions

$$\lambda(-\infty) = -\infty; \qquad \lambda(0) = 0; \qquad \lambda(+\infty) = \infty. \qquad (8.43)$$

We have, consistent with the oddness in t of λ, $\lambda(0) = 0$ at the time $t = 0$, representing the top of the barrier, where $\mathcal{A}_0$ vanishes according to our ansatz. Then, at $t = 0$ (i.e., $\lambda = 0$), the Bitar–Chang potentials reduce to the $d = 3$ trial function already used in Eq. (8.38) for the QCD sphaleron plus the specification $\beta = \pi$. The minimum QCD sphaleron barrier height at $t = 0$ is the sphaleron energy, and the saddlepoint nature of the sphaleron becomes evident because (see Eq. (8.44) below) as time increases, the energy decreases.

For the QCD sphaleron, we treat a as a variational parameter to be determined from the *massive* effective Hamiltonian. This Hamiltonian comes from inserting the full ansatz into the $d = 3 + 1$ action analogous to the $d = 4$ massive effective action S_{eff} and stripping off a time integral. It is not quite the same as the static Hamiltonian H_{eff} of Eq. (7.7) because there are contributions from the $\dot{\lambda}$ terms. The result [26] has the form

$$H_s = H_{eff} + \frac{\dot{\lambda}^2}{2g^2}\mu(\lambda, a, m) - \frac{\dot{\lambda}^2}{2g^2}\kappa(\lambda, a, m). \qquad (8.44)$$

The first term on the right is the static (potential) energy H_{eff} at $\lambda = \dot{\lambda} = 0$, and μ, κ are positive integrals [26] over the Bitar–Chang potentials and fields. The sphaleron mass M_s is simply the extremal value of H_{eff}. At $t = 0$, extremalization of H_{eff} leads to

$$a = \frac{\sqrt{3}}{2m}; \qquad M_s = \frac{4\sqrt{3}\pi m}{g^2}. \tag{8.45}$$

The saddlepoint instability of the sphaleron is evident in the negative sign for the potential coefficient κ.

The Chern–Simons number varies smoothly with t from 0 at $t = -\infty$ to 1 at $t = +\infty$. The total topological charge has the expression

$$\begin{aligned}
Q_{\text{topo}} &= -\frac{1}{4\pi^2} \int d^4x \, \text{Tr} \, \vec{\mathcal{E}} \cdot \vec{B} \\
&= \frac{24a^4}{\pi} \int_0^\infty dr \, r^2 \int_{-\infty}^\infty d\lambda \frac{1}{(\lambda^2 + r^2 + a^2)^4} = 1. \tag{8.46}
\end{aligned}$$

The integral to the top of the barrier ($\lambda = 0$) gives topological charge 1/2, as expected, consistent with Eq. (8.37).

Now change variables from λ to a new variable angular $q(t) = f(\lambda(t))$, chosen so that the kinetic energy in the Hamiltonian has the simple form $\dot{q}^2/(2I)$, with a q-independent moment of inertia and with the angular properties $q(t = -\infty) = 0$, $q(t = +\infty) = 2\pi$. This has been done numerically [26], and the resulting potential looks very much like the pendulum potential $\sim 1 - \cos q$. The sphaleron is the static but unstable point $q = \pi$ with the pendulum standing on end.

The parameter β_0 introduced in Eq. (8.39) for the trial wave function, considered a function of time, is both (approximately) the phase variable q for the upside-down pendulum and the angle to be used in N_{cs} (see Eq. (8.37)).

8.4 Chiral symmetry breakdown, nexuses, and fractional topological charge

Chiral symmetry breaking (CSB) for quarks in QCD is closely related to confinement and (via the Atiyah–Singer theorem) a condensate of topological charge. Arguments were given long ago [28, 29, 30] that confinement was sufficient for CSB. These works were based on a variety of phenomenological models of confinement, not center vortices. Later, lattice simulations showed [31] that center vortices (and nexuses) were not only sufficient but also necessary for quark CSB: there was both confinement and CSB in the presence of center vortices, but when center vortices were removed from the simulation, not only did confinement disappear, but CSB disappeared also, as shown in Figure 8.9.

Moreover, simulations show that the CSB transition temperature, at which chiral symmetry is restored, is very close to the deconfinement transition temperature,

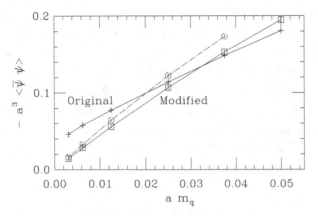

Figure 8.9. Graph of the quark condensate $\langle \bar{\psi}\psi \rangle$ versus quark mass m_q, showing CSB at $m_q = 0$ if center vortices are present (curve marked "Original") but not if they are removed (curve marked "Modified"). Reprinted with permission from P. de Forcrand and M. D'Elia, *Phys. Rev. Lett.* **82** (1999) 4582, © 1999 by the American Physical Society.

above which center vortices are unable to confine (e.g., see Chapter 9 and Cheng et al. [32]). This, too, suggests that confinement is necessary for CSB for quarks because if there were another significant mechanism, it might show up once confinement was out of the picture.

In the picture of center vortices and nexuses supported by gluon-mass generation, it is easy to see how this happens. Center vortices give confinement, as we know, and nexuses, plus vortex twist and writhe, give topological charge and CSB. Removing center vortices takes away all these effects because nexus world lines are required to live on center vortices.

The Atiyah–Singer theorem and the Banks–Casher relation [33] (showing that CSB requires a condensate of fermionic zero modes associated by the Atiyah–Singer theorem with topological charge) say that there should be fermionic zero modes (solutions of the massless Dirac equation) localized near the topological charge produced by vortex-nexus linking. Some appropriate zero modes have been found for just such linkings [34]. Clearly, when vortices are removed in lattice simulations, such zero modes, and apparently the whole fermionic condensate, should vanish.

On the other hand, confinement is not always necessary for CSB. Dirac fermions[9] in the adjoint representation show CSB [35, 36, 37] on the lattice, and of course, the adjoint representation is not confined. Some other mechanism must be at work,

---

[9] Not Majorana fermions, so supersymmetry is not an issue. In fact, Majorana fermions are impossible in Euclidean space.

which may be well approximated by a conventional gap equation based on one-gluon Feynman graphs. The gluon is coupled to the adjoint with a strength 9/4 times its coupling to quarks, so it can happen that the gap equation breaks CSB for the adjoint but not for quarks, depending on the size of $\alpha_s(0)$ [38]. The PT estimates for $\alpha_s(0)$ are in a range where just this happens [39, 40].

It is only recently that powerful lattice algorithms for chiral quarks have come into widespread use, and so there still remains much to be done in confirming the dominant role of center vortices and nexuses in CSB for quarks. However, all present indications are favorable.

References

[1] J. M. Cornwall, Nexus solitons in the center vortex picture of QCD, *Phys. Rev.* **D58** (1998) 105028.

[2] J. Ambjorn and J. Greensite, Center disorder in the 3D Georgi-Glashow model, *JHEP* **9805** (1998) 004.

[3] M. N. Chernodub, M. I. Polikarpov, A. I. Veselov, and M. A. Zubkov, Aharonov-Bohm effect, center monopoles and center vortices in $SU(2)$ lattice gluodynamics, *Nucl. Phys. B Proc. Suppl.* **73** (1999) 575.

[4] B. L. G. Bakker, A. I. Veselov, and M. A. Zubkov, The simple center projection of $SU(2)$ gauge theory, *Phys. Lett.* **B497** (2001) 159.

[5] J. M. Cornwall, Dynamic problems of baryogenesis, in *Unified Symmetry: In the Small and in the Large: Proceedings of the Conference, Coral Gables, Florida, 1994*, edited by B. Kursonoglu et al. (Plenum, New York, 1994) 243.

[6] J. M. Cornwall and B. Yan, String tension and Chern-Simons fluctuations in the vortex of $d = 3$ gauge theory, *Phys. Rev.* **D53** (1996) 4638.

[7] M. Engelhardt, Center vortex model for the infrared sector of Yang-Mills theory – topological susceptibility, *Nucl. Phys.* **B585** (2000) 614.

[8] M. Engelhardt and H. Reinhardt, Center projection vortices in continuum Yang-Mills theory, *Nucl. Phys.* **B567** (2000) 249.

[9] R. Bertle, M. Engelhardt, and M. Faber, Topological susceptibility of Yang-Mills center projection vortices, *Phys. Rev.* **D64** (2001) 074504.

[10] H. Reinhardt, Topology of center vortices, *Nucl. Phys.* **B628** (2002) 133.

[11] J. M. Cornwall and G. Tiktopoulos, Three-dimensional gauge configurations and their properties in QCD, *Phys. Lett.* **B181** (1986) 353.

[12] A. M. Polyakov, Quark confinement and topology of gauge groups, *Nucl. Phys.* **B120** (1977) 429.

[13] J. M. Cornwall, Center vortices, nexuses, and the Georgi-Glashow model, *Phys. Rev.* **D59** (1999) 125015.

[14] J. M. Cornwall, Quark confinement and vortices in massive gauge-invariant QCD, *Nucl. Phys.* **B157** (1979) 392.

[15] J. M. Cornwall, A three-dimensional scalar field theory model of center vortices and its relation to k-string tensions, *Phys. Rev.* **D70** (2004) 065005.

[16] J. M. Cornwall, Center vortices, nexuses, and fractional topological charge, *Phys. Rev.* **D61** (2000) 085012.

[17] J. M. Cornwall, On topological charge carried by nexuses and center vortices, *Phys. Rev.* **D65** (2002) 085045.

[18] L. Kaufmann, *On Knots* (Princeton University Press, Princeton, 1987).

[19] E. Witten, Current algebra theorems for the $U(1)$ "Goldstone boson," *Nucl. Phys.* **B156** (1979) 269.

[20] G. Veneziano, $U(1)$ without instantons, *Nucl. Phys.* **B159** (1979) 213.

[21] F. R. Klinkhamer and N. S. Manton, A saddle-point solution in the Weinberg-Salam theory, *Phys. Rev.* **D30** (1984) 2212.

[22] N. S. Manton, Topology in the Weinberg-Salam theory, *Phys. Rev.* **D28** (1983) 2019.

[23] J. M. Cornwall, Semiclassical physics and confinement, in *Deeper Pathways in High-Energy Physics: Proceedings of the Conference, Coral Gables, Florida, 1977*, edited by B. Kursonoglu et al. (Plenum, New York, 1977) 683.

[24] E. V. Shuryak and T. Schafer, The QCD vacuum as an instanton liquid, *Ann. Rev. Nucl. Part. Sci.* **47** (1997) 359.

[25] K. M. Bitar and S. J. Chang, Vacuum tunneling of gauge theory in Minkowski space, *Phys. Rev.* **D17** (1978) 486.

[26] J. M. Cornwall, High temperature sphalerons, *Phys. Rev.* **D40** (1989) 4130.

[27] J. M. Cornwall and N. Graham, Sphalerons, knots, and dynamical compactification in Yang-Mills-Chern-Simons theories, *Phys. Rev.* **D66** (2002) 065012.

[28] A. Casher, Chiral symmetry breaking in quark confining theories, *Phys. Lett.* **83B** (1979) 395.

[29] J. F. Donoghue and K. Johnson, The pion and an improved static bag model, *Phys. Rev.* **D21** (1980) 1975.

[30] J. M. Cornwall, Confinement and chiral symmetry breakdown: Estimates of F_π and of effective quark masses, *Phys. Rev.* **D22** (1980) 1452.

[31] P. de Forcrand and M. D'Elia, Relevance of center vortices to QCD, *Phys. Rev. Lett.* **82** (1999) 4582.

[32] M. Cheng et al., Transition temperature in QCD, *Phys. Rev.* **D74** (2006) 054507.

[33] T. Banks and A. Casher, Chiral symmetry breaking in confining theories, *Nucl. Phys.* **B169** (1980) 103.

[34] H. Reinhardt, O. Schroeder, T. Tok, and V. C. Zhukovsky, Quark zero modes in intersecting center vortex gauge fields, *Phys. Rev.* **D66** (2002) 085004.

[35] J. Engels, S. Holtmann, and T. Schulze, The chiral transition of $N_f = 2$ QCD with fundamental and adjoint fermions, *PoS* **LAT2005** (2006) 148.

[36] F. Basile, A. Pelissetto, and E. Vicari, The finite-temperature chiral transition in QCD with adjoint fermions, *JHEP* **0502** (2005) 044.

[37] F. Karsch and M. Lutgemeier, Deconfinement and chiral symmetry restoration in an $SU(3)$ gauge theory with adjoint fermions, *Nucl. Phys.* **B550** (1999) 449.

[38] J. M. Cornwall, Center vortices, the functional Schrödinger equation, and CSB, arXiv 0812.1870 [hep-ph].

[39] A. C. Aguilar, D. Binosi, and J. Papavassiliou, Infrared finite effective charge of QCD, *PoS* **LC2008** (2008) 050.

[40] A. C. Aguilar, D. Binosi, J. Papavassiliou, and J. Rodriguez-Quintero, Non-perturbative comparison of QCD effective charges, *Phys. Rev.* **D80** (2009) 085018.

9

A brief summary of $d = 3$ NAGTs

9.1 Introduction

NAGTs in three dimensions have valuable applications in their own right because they are the high-temperature limit of $d = 4$ NAGTs with infrared slavery (see Chapter 11 for more details). They also lead to important insights into $d = 4$ NAGTs at zero T, and in many ways, $d = 3$ QCD is more interesting to study to gain this insight than the far more often-invoked two-dimensional theories. It is not a free-field theory (as is a $d = 2$ pure-gauge NAGT), and it has many features strongly analogous to those of $d = 4$ NAGTs that are best understood by applying the pinch technique. In particular, although a $d = 3$ NAGT cannot be asymptotically free (because it is superrenormalizable, not possessing the usual renormalization group), it is still very much infrared unstable, with even worse singularities than those in $d = 4$. Although this $d = 3$ infrared slavery had been strongly suspected before the pinch technique on the basis of conventional Feynman graph calculations, it took the pinch techniqe to settle the issue and demonstrate the existence of infrared slavery in $d = 3$ NAGTs.

Because a $d = 3$ NAGT is the critical nonperturbative part of the high-temperature behavior of its $d = 4$ counterpart, infrared slavery prevents the use of perturbation theory (beyond $\mathcal{O}(g_3^4)$) in understanding all the phenomena of high temperature, including generation of a so-called magnetic mass, which vanishes identically to all orders of perturbation theory. Just as we have already seen at zero temperature, the magnetic mass, found from the PT Schwinger–Dyson equations, cures the otherwise intractable infrared singularities of high-temperature $d = 4$ gauge theories. We study here only the $d = 3$ NAGT part of finite-temperature $d = 4$ NAGTs,

---

In this chapter, we continue to use the notation introduced in Chapter 7. Also in the present chapter, g_3 is the $d = 3$ NAGT coupling, and g continues to be the $d = 4$ coupling.

saving the PT results for other components of finite-temperature field theories for Chapter 11.

In some respects, $d = 3$ NAGTs are somewhat easier technically than their $d = 4$ counterparts. For example, effective field theories of center vortices are fairly simple scalar field theories in $d = 3$ [1] and so are easier than in $d = 4$, where they are string theories. Unfortunately, we cannot cover these effective theories in a book of this length.

We list here a few of the many reasons for being interested in $d = 3$ QCD, most of which are really only understood with the help of the pinch technique, the gauge technique, or both:

1. It is a superrenormalizable theory, very well behaved in the ultraviolet, with corrections to the bare coupling vanishing as inverse powers of large momenta. But the pinch technique reveals infrared slavery, just as in $d = 4$, meaning that the PT propagator has unphysical singularities at finite momentum. Furthermore, a $d = 3$ gauge theory with zero bare mass (no Higgs effect or Chern–Simons (CS) term) is always strongly coupled at low momenta q, where the dimensionless expansion parameter is Ng_3^2/q for $SU(N)$. As one might by now expect, infrared slavery is resolved by generation of a gluon mass, which in turn gives rise to a $\langle \mathcal{G}_{ij}^2 \rangle$ condensate and to many of the solitons familiar in $d = 4$: center vortices, nexuses, and sphalerons.

2. In $d = 3$, we will actually prove the existence of this $\mathcal{G}_{ij}^2$ condensate and entropy dominance of the effective action, simply on the hypothesis that the full theory possesses only one mass scale (that of g_3^2 itself). We will also show that an approximation based on the pinch technique fully realizes the expected functional form of the exact effective action and the taming of all infrared-slavery singularities. The pinch technique shows that there is a direct connection between the "wrong" sign of the one-loop self-energy, responsible for infrared slavery, and the existence of a minimum in the effective action at a finite condensate VEV.

3. The vacuum wave functional of the functional Schrödinger equation (FSE) for $d = 4$ QCD is expressed in terms of a gauge-invariant effective action whose arguments are background fields given by the coordinate gauge potentials of this wave functional. Certain aspects of the form of the Schrödinger functional are governed by the pinch technique. Gauge invariance gives rise to an infinite tower of QED-like Ward identities, and the gauge technique is effective in exploiting the Ward identities. The two lowest terms in a gauge-technique-inspired expansion of the effective action around small momentum

lead approximately to $d = 3$ QCD as an effective field theory for calculating matrix elements. (This is by no means obvious because Schrödinger equation functionals depend intrinsically on square roots of operators, which are forms not encountered in conventional effective actions.) This effective field theory shows confinement (because it has a condensate of center vortices), and $d = 3$ estimates of the gluon mass actually lead to an estimate of the $d = 4$ coupling $\alpha_s(m^2) \simeq 0.4 - 0.5$, which is not too far (given the approximations) from what we found in Chapter 6 and in phenomenological evaluations.

4. Although $d = 3$ gauge theory does not have the usual $d = 4$ topological charge, it does admit topologically interesting parity-violating CS terms in the action. The coupling for this term in the action is integrally quantized and called level k. In an elegant work, Witten [2] showed that Wilson-loop expectation values in a field theory whose action was just the CS term (a so-called topological field theory) generated some deep results about knots in three dimensions. This Witten theory corresponds to very large values of k. When the conventional Yang–Mills action is included along with the CS term, it turns out that gauge bosons get mass $\sim kg_3^2$ in perturbation theory. Perhaps surprisingly, at large k, this mass does not lead to well-behaved classical solitons. The pinch technique strongly suggests that at small k ($k \simeq (1 - 2)N$), infrared slavery problems still persist, and there is a phase transition from the large-k Witten phase to a phase that also has a dynamically generated gauge-boson mass. Modified forms of the usual $k = 0$ solitons exist in this phase, which is confining.

5. The PT dynamical mass gives rise to the sphaleron, a soliton of interest purely as a $d = 3$ object. The sphaleron becomes even more interesting when it is coupled to a CS term. Because it is natural for the CS number of a sphaleron to be a half-integral, a condensate of an odd number of sphalerons challenges usual compactnesss assumptions, which suggests challenging the conventional wisdom demanding integral levels k for the CS term, as well. We show that, although noncompact theories could in principle exist, they have infinitely higher energy than the corresponding compact versions. In the process, we show that half-integrality is also related to $d = 3$ knots and to nexuses in $d = 2$. So even though there is no topological charge per se in $d = 3$ gauge theory, there are many interesting and curious topological effects.

We start next with PT perturbation theory at one loop, and then, after finding the exact form of the effective action, we show how the one-loop result realizes the exact functional form of the effective action. This illustrates how infrared slavery is directly related to condensate formation.

9.2 Perturbative infrared instability

We easily see the problems of infrared slavery in $d = 3$ by calculating the one-loop perturbative PT proper self-energy. This goes exactly as in the $d = 4$ case of Section 1.3.3, except for the values of the integrals. The result [3, 4] for the scalar part of the one-loop PT inverse propagator (as defined in Eq. (1.30)) is as follows:

$$\widehat{d}^{-1}(q) = q^2[1 - I_3(q)] = q^2 - \pi b_3 g_3^2 q, \tag{9.1}$$

where

$$I_3(q) = \frac{15N g_3^2}{4} \int \frac{d^3 k}{(2\pi)^3} \frac{1}{k^2(q-k)^2}; \qquad b_3 = \frac{15N}{32\pi}. \tag{9.2}$$

Infrared slavery is simply the fact that I_3 occurs with a negative sign in the self-energy (or equivalently, that b_3 is positive), which has the implication that there is a pole in the propagator for positive q. In our metric, where q is the magnitude of an ordinary three-momentum, this indicates a spacelike and thus *tachyonic* pole – a pole corresponding to an imaginary mass.

What could be the cure for this unphysical behavior? At first glance, it could be easy: because the coupling g_3^2 has dimensions of mass, the omitted g_3^4 term might well provide a sufficiently positive term to overcome the negative one-loop term. This is indeed what happens nonperturbatively, but not to any order of perturbation theory, where the coefficient of g_3^4 is identically zero to all orders. (If it were not zero, we could add a bare mass term to the action, which would no longer be perturbatively renormalizable.)

This is only the beginning of the bad perturbative behavior. At $\mathcal{O}(g_3^{2N})$, each perturbative integral, by simple dimensional reasoning, has the infrared behavior $g_3^4(g_3^2/q)^{N-2}$, with poles of infinitely high order in the inverse propagator. But with nonperturbative generation of a (nontachyonic) mass m, the infrared behavior of every propagator in a loop is $\sim 1/m^2$, and an easy power counting shows that q in the perturbative ordering expression is replaced by the dynamical mass $m \sim g_3^2$, so all terms are of $\mathcal{O}(m^2)$ for order $N \geq 2$.

A one-loop PT calculation only clearly shows us (i.e., gauge invariantly) the disease, not the cure – which is a dynamical gluon mass. In $d = 4$, this mass is directly related to the gluon condensate, and we now argue that this is so also in $d = 3$.

9.3 The exact form of the zero-momentum effective action

Define a condensate operator θ by

$$\theta(x) = -\frac{1}{2g_3^2} \operatorname{Tr} (\mathcal{G}_{ij})^2. \tag{9.3}$$

The key result is Eq. (9.10), giving the precise form of the effective action as a function of the zero-momentum matrix elements of θ. This equation says that this operator must have a (positive) VEV, and so there is a condensate of some sort. It further says that the condensate generates so much entropy that the entropy (a negative contribution to the effective action) overcomes the positive action from whatever is in the condensate – just what we expect for center vortices and nexuses. The condensate is important for the self-consistency of gluon mass generation because it gives [5] the coefficient of q^{-2} in the falloff of the gluon mass at large q:

$$m^2(q) \rightarrow \frac{58Ng_3^2\langle\theta\rangle}{15(N^2 - 1)q^2}. \tag{9.4}$$

Before Lavelle found this result, people were not at all sure of what was going on with the use of the OPE in gauge-boson propagators. The simple reason was that the conventional Feynman propagator was gauge dependent, meaning that not only condensates of gauge-invariant operators, such as θ, appeared in the OPE but also other condensates, such as ghost condensates of the form $\bar{c}c$ and mixed gluon-ghost condensates such as $\partial_i\bar{c}A_ic$, as explicit computations showed. But in the PT propagator, these gauge-dependent condensates drop out, leaving only Lavelle's simple result.

One can always resort to assuming the existence of a nonvanishing VEV $\langle\theta\rangle$ with no further argument. But in $d = 3$, we can actually prove [6] that there must be such a (positive) VEV by determining the exact dependence of the effective potential on the zero-momentum part of the operator θ. The answer is reminiscent of a similar one-loop result in $d = 4$ QCD [7], showing evidence for a condensate. The only assumption we need to make is that there is only one dimensional parameter in $d = 3$ QCD (without matter fields), and that is the coupling g_3^2, which has mass dimension unity. We then show that the effective action $\Gamma(\theta)$ has a minimum for a nonzero value of its argument. We can also show [8] that the exact functional form is actually found in the one-loop PT propagator in the presence of the fields constituting the condensate.

Define the generating functional for zero-momentum matrix elements of the action density θ:

$$Z(J) \equiv e^{-W(J)} = \int [dA_i] \exp\left[(1 - J)\int d^3x \frac{1}{2g_3^2}\mathrm{Tr}\,\mathcal{G}_{ij}^2\right], \tag{9.5}$$

where $W(J)$ is the space-time integral of the vacuum action density in the presence of a space-time constant source J coupled to θ:

$$W(J) = \int d^3x \,\epsilon_{\mathrm{vac}}(J). \tag{9.6}$$

In the usual way, multiple derivatives of $W(J)$, evaluated at $J = 0$, give connected matrix elements at zero momentum of the operator θ. In particular, the VEV of θ at $J = 0$ is

$$\langle\theta\rangle = -\left.\frac{d\theta}{dJ}\right|_{J=0}. \tag{9.7}$$

Given the assumption that g_3^2 is the only mass parameter, it follows that $\epsilon_{\text{vac}} \sim g_3^6$. It is now completely trivial to find $W(J)$ because it differs from $W(J = 0)$ simply by the substitution $g_3^2 \to g_3^2/(1 - J)$. So

$$\epsilon_{\text{vac}} = -\frac{\langle\theta\rangle}{3}(1 - J)^{-3}, \tag{9.8}$$

where the normalization follows from Eq. (9.7).

The next step is to make a Legendre transform to the effective action $\Gamma(\theta)$:

$$\Gamma(\theta) = W(J) + J\int d^3x\,\theta; \qquad \frac{d\Gamma}{d\theta} = J\int d^3x. \tag{9.9}$$

The effective action has the property that when the current J is turned off, it has an extremum as a function of θ, and its value at the extremum is the vacuum action $W(0)$.

The differential equation for Γ, plus Eq. (9.8), is elementary to solve:

$$\Gamma(\theta) = \int d^3x\left[\theta - \frac{4}{3}\theta^{3/4}\langle\theta\rangle^{1/4}\right]. \tag{9.10}$$

This indeed has a minimum at $\theta = \langle\theta\rangle$, and this minimum value of $-\int d^3x\,\langle\theta\rangle/3$ is negative.[1] This negative action tells us that the theory is entropy dominated, and so there are interesting nonperturbative effects.

9.3.1 The effective action and the pinch technique

Of course, Eq. (9.10) has nothing to say about what the effects are or how large $\langle\theta\rangle$ is in units of g_3^6. There are no exact results about the latter, although one can make certain approximations [8] in estimating the effective action. One of two basic approximations is to use the one-dressed-loop effective action, with the structure of the loop supplied by our preceding PT results[2]; the other, commonly used by many authors, is to replace the true condensate fields in $\theta(x)$ with a background field B that is constant in space-time. This approximation of constancy makes it possible to do the calculations but introduces an unphysical feature, noted long ago [9]:

---

[1] Could $\langle\theta\rangle$ be zero? Only if $d = 3$ gauge theory is free, which we know it is not.
[2] Or equivalently, the one-loop effective action in the background-field Feynman gauge.

a constant chromomagnetic field is *unstable* to decay into a tangle of space-time-dependent fields. This, not unexpectedly, gives an imaginary part to the effective action. We will simply ignore such features here, knowing that in the real world, the condensate is made of such a tangle of fields and that the effective action is real.

The one-loop effective action, including the classical term, is

$$\Gamma(\theta) = \int d^3x \int \frac{d^3k}{(2\pi)^3} \, \mathrm{Tr} \, \frac{-1}{2g_3^2} \mathcal{G}_{ij}(k)\mathcal{G}_{ij}(-k)[1 - I_3(k)]. \tag{9.11}$$

We can go beyond the strict one-loop form by allowing the PT function I_3 to depend on the condensate. The only feasible way to do this is to assume that the condensate is made of constant fields, so we approximate θ, needed only at zero momentum, by a constant-field condensate such that $\langle \theta \rangle \simeq \bar{B}^2/g_3^2$ for some constant magnetic field of magnitude $\bar{B} \equiv |\mathcal{B}|$.

It turns out [9] that in a constant chromomagnetic field, all gluonic fluctuation modes except one become massive, with $m^2 \sim \bar{B}$. The one exception is a tachyonic mode that carries the instability for decay of the constant field. Without going through the complicated calculations of Nielsen and Olesen [9], we can appreciate such a mass relation from Lavelle's relation in Eq. (9.4), noting[3] that for finite momenta $q^2 \sim m^2$, the mass itself obeys $m^2 \sim [g_3^2\langle\theta\rangle]^{1/2} \sim \bar{B}$. Let us replace (in the spirit of one-loop gap equations, discussed in Section 9.4.2) the free propagators in the integral I_3 used for the perturbative one-loop propagator by adding a mass term $k^2 \to k^2 + \bar{B}$, and so on. We omit detailed numerical constants that are not of interest. Then the effective action of Eq. (9.11) is

$$\Gamma(\theta) = \int d^3x \, \theta[1 - I_3(k = 0)] = \int d^3x \, \frac{1}{2g_3^2} \left(\bar{B}^2 - b_3 g_3^2 \bar{B}^{3/2}\right) + \cdots, \tag{9.12}$$

where, in calculating I_3, the propagators have been modified as discussed earlier. Because $\bar{B} \sim g_3\theta^{1/2}$, this result is a real effective action that has the correct functional form of the exact effective action in Eq. (9.10). No imaginary part shows up because we omitted the tachyonic fluctuation mode.

The minus sign in this approximate effective action is exactly the minus sign coming from infrared slavery, so as promised, infrared slavery and condensate formation are really the same thing. The reader might find it interesting to carry out the same calculation for the $d = 4$ effective action using the $d = 4$ PT self-energy. The result is the famous one-loop effective action $\sim \mathcal{G}^2 \ln \mathcal{G}^2$, which also shows condensate formation, and for the same reason.

---

[3] In this and what follows, we give explicit formulas only for $SU(2)$. For $SU(N)$, the appropriate scalings follow from Eq. (9.4); for example, Γ scales like $N^3(N^2 - 1)$.

9.4 The dynamical gauge-boson mass

The next question is estimation of the dynamical mass needed to cure infrared slavery. There are both theoretical estimates [3, 4, 10, 11, 12, 13, 14, 15, 16, 17] and lattice simulations [18, 19, 20, 21]. Some of the theoretical estimates are based on the pinch technique [3, 4, 10, 11, 15, 16] and some on conventional Feynman-graph technology [12, 13]. This is a technically difficult problem, and all authors[4] use some form of one-dressed-loop equations.

9.4.1 Early pinch technique work

The early PT papers [3, 4, 10] used the spectral form of the gauge technique to write the three-gluon vertex in terms of the PT propagator. This results in the one-loop integral equation:

$$\widehat{d}^{-1}(q) = q^2 \left[1 - \frac{2b_3 g_3^2}{\pi q} \int_0^\infty k \, dk \, \widehat{d}(k) \ln \left| \frac{2k+q}{2k-q} \right| \right] + \widehat{d}^{-1}(0), \qquad (9.13)$$

for the scalar part $\widehat{d}$ of the PT propagator. One can check that if the bare propagator $\widehat{d}(k) = 1/k^2$ and the bare value $\widehat{d}^{-1}(0) = 0$ are used on the right-hand side of this equation, the one-loop propagator of Eq. (9.16), which has a tachyonic pole (to be cured by a positive value of $\widehat{d}^{-1}(0)$), is recovered. Note that this equation, although necessarily approximate, does demand consistency in that the same propagator $\widehat{d}$ appears both on the right-hand side and the left-hand side of this equation (in contrast to the one-loop gap equations discussed next). As it stands, this equation cannot be solved for a gluon mass because the last term $\widehat{d}^{-1}(0)$ is just a placeholder for some dynamical expression. This expression has not yet been worked out because the presence of logarithmically divergent terms that are canceled at two-loop order requires working out the two-loop self-energy, and this remains to be done. However, it is possible to give a lower bound to the gluon mass, or more accurately, $\widehat{d}^{-1}(0)$, because Eq. (9.13) has no solution at all if this quantity vanishes. (If one tries to solve Eq. (9.13) by successive substitution beginning with a massless propagator, infrared singularities from the negative sign of the integral build up uncontrollably.) Numerical investigations give an approximate value of the lower limit, $\widehat{d}^{-1}(0)_{\min}$, as

$$\widehat{d}^{-1}(0)_{\min} = [1.96 b_3 g_3^2]^2, \qquad (9.14)$$

---

[4] Except Karabali et al. [17], whose methods are original and unique. The estimate of Ref. [11] is really an estimate of the ratio of the string tension to the squared mass; it is based on special methods that we will not cover here.

which is equivalent[5] to $m/(N g_3^2) \geq 0.29$. Next we will compare this lower limit to estimates based on one-loop gap equations and find some problems.

9.4.2 One-loop gap equations and lattice simulations

Since the earlier work, a number of authors [12, 13, 15, 16] have addressed the theoretical issues with one-loop gap equations, in which the internal propagators of the one-loop self-energy are approximated by a simple massive form:

$$\frac{1}{q^2} \rightarrow \frac{Z_{\text{in}}}{q^2 + m^2}. \tag{9.15}$$

The one-loop self-energy is calculated with this input, and one demands that the output mass be equal to the input mass. Here one should include the (finite in $d = 3$) input renormalization constant Z_{in} and check that it, too, is reproduced in the output, but this has not been done in the gap-equation papers, which all use $Z_{\text{in}} = 1$. It might seem reasonable to insist that not only the mass but also the residue of the output propagator agree with the input values, but this is rarely looked at. Part of the reason is that some authors [12, 13] use standard Feynman propagators in the gap equation, and with these, only the pole position, but not its residue, is gauge invariant. The specific concern of Buchmuller and Philipsen [12] and Eberlein [13] is mass generation in finite-T electroweak theory (with the $U(1)$ part dropped), and so these works include Higgs fields. But it is straightforward to suppress these Higgs fields by taking their mass to infinity [16], resulting in a form of dynamical gluon mass generation without Higgs fields.

The results are inconclusive. This happens for two reasons: the first is the use of the conventional Feynman propagator [12, 13] rather than the pinch technique. The position of the pole in the conventional propagator is gauge independent, but otherwise, the propagator, even the pole residue, is gauge dependent. The second reason is that the hypothesized input propagator – even when the one-loop PT self-energy is used [15, 8] – shows no signs of the infrared slavery that motivates the study of dynamical mass in the first place. Recall that the one-loop PT propagator $\widehat{\Delta}_{ij}(q)$ for $d = 3$ QCD is

$$\widehat{\Delta}_{ij}(q) = \left[\delta_{ij} - \frac{q_i q_j}{q^2} \right] \frac{1}{q^2 - \pi b_3 g_3^2 q} + \text{terms} \sim q_i q_j, \tag{9.16}$$

---

[5] This defines the mass in terms of the behavior of the propagator at zero momentum rather than the pole mass. This leads to minor inaccuracies, but it is still a gauge-invariant mass estimate.

Table 9.1. *Parameter values for three one-loop gap equations.*

Reference	α	β	γ
[12]	27/16	3/8	9/4
[15]	15/4	1/2	3/2
[16]	15/4	1/2	0

with $b_3 = 15N/32\pi$. Roughly speaking, with mass generation, the PT propagator would have the denominator of Eq. (9.16) replaced by something like the following:

$$q^2 - \pi b_3 g_3^2 q \rightarrow q^2 - \pi b_3 g_3^2 q + m^2. \qquad (9.17)$$

If $m > \pi b g_3^2/2$, there are no tachyonic (real) poles of the propagator.

The one-loop gap equation input of Eq. (9.15) differs from the preceding form by not having a negative term, which, of course, comes from infrared slavery. Without this infrared-slavery effect in the input propagator, the self-consistent one-loop masses are lower than they would be with this effect included. This has the effect of giving an output pole residue Z_{out} that is rather different from the input value Z_{in}, so that true self-consistency is not achieved. (This comparison of residues only makes sense for the PT gap equations because for non-PT propagators, the residues are not gauge invariant.)

All of the one-loop gap equation results have the same functional form[16]:

$$d^{-1}(q^2) = q^2 + \frac{Ng_3^2}{4\pi}\left[\left(-\alpha q + \frac{\gamma m^2}{q}\right)\arctan\frac{q}{2m} - \beta m\right], \qquad (9.18)$$

with the values shown in Table 9.1 for the parameters. Observe that in extrapolating Eq. (9.18) to Minkowski momenta ($q \rightarrow iq$), there are only normal threshholds at $-q^2 = 4m^2$. This is to be expected with the gauge-invariant pinch technique, but it only happens in the Feynman gauge otherwise, where the ghosts and Goldstone bosons all have the mass m. In other gauges, this is not so, and the self-energy of Buchmuller and Philipsen [12] would have other terms. However, these unphysical threshhold terms do not contribute to the pole mass.

Note that Alexanian and Nair [15] and Ref. [16] have the same values for α and β; this is because both use the pinch technique. The differing value of γ comes from differing treatments of the gauge-invariant mass terms used by these two sets of workers. In contrast, [12] uses the Feynman gauge and has rather different parameters.

Table 9.2. *Estimates of the SU(N) magnetic mass by various techniques*

Reference	$m/(Ng_3^2)$	Technique	Z_{out}/Z_{in}
[12]	0.14*	1-loop gap	N/A
[13]	0.17*	2-loop gap	N/A
[15]	0.19*	Pinch/gap	150
[16]	0.13*	Pinch/gap	<0
[17]	0.16	See text	N/A
[18]	0.18*	Lattice	N/A
[19]	0.24*	Lattice	N/A
[20]	0.26*	Lattice	N/A
[21]	0.19	Lattice	N/A

Evaluating this and its derivative on the mass shell (at $q = im$) yields

$$m = \frac{Ng_3^2}{4\pi}\left(\frac{\alpha + \gamma}{2}\ln 3 - \beta\right) \tag{9.19}$$

$$Z_{out} = \left(\frac{\alpha + \gamma}{2}\ln 3 - \beta\right)\left[\alpha\left(\frac{1}{4}\ln 3 - \frac{1}{3}\right) - \beta + \gamma\left(\frac{3}{4}\ln 3 - \frac{1}{3}\right)\right]^{-1}. \tag{9.20}$$

The pole masses, following from setting $d^{-1}(q^2 = -m^2) = 0$, seem reasonable when compared to lattice values, as we will see shortly. But the PT residues are not close to self-consistency, as one may easily check. It would be better to use an input propagator of the form of Eq. (9.17), or something like it, but as far as we know, this has not been done with one-loop gap equations; instead, there is the original PT calculation, which demands a self-consistent propagator at all momenta but which has only been carried (so far) to the point of estimating a lower bound for the mass.[6] Ironically, the presumably gauge-dependent parameters of [12] yield a more reasonable value of Z_{out} than do the PT parameters.

Table 9.2 shows various results for the ratio $m/(Ng_3^2)$, which should be roughly independent of N for gauge group $SU(N)$ (exactly so for one-loop gap equations). Values marked by an asterisk were calculated for $SU(2)$; the rest were calculated for $SU(3)$, and all were assumed to scale linearly in N. In Table 9.2, N/A means that no residue factors were given. Reference [17] uses a very interesting approach to $d = 3$ gauge theory that we cannot describe here; it culminates in the formula $m/(Ng_3^2) = 1/(2\pi)$.

---

[6] Another problem with all the estimates we will discuss is that they do not properly account for the fact that the magnetic mass is really a function of momentum q, vanishing like $1/q^2$ (modulo logarithms) at large momentum (see Chapter 2). This is essential for the Schwnger–Dyson equations to yield finite results.

From Table 9.2, we see that there is some spread in the ratio, with the average lattice value larger than the average gap-equation value. Note that the previously estimated lower bound of 0.29 is larger than any of the masses in the table. The lattice results vary somewhat, in part because the propagators from which the magnetic mass is extracted are in different gauges, and the extracted mass is not exactly the (gauge invariant) pole mass, which is hard to reach on the lattice because it involves extrapolation to negative values of momentum squared.

In any event, there seems to be no question that there is a finite $d = 3$ gluon mass and therefore the solitons (center vortices, nexuses) that we have already discussed.

9.5 The functional Schrödinger equation

The FSE is another way, in principle, of expressing the content of a field theory in d dimensions via functional differential equations in $d - 1$ dimensions. There is nothing in the FSE approach that could not be understood directly from the field theory, but sometimes one gains insight by looking at a hard problem in a different way.

For any field theory, the FSE is no more or less than the usual Schrödinger equation, with fields as the coordinates and functional derivatives with respect to these fields as the momenta. The fields, as coordinates, are labeled by (in $d = 3 + 1$) three spatial positions $\vec{x}$. For a gauge theory, the component $\mathcal{A}_0^a$ has no canonical momentum and is set to zero, leaving only the three magnetic potentials $\mathcal{A}_i^a(\vec{x})$ at zero time as coordinates.[7] The canonical momentum is the electric field:

$$\Pi_i^a = \mathcal{E}_i^a(\vec{x}) \rightarrow -ig^2 \frac{\delta}{\delta \mathcal{A}_i^a(\vec{x})}. \tag{9.21}$$

(Note that the commutator term is missing because we set $\mathcal{A}_0^a = 0$.) The Hamiltonian for the NAGT FSE is

$$\mathcal{H} = \int \left\{ -\frac{1}{2} g^2 \left(\frac{\delta}{\delta \mathcal{A}_i^a} \right)^2 + \frac{1}{2g^2} \left[\frac{1}{2} (\mathcal{G}_{ij}^a)^2 \right] \right\} \equiv \int \left[\frac{1}{2} (\Pi_i^a)^2 \right] + V, \tag{9.22}$$

and the Schrödinger equation is the usual $\mathcal{H}|\psi\rangle = E|\psi\rangle$. We consider the vacuum (ground state) wave functional $\psi\{\mathcal{A}_i^a(\vec{x})\}$ and the time-independent FSE that determines it. This wave functional is the matrix element

$$\psi\{\mathcal{A}_i^a(\vec{x})\} = \langle \mathcal{A}_i^a(\vec{x})|\psi\rangle, \tag{9.23}$$

---

[7] See Jackiw [22] for an elegant treatment of the fundamentals of the canonical FSE for gauge theories. We temporarily use group-component notation.

where the bra vector is an eigenvector of the field operator $\mathcal{A}_i^a$ and $|\psi\rangle$ is the vacuum-state eigenvector, whose energy we normalize to zero for the present. The vacuum wave functional has the form

$$\psi\{\mathcal{A}_i^a(\vec{x})\} = \exp[-S\{\mathcal{A}_i^a(\vec{x})\}], \tag{9.24}$$

where S can be written as a formal power series with infinitely many terms:

$$g^2 S = \frac{1}{2!} \iint \mathcal{A}_i^a \Omega_{ij} \mathcal{A}_j^a + \frac{1}{3!} \iiint \mathcal{A}_i^a \mathcal{A}_j^b \mathcal{A}_k^c \Omega_{ijk}^{abc} + \cdots. \tag{9.25}$$

The Ω functions relate their associated gauge potentials nonlocally and may have derivatives of high order.

The exponent S is real, bounded below for finite arguments (vacuum wave functionals do not have nodes), and positive for sufficiently large arguments (it is normalizable). Most important, it is a gauge-invariant functional of its arguments. These properties of S, plus the usual rules for constructing vacuum matrix elements, allow us to interpret $2S$ as an effective $d = 3$ action. The vacuum matrix elements are of the type[8]

$$\langle \psi | \cdot | \psi \rangle = \int \left[d\mathcal{A}_i^a \right] e^{-2S}(\cdot). \tag{9.26}$$

Define the effective $d = 3$ action by

$$I_{d=3} = 2S. \tag{9.27}$$

So FSE matrix elements such as $\langle \psi | W | \psi \rangle$, where W is a spacelike Wilson loop, are expectation values in the $d = 3$ theory with effective action $I_{d=3}$.

The question is: how do we solve for this effective action, and what does it look like? This is not an easy question. In the first place, it is not possible to solve the FSE exactly,[9] so there is little guidance from existing solutions. (Often workers simply postulate what seems to be a reasonable approximate form for ψ – typically Gaussian – to be used for variational estimates, but often, in the process, gauge invariance is lost.) A few low-N terms of the N-point coefficients $\Omega_{ijk...}$ can be found order by order in perturbation theory, but that is not very interesting; it is analogous to a bare-loop expansion of the effective action. Much more interesting is the dressed-loop expansion, in which the three-point and higher functions $\Omega_{ijk...}$ are expressed in terms of the dressed two-point function Ω_{ij}; then the FSE (or, equivalently, extremalization of the effective action S) yields a nonlinear equation

---

[8] As usual, we do not explicitly indicate ghost and gauge-fixing terms.

[9] To forestall confusion, there is an exact zero-energy formal solution [23, 24, 25, 26] to the vacuum FSE, which is $\psi \sim \exp(-(8\pi^2/g^2)N_{CS})$, with N_{CS} being the CS integral (see Eq. (9.51)). Although this solution is not normalizable because the CS integral does not have a definite sign, it is applicable for certain high-energy, few-to-many scattering processes; see [26].

for the two-point function, quite analogous to the Schwinger–Dyson equation for the PT propagator. This approach is quite successful [27] for the anharmonic oscillator in one dimension (ordinary quantum mechanics) even in the limit where the quadratic term in the potential vanishes and perturbation theory completely fails. Because $I_{d=3}$ is gauge invariant under gauge transformations of its background-field arguments $\mathcal{A}_i(\vec{x})$, it is natural to use the pinch technique and gauge technique to approximate it, along with the dressed-loop expansion. Whether one uses a dressed-loop approximation or a bare-loop expansion, the solution to the FSE always involves the square root of operators. For example, a little experimentation shows that a perturbative expansion involves unfamiliar operators such as $\sqrt{-\nabla^2}$. Because we know that NAGTs show dynamic gluon mass generation, we expect that square-root operators of the form $\sqrt{m^2 - \nabla^2}$ are what turn up in the dressed-loop expansion.

9.5.1 The gauge technique and the FSE

The generator of infinitesimal gauge transformations is $\mathcal{D}_j^{ab} \times (-i\delta/\delta \mathcal{A}_j^b)$, and this must annihilate ψ or, equivalently, S. Invariance of S under infinitesimal gauge transformations is trivial for the two-point function Ω_{ij}; this quantity must be conserved (as in an Abelian gauge theory) so that in Fourier space,

$$\Omega_{ij}(k) = \Omega(k)P_{ij}(k) \qquad P_{ij} = \delta_{ij} - \frac{k_i k_j}{k^2}. \tag{9.28}$$

For the free theory, $\Omega_0(k) = k$, but for the dressed theory, we expect something like $\Omega(k) = \sqrt{k^2 + m^2}$.

Gauge invariance is more complicated for higher-point functions. Annihilating ψ with the generator of gauge transformations yields a set of ghost-free Ward identities, just as in the pinch technique. For example, the Ward identity for the three-point function is

$$k_{1i}\Omega_{ijk}^{abc}(k_1, k_2, k_3) = f^{abc}\left[\Omega_{jk}(2) - \Omega_{jk}(3)\right], \tag{9.29}$$

where $\Omega_{jk}(2) \equiv \Omega_{jk}(k_2)$, and so on.

Further information comes from the FSE, where one finds that the equation determining the three-point function has the general form

$$\Omega_{il}(1)\Omega_{ljk}^{abc} + \Omega_{jl}(2)\Omega_{lik}^{bac} + \Omega_{kl}(3)\Omega_{lij}^{cab} = f^{abc}\Gamma_{ijk}. \tag{9.30}$$

The right-hand side Γ_{ijk} comes from the cubic term in $\mathcal{H}$ plus another term from the five-point function. The Ward identity for Γ_{ijk} is determined by the preceding equation plus the Ward identities for the two- and three-point functions, as already given, and multiplying both sides of Eq. (9.30) by k_{1i} yields

$$k_{1i}\Gamma_{ijk} = \Omega_{jk}^2(3) - \Omega_{jk}^2(2). \tag{9.31}$$

For free particles with $\Omega = \Omega_0$, this is satisfied by the usual free three-point vertex

$$\Gamma_{ijk}^0 = i(k_1 - k_2)_k\delta_{ij} + \text{c.p.} \tag{9.32}$$

The reader can verify that Eq. (9.30) has a solution of the form

$$\Omega_{ijk}^{abc}(k_1, k_2, k_3) = f^{abc}[\Omega(1) + \Omega(2) + \Omega(3)]^{-1}$$

$$\times \left\{ \Gamma_{ijk} + \left[\Omega(1)\frac{k_{1i}}{k_1^2} \left(\Omega_{jk}(2) - \Omega_{jk}(3) \right) + \text{c.p.} \right] \right\}, \tag{9.33}$$

which respects the Ward identity of Eq. (9.29) by virtue of the massless pole terms of Eq. (9.33). It should now be clear that these longitudinally coupled massless excitations will occur, as a result of enforcing gauge invariance, for every n-point function. We will shortly identify these with couplings of the gauged nonlinear sigma (GNLS) field introduced in our conjecture for the infrared-effective action.

So far, the vertex function Γ_{ijk} is undetermined, but we will find an approximation to it, useful in the infrared, with the gauge technique. We can read off from Chapter 5 the needed relation

$$\Gamma_{ijk} = \delta_{ij}(k_1 - k_2)_k - \frac{k_{1i}k_{2j}}{2k_1^2k_2^2}(k_1 - k_2)_l \Pi_{lk}(k_3)$$

$$- [P_{il}(k_1)\Pi_{lj}(k_2) - P_{jl}(k_2)\Pi_{li}(k_1)]\frac{k_{3k}}{k_3^2} + \text{c.p.}, \tag{9.34}$$

where the first term on the right-hand side is the free vertex Γ_{ijk}^0, and $\Pi_{ij}(k) \equiv P_{ij}(k)\Pi(k)$ is the transverse PT self-energy, related to Ω_{ij} by

$$\Omega_{ij}^2 = P_{ij}[\Omega_0^2 + \Pi\{\Omega\}], \tag{9.35}$$

where $\Omega_0^2 = k^2$ is the free gluon contribution.

In the simple case studied here, $\Pi = m^2$, and the resulting expression for Γ_{ijk} is

$$\Gamma_{ijk} = \delta_{ij}(k_1 - k_2)_k + \frac{m^2}{2}\frac{k_{1i}k_{2j}(k_1 - k_2)_k}{k_1^2k_2^2} + \text{c.p.} \tag{9.36}$$

Combining the pinch technique and the gauge technique by solving the Ward identities ensures exact gauge invariance but is nonetheless an approximation (expected to be valid in the infrared regime). Ultimately, it yields a dressed-loop equation for a single transverse operator $\Omega_{ij}(k) \equiv P_{ij}(k)\Omega(k)$. We will not explore this difficult program further here.

The order-by-order appearance of massless longitudinal poles in the gauge-completion process is directly mirrored in the order-by-order solution of the classical GNLS model. Because the notation is more compact, we now switch to anti-Hermitean matrix notation. The local GNLS model, normalized appropriately, has the action

$$I_{\text{GNLS}} = -\frac{m}{g^2} \int d^3x \, \text{Tr} \, [U^{-1} \mathcal{D}_i U]^2, \tag{9.37}$$

where U is a unitary matrix transforming as $U \to V U$ under the gauge transformation

$$\mathcal{A}_i \to V \mathcal{A}_i V^{-1} + V \partial_i V^{-1}. \tag{9.38}$$

The classical equations for U express this quantity in terms of the $\mathcal{A}_i$ (see Chapter 7), with the result

$$U = e^\omega$$

$$\omega = -\frac{1}{\nabla^2} \partial \cdot \mathcal{A} + \frac{1}{\nabla^2} \left\{ \left[\mathcal{A}_i, \partial_i \frac{1}{\nabla^2} \partial \cdot \mathcal{A} \right] + \frac{1}{2} \left[\partial \cdot \mathcal{A}, \frac{1}{\nabla^2} \partial \cdot \mathcal{A} \right] + \cdots \right\}, \tag{9.39}$$

showing the appearance of massless scalars. More generally, because $U^{-1} \mathcal{D}_i U$ is a gauge transformation of $\mathcal{A}_i$, functional integration over U is equivalent to projecting the gauge-invariant part of the mass term. Note that the linear term in $\mathcal{A}_i$ of the GNLS model field $U^{-1} \mathcal{D}_i U$ is the transverse part of $\mathcal{A}_i$. This linear term is the Abelian mass term that began our investigations. All higher-order terms of ω in Eq. (9.39) are non-Abelian. One can straightforwardly verify that the three-point function of Eq. (9.36) corresponds precisely to the three-point term found by using the expansion of Eq. (9.39) in the GNLS model action. Because the GNLS action is fully gauge invariant, it gives one solution to the all-orders ghost-free Ward identities, and this solution is what is emerging from direct calculations using the gauge technique.

9.5.2 The proposed infrared-effective action

Our proposed form of the effective action answer [28] is that in the infrared regime, where no momenta are large compared to the gluon mass m, this action is reasonably well approximated by a $d = 3$ action that is essentially the $d = 3$ massive effective action already studied in the last chapter. This action consists of a gauged, nonlinear sigma model mass term, giving the gluon a mass m and the usual Yang–Mills term

$$I_{d=3} = -\int d^3x \left\{ \frac{m^2}{g_3^2} \text{Tr} \, [U^{-1} \mathcal{D}_i U]^2 + \frac{1}{2g_3^2} \text{Tr} \mathcal{G}_{ij}^2 \right\}. \tag{9.40}$$

One finds this form by saving the first two terms in an expansion of a certain approximation to S in powers of the operator $-\nabla^2/m^2$ or, equivalently, k^2/m^2 (k is a momentum). The leading term gives the gauged, nonlinear sigma model, and the next leading term gives the conventional Yang–Mills action. As momenta get larger, correction terms with more and more derivatives enter, and finally, in the region of large momenta, expanding in powers of k^2/m^2 is useless. Fortunately, because $d = 3 + 1$ QCD is asymptotically free, perturbation theory determines the leading large-momentum terms in ψ, but this is not of interest here.

Because this action must describe the same phenomena as, for example, the Schwinger–Dyson equations do, it must be that the gluon mass described by the effective action is the same in $d = 3, 4$. As for the $d = 3$ coupling g_3^2, we have already seen that $d = 3$ gauge dynamics determine the dimensionless ratio m/g_3^2, so knowing m gives the $d = 3$ coupling.

To understand how such an action might arise, consider just the two-point term in S of Eq. (9.25), called S_2. In perturbation theory, one can easily check that choosing

$$\Omega_{ij} = \Omega_0 P_{ij}, \quad \text{with } P_{ij} = \delta_{ij} - \frac{\partial_i \partial_j}{\nabla^2} \quad \text{and } \Omega_0 = \sqrt{-\nabla^2}, \quad (9.41)$$

solves the FSE for the free part of the action and also has the crucial property of gauge invariance (in this case, under Abelian $U(1)^{N^2-1}$ gauge transformations). To describe mass generation, make the simple replacement

$$\Omega_0 \to \Omega \equiv \sqrt{m^2 - \nabla^2} \quad (9.42)$$

so that S_2 is

$$S_2 = \frac{1}{2g^2} \int A_i^a \sqrt{m^2 - \nabla^2} \, P_{ij} A_j^a. \quad (9.43)$$

This S_2 is an exact solution of an FSE with an Abelian gauge Hamiltonian with gauge-invariant mass generation put in by hand:

$$H = \int \left\{ -\frac{1}{2} g^2 \left(\frac{\delta}{\delta A_i^a} \right)^2 + \frac{1}{2g^2} \left[\frac{1}{2} (\mathcal{F}_{ij}^a)^2 + m^2 A_i^a P_{ij} A_j^a \right] \right\}$$

$$\equiv \int \left[\frac{1}{2} (\Pi_i^a)^2 \right] + V, \quad (9.44)$$

where $\mathcal{F}_{ij}^a = \partial_i A_j^a - \partial_j A_i^a$ are the Abelian field strengths.

Although this is a familiar Hamiltonian, closely related to that of the Abelian Higgs model, the action $I_{d=3} = 2S_2$ is not familiar, involving as it does a square root of an operator. Try the infrared expansion

$$\sqrt{m^2 - \nabla^2} \to \frac{1}{m}\left(m^2 - \frac{1}{2}\nabla^2\right) + \cdots. \tag{9.45}$$

Now we do see familiar operators, and saving these two terms in the infrared expansion of $I_{d=3} = 2S_2$ is almost a repeat of the Hamiltonian of Eq. (9.44), divided by m:

$$I_{d=3} \to \frac{m}{g^2}\int A_i^a P_{ij} A_j^a + \frac{1}{4mg^2}\int [\mathcal{F}_{ij}^a]^2 + \cdots. \tag{9.46}$$

Unfortunately, this action describes gluons of mass $\sqrt{2}m$ and not m because of the $1/2$ in the expansion of the square root. The problem is in trying to make a strict expansion around zero momentum when, in fact, momenta of $\mathcal{O}(m)$ are important. Reference [28] describes a least-squares operator approximation, intended to be more or less accurate over the range of momenta from 0 to $\mathcal{O}(m)$, of the form

$$\sqrt{m^2 - \nabla^2} \to \frac{Z}{m}\left(m^2 - \nabla^2 + \cdots\right), \tag{9.47}$$

where Z is perhaps 1.1 or 1.2. This new approximation does describe gluons of mass m, as required.

The next step is to make a gauge completion of this Abelian form by adding the infinitely many terms in the expansion of S in Eq. (9.25) that are required by gauge invariance. It turns out that for any given Ω, there are infrared-useful approximations to all these terms that exactly preserve gauge invariance using the techniques of [11]. The first term in a large-mass expansion of this gauge completion is, as might be expected, equivalent to a gauged nonlinear sigma model mass term. *Equivalent* means that what one actually finds is the $d = 3$ version of the perturbative expansion of this model, as given in Eq. (9.39). The second is the usual Yang–Mills term involving the full field strengths $\mathcal{G}_{ij}^a$, as described in Eq. (9.40). The same problem arises as in the Abelian case – the free mass is $\sqrt{2}M$ – and is approximately resolved in the same way as indicated for the Abelian theory with a modified infrared expansion. The final result is the obvious modification of the Abelian S_2:

$$-2S = -I_{d=3}$$
$$\to \frac{2mZ}{g^2}\int d^3x \, \mathrm{Tr}\,[U^{-1}\mathcal{D}_i U]^2 + \frac{Z}{mg^2}\int d^3x \, \mathrm{Tr}\,\mathcal{G}_{ij}^2 + \mathcal{O}(M^{-3}). \tag{9.48}$$

The next question is: what are the consequences of this action? Three are worth mentioning [28]. First, we know already that it has center vortices and nexuses. The center vortices are closed strings, corresponding to the projection of closed surfaces in $d = 4$ onto $d = 3$; similarly, the nexuses are points on these strings. Given an entropy-driven condensate of vortices, these will describe confinement through matrix elements of the FSE.

Second, given values of Ng_3^2/m (see Section 9.4), as found strictly in three dimensions, we can actually estimate the on-shell value of the $d = 4$ coupling $\alpha_s(m^2) \equiv g^2/4\pi$. Just compare the two forms of the conjecture, as stated in Eqs. (9.40) and (9.48), and find the equation:

$$g^2 = \frac{2Zg_3^2}{m}. \tag{9.49}$$

This equation expresses a $d = 4$ quantity, g^2, in terms of the $d = 3$ ratio m/g_3^2, estimates for which we summarized in Table 9.2. Using $Z \simeq 1.2$ and the estimate [16] $Ng_3^2/m \simeq 7.7$ for $N = 3$ gives $\alpha_s(M^2) \simeq 0.5$, a value holding for no quarks. This is close both to an early estimate [4] that comes out of the first PT attempt to find the gluon mass and to modern estimates given in Chapter 6. The early analytic PT estimate is

$$\alpha_s(M^2) = \frac{g^2}{4\pi} = \frac{12\pi}{[11N - 2N_f]\ln[5M^2/\Lambda^2)]} \simeq 0.4, \tag{9.50}$$

where the numerical value comes from $m = 0.6$ GeV, $\Lambda = 0.3$ GeV, and no quarks ($N_f = 0$). Chapter 6 gives a value $\simeq 0.5$. We can also compare this result to phenomenological determinations [29, 30, 31] of $\alpha_s(q^2 \simeq 0) \simeq 0.7 \pm 0.3$ coming from studies of infrared-sensitive scattering data. But in the real world to which these data apply, there are three families of light quarks, so we have to modify the FSE estimate. Assuming that the PT formula of Eq. (9.50) applies, we should multiply the result of Eq. (9.49) by 11/9, ending up with an FSE estimate of $\alpha_s(m^2)$ of about 0.6 – near the lower end of the phenomenological range.

The third consequence of this FSE work comes from using it in one less dimension. The same steps go through, yielding an effective action in two dimensions, $I_{d=2}$, that is once again the sum of a gauged nonlinear sigma model and a Yang–Mills term. This action has center vortex solutions, and if they condense, they give confinement as usual. Note the big difference with the usual confinement mechanism in $d = 2$ QCD, which just has the Yang–Mills term. The Yang–Mills action by itself is a free-field theory with a confining propagator, and so all nontrivial group representations of $SU(N)$ are confined. But this is not correct for $d = 3$, the dimensionality where $I_{d=2}$ is supposed to apply; the adjoint and similar representations are not confined but screened. Therefore, it is essential to have the mass term in $I_{d=2}$.

There is another way of creating gluon mass in $d = 3$, which works even in the classical theory: add a CS term to the action.

9.6 Dynamical gluon mass versus the Chern–Simons mass: Two phases

In Chapter 8, we already encountered the CS integral as a time slice of the topological charge density. We repeat the definition of the CS number N_{cs}:

$$N_{\text{cs}} = -\frac{1}{8\pi^2} \int d^3x \, \epsilon_{ijk} \text{Tr}\left(\mathcal{A}_i \partial_j \mathcal{A}_k + \frac{2}{3}\mathcal{A}_i \mathcal{A}_j \mathcal{A}_k \right). \tag{9.51}$$

Form the Yang–Mills–Chern–Simons (YMCS) action by adding the CS action $2\pi i k N_{\text{cs}}$ to the Yang–Mills action. The $d = 3$ functional integral with a CS term (as always, omitting the gauge-fixing and ghost terms) with a new coupling k, chosen to be real, is

$$Z = \int [d\mathcal{A}_i] \exp\left[2\pi i k N_{\text{cs}} + \int d^3x \frac{-1}{2g_3^2}\text{Tr}\, \mathcal{G}_{ij}^2 \right]. \tag{9.52}$$

We can and will always choose the level k to be positive by changing the sign of the gauge potential, that is, by making a parity transformation. In fact, because either sign contributes equally in Z, the partition function is an even function of k. The factor of i in the Euclidean action comes about because in transforming from Minkowski space (where all actions have an i factor in the path integral) to Euclidean space, no extra factor of i arises, as it usually does. So for real gauge potentials, the resulting pure imaginary action contributes a phase factor to the path integral. But once there is an imaginary part to the Euclidean action, there is no longer any requirement that the dominant contributions to the path integral come from real gauge potentials, and the CS action is generally complex. The partition function is real because there are equal contributions from a complex CS action and its complex conjugate.

The CS term is not gauge invariant under a so-called large-gauge transformation, the kind that carries topological charge. According to Eq. (8.32), the CS term changes by the integer $\mathcal{N}$ of that gauge transformation. We do not need to require that the action itself be gauge invariant; just as with Dirac monopoles, only the functional integrals created from it, such as Z, must be gauge invariant. That requires [32, 33] the coupling k to be an integer.[10] This integer is called the *level* of the CS action.

In perturbation theory, the main effect of adding the CS term is that the gluon acquires a mass m_{cs}. This follows from the easily established Euler–Lagrange variation of the CS term:

$$\frac{\delta N_{\text{cs}}}{\delta \mathcal{A}_i(x)} = \frac{1}{16\pi^2}\epsilon_{ijk}\mathcal{G}_{jk}(x) \equiv \frac{1}{8\pi^2}\mathcal{B}_i, \tag{9.53}$$

[10] One might wonder what happens if k is other than integral. If the only large gauge transformations allowed in the path integrals change the CS number by an integer, then the partition function Z vanishes for k nonintegral. We discuss in Section 9.7 [34] that there is no absolute prohibition on considering nonintegral CS numbers, but a theory accommodating nonintegral CS numbers is energetically disfavored.

from which the classical equations of motion for YMCS theory are

$$[\mathcal{D}_i, \mathcal{G}_{ij}] - \frac{ikg_3^2}{8\pi}\epsilon_{jkl}\mathcal{G}_{kl} = 0. \tag{9.54}$$

This is a peculiar equation because it is complex. However, it has a perfectly fine perturbative expansion. The linearized version of Eq. (9.54) is

$$\epsilon_{ijk}\partial_j\mathcal{B}_k = im_{\text{cs}}\mathcal{B}_i, \tag{9.55}$$

where the CS mass is

$$m_{\text{cs}} = \frac{kg_3^2}{4\pi}, \tag{9.56}$$

and $\mathcal{B}_i \equiv (1/2)\epsilon_{ijk}\mathcal{G}_{jk}$ is the magnetic field. Taking the curl of Eq. (9.55) gives

$$(\nabla^2 - m_{\text{cs}}^2)\mathcal{B}_i = 0. \tag{9.57}$$

Precisely because there is an i in front of the CS term in the equations of motion, this linear propagation equation is nontachyonic and corresponds to a gluon of physical mass m_{cs}, although it is associated with the peculiarities of complexness and parity violation.

We cannot immediately conclude that this mass, present in perturbation theory, removes the infrared instability of ordinary Yang–Mills theory. It turns out, as we will see using the pinch technique, that if the mass is large enough – that is, if k is large enough – infrared slavery is indeed gone. The perturbative expansion parameter Ng_3^2/m_{cs} behaves like N/k, and so large-k perturbation theory should be well defined, as it is in QED. If perturbation theory is to work, there should be no classical solitons – a result proven long ago [35]. This large-k theory, which is in effect the theory without the Yang–Mills term, is a particularly beautiful and mathematically powerful theory that is exactly soluble and beautifully organizes some of the mathematics of $d = 3$ knots [2].

However, if k is small enough, we will show that the CS mass is not large enough to remove infrared-slavery tachyons and that a dynamical gluon mass is required as well. Quantum solitons return [36] along with the expected nonperturbative effects, including confinement. There is a phase transition in YMCS theory at a value $k = k_c$, with $k_c \simeq (1-2)N$. For $k > k_c$, perturbation theory and Witten's results hold, whereas for smaller k, infrared slavery must be solved the way we have presented in this book. Of course, this nonperturbative phase is different from that of QCD because of the parity-violating CS term, and the solitons differ in detail. But there are still center vortices, nexuses, and sphalerons.

It is natural, from a physics point of view, to start with the Yang–Mills action as fundamental – something to which we add the CS term. But Witten [2] showed that the theory defined by dropping the Yang–Mills action and keeping only the CS term is not only sensible but is an example of a particularly interesting class of field theories called *topological field theories*. The theory with only the CS term – called CS theory – has no propagating gluonic modes and in fact can be solved exactly. Its observables are completely characterized by the topologically invariant CS numbers and the VEVs of other topological invariants such as Wilson loops. Ultimately it turns out that the phase space of CS theory is finite. Witten showed how the VEVs of multiple, knotted Wilson loops can be calculated to yield the Jones polynomials that characterize the linkings and knottings of the loops. For large k, Witten looks at a semiclassical expansion around the classical extrema of the action that, according to Eq. (9.53), comes from configurations of vanishing field strength or pure-gauge potentials. Canonical quantization of the theory requires a choice of gauge; the choice $\mathcal{A}_3 = 0$ reduces the action to a quadratic form, and that is, in part, why the theory is exactly soluble. We will not pursue this fascinating topic any further, except to say that to define rigorously pure CS theory requires a regulator, and the obvious one at large k is the Yang–Mills term. The next sections will focus on the infrared-unstable phase.

9.6.1 The nonperturbative phase uncovered by the pinch technique

First, we make a heuristic argument. Because Z of Eq. (9.52) is even in k, we can write it as

$$Z = \int [\mathrm{d}\mathcal{A}_i] \cos\left(2\pi i k N_{\mathrm{cs}}\right) \exp\left[-\int \mathrm{d}^3 x \frac{1}{2g_3^2} \mathrm{Tr}\mathcal{G}_{ij}^2\right] \equiv \mathrm{e}^{-\Gamma(\theta,k)}. \tag{9.58}$$

In the formal limit of small k, this equation says that

$$\Gamma(\theta, k) \simeq \Gamma(\theta) + 2\pi^2 k^2 \langle N_{\mathrm{cs}}^2 \rangle; \tag{9.59}$$

the CS term increases the effective action and at some point can be expected to overcome the entropic effects that tend to make Γ negative. So there might be a phase transition at some value of $k = k_c$ of $\mathcal{O}(N)$ separating a nonperturbative phase with all the usual phenomena (gluon mass, a condensate, solitons) from a perturbative phase.

The first step is to calculate the one-loop PT propagator $\widehat{\Delta}(p)_{ij}$ for YMCS theory. The corresponding bare propagator has a new parity-violating term:

$$\widehat{\Delta}_0^{-1}(p)_{ij} = (p^2 \delta_{ij} - p_i p_j) + m_{\mathrm{cs}} \epsilon_{ija} p_a + \frac{1}{\xi} p_i p_j; \tag{9.60}$$

here $m_{\mathrm{cs}} = kg_3^2/4\pi$ is the classical CS mass and ξ is a gauge-fixing parameter. The PT self-energy, which enters the full propagator through

$$\widehat{\Delta}^{-1}(p)_{ij} = \Delta_0^{-1}(p)_{ij} - \widehat{\Pi}(p)_{ij}, \qquad (9.61)$$

also has two conserved terms, a parity-conserving term and a parity-violating term:

$$\widehat{\Pi}(p)_{ij} = (p^2\delta_{ij} - p_i p_j)\widehat{A}(p) + m_{\mathrm{cs}}\epsilon_{ija}p_a\widehat{B}(p). \qquad (9.62)$$

These equations yield the propagator

$$\widehat{\Delta}(p)_{ij} = \left(\delta_{ij} - \frac{p_i p_j}{p^2}\right)\frac{1}{(1 - \widehat{A})(p^2 + m_{\mathrm{R}}^2)}$$

$$- m_{\mathrm{R}}\epsilon_{ija}p_a\frac{1}{p^2(1 - \widehat{A})(p^2 + m_{\mathrm{R}}^2)} + \xi\frac{p_i p_j}{p^4} \qquad (9.63)$$

in terms of a running mass

$$m_{\mathrm{R}}(p) = m_{\mathrm{cs}}\left(\frac{1 - \widehat{B}}{1 - \widehat{A}}\right). \qquad (9.64)$$

A lengthy calculation [36] gives equally lengthy results for $\widehat{A}(m_{\mathrm{cs}}/p)$, $\widehat{B}(m_{\mathrm{cs}}/p)$, and we will not quote them in full here. Both positive and negative powers of m_{cs}/p appear, but owing to cancellations, the propagator is finite both in the $m_{\mathrm{cs}} = 0$ limit and in the $p = 0$ limit. One simple-looking result is the one-loop PT propagator in the limit of no CS mass. One might expect this to reduce to the usual QCD expression of Eq. (9.16), but in the limit $m_{\mathrm{cs}} = 0$, there is a term $\sim 1/m_{\mathrm{cs}}$ that leads to a cancellation:

$$\widehat{\Delta}^{-1}(p)_{ij} = (p^2\delta_{ij} - p_i p_j)(1 - \pi bg_3^2 p) + \epsilon_{ija}p_a g_3^2\left(\frac{k + N}{4\pi}\right). \qquad (9.65)$$

Note the replacement of k by $k + N$, which happens also in Witten's pure CS topological theory. The N here arises from the mass cancellation. There is also the infrared-slavery term already uncovered in Eq. (9.16), with $b_3 = 15N/(32\pi)$.

So the question now is whether some finite value of m_{cs} can overcome the infrared slavery problem. By looking at the full expressions for $\widehat{A}$, $\widehat{B}$, one can check that the relevant self-energy $\widehat{A}$ is positive and monotone decreasing in momentum p, vanishing like $1/p$ at large momentum. So the infrared-slavery problem is solved if $\widehat{A}(p = 0)$ is less than 1. The result for this quantity is as follows:

$$1 - \widehat{A}(p = 0) = 1 - \frac{29N}{12k}. \qquad (9.66)$$

It then follows that, at the one-loop level, YMCS theory is consistent and free of tachyons only if k is larger than a critical value k_c, where

$$k_c = \frac{29N}{12}.$$
(9.67)

What happens for higher loops has not yet been studied, but a fair guess is that, for example, the denominator k in Eq. (9.66) would be replaced by $k + N$, in which case k_c would be $17N/12$. If so, these two values of k_c suggest that we know the critical value of k_c to within a factor of 2 and that higher loops do not change the fact that there is a critical value.

So the infrared slavery problem persists if $k < k_c$, in which case, we solve it just as before: there has to be a dynamical gluon mass m generated, and this mass generation is self-consistently supported by condensates of solitons of the massive effective action. If so, the resulting infrared-effective action has both a CS term and a GNLS term (see Eq. (9.71)). In the following, we argue that as $k \to k_c$ from below, the dynamical gluon mass along with the condensate supporting it must vanish, and the solitons composing the condensate no longer exist. There are [36] qualitative arguments suggesting that the exact form, given in Eq. (9.10), of the zero-momentum effective action as a function of the operator θ of Eq. (9.3) gets modified in a certain way by a CS term. This modified effective action $\Gamma(\theta, k)$ has all the right qualitative properties, including correct scaling in N and g for all quantities appearing in it, a dynamical gluon mass consistent with the operator product expansion of Eq. (9.4), a phase transition at a critical value of k at which the condensate vanishes, a quadratic increase in Γ as a function of k for small k, and the correct zero-k limit. We will not detail the arguments, all based on the one-loop equations given so far, but will simply state the result here:

$$\Gamma(\theta, k) = \int \theta \left\{ 1 - k_c \left[k^2 + \left(\frac{4\pi}{g_3^2} \right)^2 (a_3 g_3^2 \theta)^{1/2} \right]^{-1/2} \right\};$$
(9.68)

here a_3 is the Lavelle constant of Eq. (9.4), and the critical value k_c is proportional to N. The pure numbers in this expression are not to be taken seriously.

Now let us check the properties of this modified effective action. First, in the limit $k = 0$, it is of the necessary form given in Eq. (9.10), with $\langle \theta \rangle \simeq (Ng_3^2)^3(N^2 - 1)$. Second, for small k, the leading correction term is positive and quadratic, as expected from Eq. (9.59). Third, at $k = k_c$, the minimum of Γ moves to $\theta = 0$ so there is no condensate, whereas for positive θ, the effective action Γ is also positive, indicating that entropy effects are no longer dominant.

The order parameter for this phase transition is the dynamical gluon mass $m(k)$, which now depends on k. When $k \lesssim k_c$, some simple algebra shows that the minimum of Γ in Eq. (9.68) obeys the following:

$$\theta^{1/2} \sim (k_c - k), \tag{9.69}$$

and so

$$m(k) \sim \theta^{1/4} \sim (k_c - k)^{1/2}, \tag{9.70}$$

characteristic of a second-order phase transition. If there really is a phase transition at $k = k_c$, then one would expect solitons to appear for smaller k.

9.6.2 YMCS solitons

D'Hoker and Vinet [35] long ago looked for classical solitons of YMCS theory. Their result is that there are no finite-action classical solitons of the vortex or sphaleron type in classical YMCS theory. There are solitons, but they have a curious instability that creates a singularity, preventing them from having finite action.

The remaining questions are as follows: are there any finite-action solitons when there is both a dynamical mass m and a CS mass in the effective action? How do these solitons behave when $m \to 0$? For $k < k_c$, the full effective action with dynamical mass term is

$$\int d^3x \left[2\pi i k N_{\text{cs}} - \frac{1}{2g_3^2} \int \text{Tr}\, \mathcal{G}_{ij}^2 - \frac{m^2}{g_3^2} \int \text{Tr}[U^{-1}\mathcal{D}_i U]^2 \right]. \tag{9.71}$$

First, let us look for center vortices using the classical equations from Eq. (9.71). Center vortices are Abelian, and this action leads to the Abelian solution [36]:

$$\mathcal{A}_i^a(x) = \frac{2\pi Q^a}{\mu} \oint dz_k \left\{ \epsilon_{ijk}\partial_j \left[\mu_-(\Delta_+(x-z) - \Delta_0(x-z)) + (+ \leftrightarrow -) \right] \right.$$
$$\left. + i\delta_{ik}\mu_+\mu_- \left[\Delta_+(x-z) - \Delta_-(x-z) \right] \right\}, \tag{9.72}$$

where the masses μ, $\mu_\pm$ are

$$\mu_\pm = \frac{1}{2}[\pm m_{\text{cs}} + (m_{\text{cs}}^2 + 4m^2)^{1/2}] \qquad \mu = \mu_+ + \mu_-, \tag{9.73}$$

and $\Delta_\pm$ is the free Feynman propagator for mass $\mu_\pm$. This is a peculiar soliton because it has a (parity violating) imaginary part. But the total action, including the CS term, is real, and the action per unit length is finite (given that the running mass $m(q)$ decreases as in Eq. (9.4)). This soliton is twisted and has a nonzero $\langle N_{\text{cs}} \rangle$. The twist (or N_{cs}) changes sign when k changes sign, and so the action is even in

k. In the limit $m_{cs} \to 0$, this vortex reduces to the usual center vortex, and in the limit $m \to 0$, the soliton vanishes.[11]

Next, look for spherical solitons in $SU(2)$, which should resemble modified sphalerons. The usual spherical decomposition of the gauge potential is

$$2i\mathcal{A}_i = \epsilon_{iak}\tau_a \widehat{x}_k \left[\frac{\phi_1(r) - 1}{r} \right] - (\tau_i - \widehat{x}_i\widehat{x} \cdot \vec{\tau}) \frac{\phi_2(r)}{r} + \widehat{x}_i\widehat{x} \cdot \vec{\tau} H_1(r) \quad (9.74)$$

$$U = \exp\left[i\beta(r)\frac{\vec{\tau} \cdot \widehat{x}}{2} \right]. \quad (9.75)$$

Inserting these into the effective YMCS plus mass action of Eq. (9.71) gives [36, 34] what looks like four equations of motion, one for each of the four functions in Eq. (9.74):

$$0 = (\phi_1' - H_1\phi_2)' + \frac{1}{r^2}\phi_1(1 - \phi_1^2 - \phi_2^2)$$
$$+ (im_{cs} - H_1)(\phi_2' + H_1\phi_1) - m^2(\phi_1 - \cos\beta) \quad (9.76)$$

$$0 = (\phi_2' + H_1\phi_1)' + \frac{1}{r^2}\phi_2(1 - \phi_i^2 - \phi_2^2)$$
$$- (im_{cs} - H_1)(\phi_1' - H_1\phi_2) - m^2(\phi_2 + \sin\beta) \quad (9.77)$$

$$0 = \phi_1\phi_2' - \phi_2\phi_1' + H_1(\phi_1^2 + \phi_2^2)$$
$$+ \left(im_{cs}(1 - \phi_1^2 - \phi_2^2) + \frac{1}{2}m^2r^2(H_1 - \beta') \right) \quad (9.78)$$

$$0 = \frac{1}{r^2}[r^2(\beta' - H_1)]' - \frac{2}{r^2}(\phi_1\sin\beta + \phi_2\cos\beta). \quad (9.79)$$

Here primes indicate radial derivatives. In fact, there are only three independent equations; Eq. (9.78), which comes from varying the matrix U, is (as we already know) not independent and can be derived from the other three.

Why are there no classical solitons (at $m = 0$) but there are (quantum) solitons for finite m? To a large extent, the answer to this question appears in Eq. (9.77) for the amplitude H_1 or, equivalently, A of Eq. (9.82). This equation is algebraic, not differential, and has the solution

$$A = \frac{1}{\phi_1^2 + \frac{m^2r^2}{2} - \frac{m_{cs}^2 B^2}{m^2}} \left[\frac{1}{m}(B\phi_1' - B'\phi_1) + 1 - \phi_1^2 + \frac{m_{cs}^2}{m^2}B^2 \right]. \quad (9.80)$$

---

[11] Because the Abelian equations are linear, there are other linear combinations of the preceding $\pm$ solutions without this property, but they do not have finite action per unit length.

D'Hoker and Vinet [35] have an analogous equation, but in the gauge $B = 0$ and with no dynamical mass, so their equation is recovered by setting B, B/m and m to zero. Their denominator, then, is just ϕ_1^2. They show that there is at least one zero of this denominator and that the existence of one zero leads to an infinite number of zeros and a "soliton" having an accumulation point of zeroes at $r = 0$. In our case, if m is large enough, it is possible that the $m^2 r^2/2$ term in the denominator prevents the denominator from vanishing, and this does happen, at least numerically [36]. The numerics show that for small enough m, there is at least one zero, and the D'Hoker–Vinet disease arises: there are no sphaleron-like solitons.

So several different lines of investigation, all of them qualitative, lead to the same conclusion: for large k, YMCS theory is in the Witten phase and can be solved exactly, but for $k < k_c$, with $k_c \simeq (1 - 2)N$, there is a second-order phase transition to a nonperturbative phase with a dynamical gluon mass in addition to the CS mass.

9.7 Compactness and the Chern–Simons number of YMCS solitons

The developments so far provide a setting for investigating whether the assumption of compactness, which quantizes various topological indices, is physically necessary [34]. We know already that topological charge may consist of localized lumps of nonintegral charge whose sum over all Euclidean space-time is integral, but this does not challenge the notion of compactness, which is only needed for infinite spaces and their boundaries at infinity. Compactness requires that quantum numbers defined on such boundaries be integral, but here we assume otherwise and look for the consequences, using a model of a dilute condensate of YMCS sphalerons. The result is that the noncompact model has a vacuum energy density higher than that of the compact theory by a finite amount and hence a vacuum energy higher by an infinite amount after integrating over all three-space [34]. So compactness is energetically preferred.

The model begins with sphalerons from Eqs. (9.76), (9.76), (9.77), and (9.78), with boundary conditions

$$r = 0: \quad \phi_1(0) = 1, \qquad \phi_2(0) = H_1(0) = \beta(0) = 0;$$

$$r = \infty: \quad \phi_1(\infty) = \cos \beta(\infty), \qquad \phi_2(\infty) = -\sin \beta(\infty), \qquad H_1 \to \beta'. \quad (9.81)$$

These equations are again complex, but in one case, they can be reduced to real equations for real functions, and this is the case of interest. Set $\beta = \pi$ and $\alpha = 0$. Then one can choose H_1, ϕ_2 to be pure imaginary, with all other functions real:

$$H_1 = im_{cs}A(r); \qquad \phi_2 = i\frac{m_{cs}}{m}B(r). \quad (9.82)$$

The equations become real, and so ϕ_1, A, B are all real. Generally, solitons with both real and imaginary field components have conjugate solitons found by complex conjugation (which changes the sign of the CS number), but the soliton here is self-conjugate, and its ordinary CS number vanishes. For this choice of boundary conditions, the CS action $2\pi i k N_{\mathrm{cs}}$ is real (the integral in Eq. (9.51) is imaginary). The general form of N_{cs} for a spherical soliton is

$$N_{\mathrm{cs}} = \frac{1}{8\pi^2} \int \frac{d^3 x}{r^2} [\phi_1 \phi_2' - \phi_2 \phi_1' - \phi_2' - H_1(1 - \phi_1^2 - \phi_2^2)]. \qquad (9.83)$$

Substitute the forms of Eq. (9.82) to find a purely imaginary N_{cs} and so a purely real CS action. (There is no reason that the contribution of solitons to N_{cs} should be integral or even real.) This action is $\mathcal{O}(k^2)$ and positive for small k, as we argued earlier on general grounds.

The only interpretation we can make of a pure imaginary N_{cs} is that what we usually think of as the (topological) CS number vanishes. This soliton is very much like the QCD sphaleron of Chapter 7, which, considered only as a $d = 3$ soliton, has no CS number. However, by making a special gauge transformation on this solution, we can endow it with a genuine (and nonintegral) CS number. The spherical equations of motion have a residual $U(1)$ gauge invariance that preserves spherical symmetry:

$$\phi_1(r) \rightarrow \phi_1(r) \cos \alpha(r) + \phi_2(r) \sin \alpha(r)$$

$$\phi_2(r) \rightarrow \phi_2(r) \cos \alpha(r) - \phi_1(r) \sin \alpha(r)$$

$$\beta(r) \rightarrow \beta(r) + \alpha(r)$$

$$H_1(r) \rightarrow H_1(r) + \alpha'(r). \qquad (9.84)$$

These transformations can be read off from the gauge transformation:

$$\mathcal{A}_i^a \rightarrow V \mathcal{A}_i^a V^{-1} + V \partial_i V^{-1}, \qquad (9.85)$$

with[12]

$$V(\alpha) = \exp\left[\frac{i}{2} \vec{\tau} \cdot \hat{r} \alpha(r)\right]. \qquad (9.86)$$

Of course, the subgroup generated by all group elements of the form in Eq. (9.86) is Abelian, but it does not commute with the general vector potential. We call gauge transformations of the type in Eq. (9.86) spherical gauge transformations.

For any gauge transformation, as in Eq. (9.85), N_{cs} changes according to Eq. (8.33). With the assumption of compactness, the gauge transformation V approaches the identity on the sphere at infinity, and the gauge potential $\mathcal{A}_i$ vanishes at least as

---

[12] To avoid singularities at the origin, choose $\alpha(0) = 0$.

fast as $1/r$, and so the change in N_{cs} reduces to the winding number of Eq. (9.87), with the added requirement that $\alpha(r = \infty) = 2\pi L$ for an integer L. When V approaches I on the sphere at infinity, the space of gauge transformations is really defined on the three-sphere S^3 rather than on $\mathcal{R}_3$ because all the points at infinity are mapped to a single point. This integral winding number is that of the map of the group space S^3 onto the spatial S^3 that we just identified, or in other words, the homotopy $\Pi^3(S^3) \simeq Z$, and this winding number is L. The winding number is topological, which means two things: it is independent of a choice of metric, and it can be expressed as a boundary-value integral. It is not completely elementary to find the function whose divergence is the winding-number integrand; the answer is in the work of Deser et al. [32] for general gauge transformations.[13] For the spherical gauge transformation of Eq. (9.86), a straightforward calculation (easiest with Eq. (9.83)) gives

$$\frac{1}{8\pi^2} \int \epsilon_{ijk} \operatorname{Tr}\frac{1}{3} V(\alpha)^{-1}\partial_i V(\alpha)V^{-1}(\alpha)\partial_j V(\alpha)V^{-1}(\alpha)\partial_k V(\alpha)$$

$$= \frac{1}{2\pi} \int_0^\infty dr\, \alpha'(r)[1 - \cos\alpha(r)] = \frac{1}{2\pi} [\alpha(r) - \sin\alpha(r)]|_0^\infty . \qquad (9.87)$$

The answer depends only on the boundary value $\alpha(r = \infty)$ because we choose $\alpha(0) = 0$.

Now we abjure compactness and let $\alpha(r = \infty)$ be arbitrary. There is a (real) CS number that is not integral. A particularly interesting case removes an integrable singularity arising from $\beta(0) \neq 0$; this singularity can be removed by invoking a spherical gauge transformation with $\alpha(0) = -\pi$, $\alpha(\infty) = 0$. The CS number is $1/2$, just as we would expect for a sphaleron.

We assume that for entropic reasons, there is a dilute (noninteracting) condensate of sphalerons in the vacuum so that all solitons are essentially independent. When a CS term is present in the action, the dilute-gas partition function Z is the usual expansion as a sum over sectors of different sphaleron number:

$$Z(k) = \sum_J Z_J \qquad Z_J(k) = \sum_{c.c.} \frac{1}{J!}e^{-\sum I_c} + \cdots, \qquad (9.88)$$

where $Z_J(k)$ is the partition function in the sector with J sphalerons; the subscript c.c. indicates a sum over collective coordinates of the sphalerons; I_c is the action (including CS action) of a sphaleron, and the omitted terms indicate corrections to the dilute-gas approximation. To be explicit, separate the sum over collective

---

[13] We know without calculation that the winding number for spherical gauge transformations has to be a total divergence because it changes the action without changing the equations of motion.

coordinates into kinematic coordinates, such as spatial position (the ath soliton is at position $\vec{r} - \vec{a} \equiv \vec{r}(a)$) and gauge-collective coordinates. The former we represent in the standard dilute-gas way; the latter, we indicate as a functional integral over spherical gauge transformations U:

$$Z(k) = \int [dU] \sum \frac{1}{J!} \left(\frac{V}{V_c} \right)^J \exp \left\{ -J \Re I_c - 2\pi i k \left[J N_{cs}(\mathcal{A}) + N_{cs}(U) \right] \right\}. \quad (9.89)$$

Here V is the volume of all three-space; V_c is a finite collective-coordinate volume; $\Re I_c$ is the real part of the action; $N_{cs}(\mathcal{A})$ is the CS number of each individual soliton of gauge potential $\mathcal{A}$; and $N_{cs}(U)$ is the CS number of the large gauge transformation.

Even when we consider the apparently innocuous case of sphalerons of CS number $1/2$, choose k integral, and allow only compact gauge transformations, problems arise. The N_{cs} contribution to a term in the sum in the partition function is a phase factor $\exp(i\pi k J)$, which is -1 if both k and J are odd. When k is odd, the odd J terms in $\ln Z$ have the opposite sign to those of a normal dilute-gas condensate, which means that the free energy, which for a normal dilute-gas condensate is negative, has turned positive. So the noncompactified theory splits into two sectors, one with even numbers of sphalerons and the other with odd numbers, and the odd-number sector has infinitely higher free energy than the (compactified) even-number sector. (Noncompactification also leads to a number of other unphysical results in the dilute-gas approximation not considered here.)

Now generalize to arbitrary noncompact gauge transformations. Begin with potentials $\mathcal{A}$ (all indices suppressed) of the self-conjugate soliton given earlier, satisfying the equations of motion with boundary conditions of Eq. (9.81) and fixed to a standard spherical gauge (to be specific, we use the self-conjugate soliton with zero CS number). Introduce a different CS number for each soliton (labeled by a set of indices a) by making a spherical gauge transformation characterized by $\alpha(a; \infty)$ for the ath soliton. If we assume compactness, these gauge transformations, with integral CS numbers, have no effect (as long as the CS index k is integral). But we give up compactness, so these gauge transformations do have an effect, and each $\alpha(a; \infty)$ is a collective coordinate.

Because the total CS number of all J sphalerons comes from a surface contribution, we can immediately write the phase factor in the action by using Eq. (9.87):

$$Z(k) = \sum_J \frac{1}{J!} \left(\frac{V}{V_c} \right)^J \exp[-J \Re I_c] \exp[ik(\alpha - \sin \alpha)], \quad (9.90)$$

where

$$\alpha = \sum_{a=1}^{J} \alpha(a; \infty). \tag{9.91}$$

The $\alpha(a; \infty)$ are collective coordinates, and we integrate over them:

$$Z(k) = \sum_{J} Z_{RJ} \times \left\{ \prod_{a} \int_{0}^{2\pi} \frac{d\alpha(a; \infty)}{2\pi} \right\} \exp[ik(\alpha - \sin \alpha)], \tag{9.92}$$

where Z_{RJ} indicates the explicitly real terms in the summand of Eq. (9.90). This integral is reduced to a product by using the familiar Bessel identity

$$e^{iz \sin \theta} \equiv \sum_{-\infty}^{\infty} J_{N}(z) e^{iN\theta}, \tag{9.93}$$

and the integral becomes, for integral k, $[J_{k}(k)]^{J}$. So the dilute-gas partition function is

$$Z(k) = \sum_{J} \frac{1}{J!} \left(\frac{V}{V_{c}} \right)^{J} \exp \left\{ J[-\Re e\, I_{c} + \ln J_{k}(k)] \right\} = \exp \left[\frac{V}{V_{c}} e^{-\Re e\, I_{c}} J_{k}(k) \right]. \tag{9.94}$$

Because $1 \geq J_{k}(k) > 0$ for all levels k, integrating over the collective coordinates has increased the free energy (the negative logarithm of Z). This shows that by compactifying the sphalerons, we lower the free energy, yielding something like the usual dilute-gas partition function (which is Eq. (9.94) without the $J_{k}(k)$ factor). This simply requires that the total CS number α be an integer, not that each contributing soliton have integral CS number.

There are a number of other issues concerning compactness that we will not discuss here; for example, what happens if the CS level k is nonintegral [34]? The upshot is that compactness is more than just a mathematical assumption because compact theories always have infinitely less free energy than noncompact theories.

9.7.1 Sphalerons, knots, and compactness

We conclude this chapter by noting [34] that sphalerons of CS number $1/2$ can be topologically mapped onto the linkages of $d = 3$ knots (see Kaufmann [37]) – the same sort of linkages that occur between center vortices and Wilson loops. For compact knots (knots whose links are closed strings), the total link number is integral and is composed of a sum of an even number of terms, each $\pm 1/2$, one for each crossing (defined subsequently). But noncompact knots, involving nonclosed strings, may have half-integral link numbers.

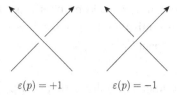

$$\varepsilon(p) = +1 \qquad \varepsilon(p) = -1$$

Figure 9.1. Overcrossings or undercrossings of knot components; the sign $\varepsilon(p)$ distinguishes the two possibilities shown.

The connection between the non-Abelian gauge potentials of a sphaleron and link numbers is an Abelian gauge potential formed from the sphaleron gauge potential, whose Abelian CS integral describes the linkages in terms of knots that occur in the Abelian magnetic field lines. For gauge group $SU(2)$, there is a deep relation between the CS integral and these Abelian gauge potentials and field strengths. This turns the non-Abelian CS form, with its characteristic cubic term, into an Abelian CS form that measures the linkages of the closed lines of Abelian field strength. The Abelian CS form, when described in terms of its Dirac string as in the confinement picture, has [34] exactly the form of the integral used in knot theory to describe over- and undercrossings, as shown in Figure 9.1. Aside from the topological characteristics we briefly note here, there is no particular physical meaning to this Abelian gauge potential.

The simplest way to think of knots in three dimensions is to project them onto a $d = 2$ plane, carefully distinguishing the various overcrossings and undercrossings that arise (see Kaufmann [37] for details). Knots are made of links or closed oriented loops (such as those occurring in $d = 3$ center vortices). A single link can be self-knotted, but the description of such knots is ambiguous until either the twist or the writhe of the single link is prescribed. Spread a system of linked closed strings out on a table to see undercrossings and overcrossings, such as idealized in Figure 9.1.

For the crossings of two distinct curves, the link number Lk, which is an integer, is defined as

$$Lk = \frac{1}{2} \sum_{p \in C} \epsilon(p), \tag{9.95}$$

where C is the set of crossing points of one curve with the other. It turns out that a single sphaleron corresponds (through the knot structure of its associated Abelian field lines) to a single term in this sum for N_{Lk}, and the $1/2$ for every term is precisely the CS number for the sphaleron.

The projection of two links formed from (nonpathological) closed loops has to have an even number of terms in the sum for N_{Lk}, so the sum yields an integer. But if the links are open, there can be an odd number (e.g., either of the crossings in Figure 9.1). An open link is equivalent to a closed link with one closure at infinity or, in other words, a noncompact link. So compactness of the knotted links implies integrality of N_{cs}.

The underlying topology comes from the perhaps surprising result that there is a topological map $S^3 \to S^2$, with the homotopy $\Pi^3(S^2) \simeq Z$. (As usual, that this map be described by integral indices requires compactness.) This map and homotopy were found by Hopf, and the map is called the Hopf fibration. The simplest way to look at the Hopf fibration (which describes, in essence, a total bundle space S^3 as the fibration of the base space S^2 by a fiber in S^1) is to begin with an $SU(2)$ matrix $U(\vec{r})$ in the fundamental representation and from it construct a unit vector $\widehat{n}$ by

$$U\tau_3 U^{-1} = \vec{\tau} \cdot \widehat{n} \qquad (9.96)$$

(the $\vec{\tau}$ are the Pauli matrices). The unit vector lives on S^2, and the group space of $SU(2)$ is S^3. Something must be redundant in such a map, and it is that U can be right multiplied by $\exp[i\alpha(\vec{r})\tau_3/2]$ without changing $\widehat{n}$; the unit vector field corresponds to the coset $SU(2)/U(1)$. This redundancy, parametrized by $\alpha(\vec{r})$, will turn out to be a change of gauge for the fictitious Abelian potential (a shift by $\partial_i\alpha$). The fictitious Abelian gauge potential is

$$A_i = i \, \text{Tr} \left(\tau_3 U \partial_i U^{-1} \right), \qquad (9.97)$$

and its field strength is

$$B_i = \epsilon_{ijk}\partial_j A_k = -i\epsilon_{ijk} \text{Tr} \left(\tau_3 U \partial_j U^{-1} U \partial_k U^{-1} \right) = \frac{1}{2}\epsilon_{ijk}\epsilon_{abc}n^a\partial_j n^b \partial_k n^c. \quad (9.98)$$

Some manipulations using the antisymmetric property of the ϵ symbol lead to a very elegant formula, expressing the pure-gauge form of N_{cs} as an Abelian CS integral:

$$\frac{-1}{12\pi^2} \int d^3r \, \epsilon_{ijk} \text{Tr} \left(U\partial_i U^{-1} U\partial_j U^{-1} U\partial_k U^{-1} \right) = \frac{1}{16\pi^2} \int d^3r \, A_i B_i. \quad (9.99)$$

We have already encountered such an Abelian CS integral in interpreting the link number of a center vortex and a Wilson loop in Eq. (7.41). For the Hopf map, the link number is the integer in the homotopy $\Pi^3(S^2) \simeq Z$. However, for the standard sphaleron, this link number is half-integral because the standard sphaleron is noncompact; the field lines of B_i terminate only at spatial infinity. Half-integrality of the link number of Eq. (9.95) also occurs when the links are noncompact. Reference [34] gives full details of this relation between sphalerons and knots.

Each 1/2, with appropriate sign, marks an overcrossing or undercrossing of knots in its $d = 2$ projection. Although knots are uniquely a property of Euclidean three-space, because $d = 1$ strings do not link in any other space,[14] a great deal of knot theory is essentially two-dimensional, based on projecting the knots' overcrossings and undercrossings onto a $d = 2$ space (see Figure 9.1).

This leads to a useful $d = 2$ interpretation of knots and sphalerons that is quite analogous to the $d = 4$ interpretation of topological charge as counting, through a nonoriented intersection number, the linkages of center vortices and nexuses. An analogous nonoriented $d = 2$ intersection-number integral of closed $d = 1$ strings, with $SU(2)$ nexuses put by hand on the strings, yields some elementary knot properties.

A special ribbon framing is often used. In this, called the *Frenet–Serret* framing, the unit vector field of Eq. (8.28) is the principal normal vector $\widehat{e}_2$, lying in the direction of the curve's curvature vector (or derivative of the tangent vector $\widehat{e}_1$). With this framing in the $\epsilon \to 0$ limit, the twist is

$$Tw = \frac{1}{2\pi} \oint_\Gamma ds\, \widehat{e}_2 \cdot \frac{d\widehat{e}_3}{ds}, \tag{9.100}$$

where $\widehat{e}_3$ is the principal binormal vector, given by $\widehat{e}_3 = \widehat{e}_1 \times \widehat{e}_2$. This expression for twist is well defined, provided that the curvature of the curve is not zero somewhere (in which case, $\widehat{e}_2$ is not defined).

References

[1] J. M. Cornwall, A three-dimensional scalar field theory model of center vortices and its relation to k-string tensions, *Phys. Rev.* **D70** (2004) 065005.
[2] E. Witten, Quantum field theory and the Jones polynomial, *Commun. Math. Phys.* **121** (1989) 351.
[3] J. M. Cornwall, Non-perturbative mass gap in continuum QCD, in *Proceedings of the French–American Seminar on Theoretical Aspects of Quantum Chromodynamics*, Marseille, France (1981).
[4] J. M. Cornwall, Dynamical mass generation in continuum QCD, *Phys. Rev.* **D26** (1982) 1453.
[5] M. Lavelle, Gauge invariant effective gluon mass from the operator product expansion, *Phys. Rev.* **D44** (1991) R26.
[6] J. M. Cornwall, Exact zero-momentum sum rules in $d = 3$ gauge theory, *Nucl. Phys.* **B416** (1994) 335.
[7] V. A. Novikov, M. A. Shifman, A. I. Vainshtein, and V. I. Zakharov, Are all hadrons alike? *Nucl. Phys.* **B191** (1981) 301.
[8] J. M. Cornwall, How $d = 3$ QCD resembles $d = 4$ QCD, *Phys. A* **158** (1989) 97.

---

[14] A $d = 1$ string links to a $d - 2$-dimensional closed hypersurface, so in any dimension, a Wilson loop links to a center vortex.

[9] N. K. Nielsen and P. Olesen, An unstable Yang-Mills field mode, *Nucl. Phys.* **B144** (1978) 376.

[10] J. M. Cornwall, W. S. Hou, and J. E. King, Gauge-invariant calculations in finite-T QCD: Landau ghost and magnetic mass, *Phys. Lett.* **B153** (1985) 173.

[11] J. M. Cornwall and B. Yan, String tension and Chern-Simons fluctuations in the vortex of $d = 3$ gauge theory, *Phys. Rev.* **D53** (1996) 4638.

[12] W. Buchmuller and O. Philipsen, Magnetic screening in the high temperature phase of the standard model, *Phys. Lett.* **B397** (1997) 112.

[13] F. Eberlein, Two-loop gap equations for the magnetic mass, *Phys. Lett.* **B439** (1998) 130.

[14] F. Eberlein, The gauge-Higgs system in three dimensions to two-loop order, *Nucl. Phys.* **B550** (1999) 303.

[15] G. Alexanian and V. P. Nair, A self-consistent inclusion of magnetic screening for the quark-gluon plasma, *Phys. Lett.* **B352** (1995) 435.

[16] J. M. Cornwall, On one-loop gap equations for the magnetic mass in $d = 3$ gauge theory, *Phys. Rev.* **D57** (1998) 3694.

[17] D. Karabali, C. J. Kim, and V. P. Nair, On the vacuum wavefunction and string tension of Yang-Mills theories in (2 + 1) dimensions, *Phys. Lett.* **B434** (1998) 103, and references therein.

[18] F. Karsch, T. Neuhaus, A. Patkos, and J. Rank, Gauge boson masses in the 3D, SU(2) gauge-Higgs model, *Nucl. Phys.* **B 474** (1997) 217.

[19] U. M. Heller, F. Karsch, and J. Rank, Gluon propagator at high temperature: Screening, improvement and nonzero momenta, *Phys. Rev.* **D57** (1998) 1438, and references therein.

[20] A. Cucchieri, F. Karsch, and P. Petreczky, Propagators and dimensional reduction of hot $SU(2)$ gauge theory, *Phys. Rev.* **D64** (2001) 036001.

[21] A. Nakamura, T. Saito, and S. Sakai, Lattice calculation of gluon screening masses, *Phys. Rev.* **D69** (2004) 014506.

[22] R. Jackiw, Introduction to the Yang-Mills quantum theory, *Rev. Mod. Phys.* **52** (1980) 661.

[23] H. G. Loos, Canonical gauge- and Lorentz-invariant quantization of the Yang-Mills field, *Phys. Rev.* **188** (1969) 2342.

[24] J. P. Greensite, Calculation of the Yang-Mills vacuum wave functional, *Nucl. Phys.* **B158** (1979) 469.

[25] R. Jackiw, in *Current Algebras and Anomalies*, ed. S. B. Trieman, R. Jackiw, B. Zumino, and E. Witten (Princeton University Press, Princeton, 1985), 258.

[26] J. M. Cornwall and G. Tiktopoulos, Functional Schrödinger equation approach to high-energy multileg amplitudes, *Phys. Rev.* **D47** (1993) 1629.

[27] J. M. Cornwall, Nonperturbative treatment of the functional Schrödinger equation in QCD, *Phys. Rev.* **D38** (1988) 656.

[28] J. M. Cornwall, Conjecture on the infrared structure of the vacuum Schrodinger wave functional of QCD, *Phys. Rev.* **D76** (2007) 025012.

[29] E. G. S. Luna, Diffraction and an infrared finite gluon propagator, *Braz. J. Phys.* **37** (2007) 84.

[30] A. A. Natale, Phenomenology of infrared finite gluon propagator and coupling constant, *Braz. J. Phys.* **37** (2007) 306.

[31] A. C. Aguilar, A. Mihara, and A. A. Natale, Freezing of the QCD coupling constant and solutions of Schwinger-Dyson equations, *Phys. Rev.* **D65** (2002) 054011.

[32] S. Deser, R. Jackiw, and S. Templeton, Three-dimensional massive gauge theories, *Phys. Rev. Lett.* **48** (1982) 975.

[33] S. Deser, R. Jackiw, and S. Templeton, Topologically massive gauge theories, *Ann. Phys.* **140** (1982) 204 [Erratum, ibid. **185** (1988) 406, **281** (2000) 409].

[34] J. M. Cornwall and N. Graham, Sphalerons, knots, and dynamical compactification in Yang-Mills-Chern-Simons theories, *Phys. Rev.* **D66** (2002) 065012.

[35] E. D'Hoker and L. Vinet, Classical solutions to topologically massive Yang-Mills theory, *Ann. Phys.* **162** (1985) 413.

[36] J. M. Cornwall, Phase transition in $d = 3$ Yang-Mills-Chern-Simons gauge theory, *Phys. Rev.* **D54** (1996) 1814.

[37] L. Kaufmann, *On Knots* (Princeton University Press, Princeton, 1987).

10

The pinch technique for electroweak theory

In this chapter, we give a general overview of how the pinch technique (PT) is modified in the case of a theory with spontaneous (tree level) symmetry breaking (Higgs mechanism) [1, 2, 3], using the electroweak sector of the standard model as the reference theory.

The application of the pinch technique in the electroweak sector brings about significant conceptual and practical advantages. First, from the purely theoretical point of view, it is important to know that the PT construction is sufficiently general to encompass theories other than massless Yang–Mills. Though the required technical manipulations in the electroweak sector turn out to be fairly cumbersome, the basic underlying principles are practically the same; what increases the complexity is not the principle itself but rather the proliferation of fields and vertices involved. In addition, the PT algorithm exposes systematically the vast number of cancellations that take place when the (nonrenormalizable) Green's functions of the unitary gauge are put together to form physical amplitudes. In fact, one may start directly from the unitary gauge and derive the same PT Green's function constructed in the context of the (renormalizable) R_ξ gauges. The application of the pinch technique provides a deeper understanding of the connection between the unitary gauge and the optical theorem and analyticity. Moreover, in the context of a theory with symmetry breaking, one gains new, important insights on the connection between the pinch technique and the background field method (BFM). Specifically, the BFM Feynman gauge is uniquely and unambiguously singled out by the powerful physical requirement of having Green's functions that display only *physical thresholds*. Last, but not least, one may derive crucial Ward identities, relating the various propagators of the theory from the sole requirement of the complete gauge independence of S-matrix elements.

10.1 General considerations

The application of the pinch technique in theories with tree-level symmetry break-ing in general, and in the electroweak sector of the standard model in particular, is significantly more involved than in the QCD case. The reasons are both bookkeep-ing related, because of the proliferation of particles and Feynman diagrams, and conceptual, related to the correct allocation of the various pinch terms among the self-energies and vertices under construction:

1. There is a considerable increase in the number of sources of gauge-fixing parameter-dependent terms. In particular, in the R_ξ gauges, the tree-level gauge-boson propagators – three massive gauge bosons ($W^\pm$ and Z) and a massless photon (A) – are given by

$$\Delta_i^{\mu\nu}(q) = \left[g^{\mu\nu} - \frac{(1 - \xi_i)q^\mu q^\nu}{q^2 - \xi_i M_i^2} \right] d_i(q^2)$$

$$d_i(q^2) = \frac{-i}{q^2 - M_i^2}, \tag{10.1}$$

where $i = W, Z, A$ and $M_A^2 = 0$. In general, the gauge-fixing parameters ξ_w, ξ_z, and ξ_A will be considered different from one another. The inverse of the gauge-boson propagators, to be denoted by $\Delta_{i,\mu\nu}^{-1}$, is given by

$$\Delta_{i,\mu\nu}^{-1}(q) = i \left[(q^2 - M_i^2)g_{\mu\nu} - q_\mu q_\nu + \frac{1}{\xi_i}q_\mu q_\nu \right]. \tag{10.2}$$

Three unphysical (would-be) Goldstone bosons are associated with the three massive gauge bosons, to be denoted by $\phi^\pm$ and χ. Their tree-level propaga-tors are ξ dependent, and given by

$$D_i(q) = \frac{i}{q^2 - \xi_i M_i^2}, \tag{10.3}$$

with $i = W, Z$ (no Goldstone boson is associated with the photon). Note that

$$\Delta_i^{\mu\nu}(q) = U_i^{\mu\nu}(q) - \frac{q^\mu q^\nu}{M_i^2} D_i(q), \tag{10.4}$$

where

$$U_i^{\mu\nu}(q) = \left(g^{\mu\nu} - \frac{q^\mu q^\nu}{M_i^2} \right) d_i(q^2) \tag{10.5}$$

is the corresponding propagator in the so-called *unitary gauge* ($\xi_i \to \infty$). In addition, the ghost propagators are also given by $D_i(q)$, with $i = W, Z, A$

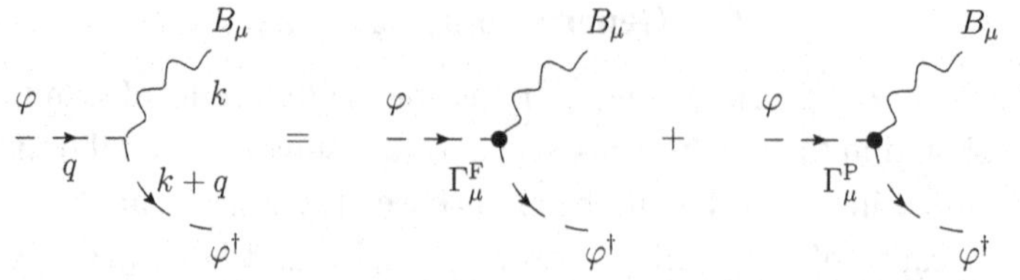

Figure 10.1. The PT decomposition of the generic elementary gauge-boson-scalar vertex $\Gamma_{B_\mu \varphi \varphi^\dagger}$.

(there is a massless ghost associated with the photon). Finally, the bare propagator of the physical Higgs boson is gauge-fixing parameter-independent at tree level, and given by $\Delta_H(q) = i/(q^2 - M_H^2)$.

2. In addition to the longitudinal momenta coming from the propagators of the gauge bosons (proportional to $\lambda_i = 1 - \xi_i$) and the PT decomposition of the vertices involving three gauge bosons, a new source of pinching momenta appears, originating from graphs having an external (i.e., carrying the physical momentum q) would-be Goldstone boson. Specifically, interaction vertices, such as $\Gamma_{A_\alpha \phi^\pm \phi^\mp}$, $\Gamma_{Z_\alpha \phi^\pm \phi^\mp}$, $\Gamma_{W_\alpha^\pm \phi^\mp \chi}$, and $\Gamma_{W_\alpha^\pm \phi^\mp H}$, also furnish pinching momenta when the gauge boson is inside the loop carrying (virtual) momentum k. Such a vertex will then be decomposed as (see Figure 10.1)

$$\Gamma_\alpha^{(0)}(q, k, -q-k) = \Gamma_\alpha^F(q, k, -q-k) + \Gamma_\alpha^P(q, k, -q-k) \quad (10.6)$$

with

$$\Gamma_\alpha^{(0)}(q, k, -q-k) = (2q+k)_\alpha$$
$$\Gamma_\alpha^F(q, k, -q-k) = 2q_\alpha$$
$$\Gamma_\alpha^P(q, k, -q-k) = k_\alpha, \quad (10.7)$$

which is the scalar case analog of Eqs. (1.41), (1.42), and (1.43).

3. When the fermions involved (external or inside loops) are massive, the Ward identity of Eq. (1.45) receives additional contributions, which correspond precisely to the tree-level coupling of the would-be Goldstone bosons to the fermions. To see this concretely, let us consider the analog of the fundamental pinching Ward identity of Eq. (1.45), (e.g., in the case in which the incoming boson is a W). Contracting the $\Gamma_{W_\mu^+ \bar{u} d}$ vertex with k^μ (the fermions u and d are isodoublet partners), we have

$$\not{k} P_L = P_R S_d^{-1}(k+p) - S_u^{-1}(p) P_L + [m_d P_R - m_u P_L], \quad (10.8)$$

where the chirality projection operators are defined according to $P_{L,R} = (1 \mp \gamma_5)/2$. The first two terms will pinch and vanish on shell, respectively, as they did in the case of QCD; the leftover term in the square bracket corresponds precisely to the coupling $\phi^+\bar{u}d$ (the case involving the $\Gamma_{W_\mu^-\bar{d}u}$ is identical). A completely analogous Ward identity is obtained when the incoming boson is a Z. Again, contraction with the vertex $\Gamma_{Z\bar{f}f}$ furnishes a Ward identity similar to Eq. (10.8), with the additional term proportional to $m_f\gamma_5$, which corresponds to the coupling $\Gamma_{\chi\bar{f}f}$.

4. After the various pinch contributions have been identified, particular care is needed when allocating them among the PT quantities that one is constructing. So unlike the QCD case, in which all propagator-like pinch contributions were added to the only available self-energy, $\Pi_{\alpha\beta}$ (to construct $\widehat{\Pi}_{\alpha\beta}$), in the electroweak case, such pinch contributions must in general be split among various propagators. Thus, in the case of the charged channel, they will be shared in general between the self-energies $\Pi_{W_\alpha W_\beta}$, $\Pi_{W_\alpha\phi}$, $\Pi_{\phi W_\beta}$, and $\Pi_{\phi\phi}$. This is equivalent to saying that when forming the inverse of the W propagator, in general, the longitudinal parts may no longer be discarded from the four-fermion amplitude because the external current is not conserved up to terms proportional to the fermion masses. The correct way of treating the longitudinal terms is to employ identities such as [3]

$$ig_\alpha^\nu = q^\nu q_\alpha D_i(q) - \Delta_i^{\nu\mu}(q)\left[(q^2 - M_i^2)g_{\mu\alpha} - q_\mu q_\alpha\right]$$
$$iq^\mu = q^2 D_i(q)q^\mu + M_i^2 q_\nu \Delta_i^{\mu\nu}(q). \tag{10.9}$$

The neutral channel is even more involved; one has to split the propagator-like pinch contributions among the self-energies $\Pi_{Z_\alpha Z_\beta}$, $\Pi_{A_\alpha A_\beta}$, $\Pi_{Z_\alpha A_\beta}$, $\Pi_{A_\alpha Z_\beta}$, $\Pi_{Z_\alpha\chi}$, $\Pi_{\chi Z_\beta}$, $\Pi_{\chi\chi}$, and Π_{HH}.

We emphasize that the preceding four points are tightly intertwined. The extra terms appearing in the Ward identity are precisely needed to cancel the gauge dependence of the corresponding graph in which the gauge boson is replaced by its associated Goldstone boson. In addition, as we will see in Section 10.3, when the external currents are not conserved, the appearance of the scalar-scalar or scalar-gauge-boson self-energies is crucial for enforcing the gauge-fixing parameter independence of the physical amplitude.

10.2 The case of massless fermions

We will now study the application of the pinch technique in the case where all fermions involved are massless. This simplification facilitates the PT procedure considerably because no scalar particles (Higgs and would-be Goldstone bosons) couple to the massless fermions. As a result, (1) the scalars can appear only

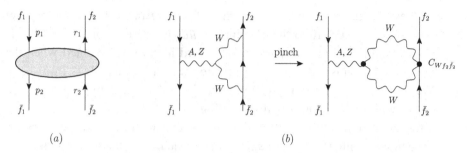

Figure 10.2. (a) The general process $f_1(p_1)\bar{f}_1(p_2) \rightarrow f_2(r_1)\bar{f}_2(r_2)$ mediated at tree level by a Z-boson and a photon and (b) the basic pinching and one of the unphysical vertices produced at the one-loop level.

inside the self-energy graphs, where they obviously cannot pinch; (2) Eq. (10.8) is practically reduced to its QCD equivalent; and (3) there are no self-energies with incoming scalars (i.e., no $\Pi_{W_\alpha\phi}$, $\Pi_{Z_\alpha\chi}$, $\Pi_{\phi\phi}$, etc.).

We now focus for concreteness on the the process $f_1(p_1)\bar{f}_1(p_2) \rightarrow f_2(r_1)\bar{f}_2(r_2)$, mediated at tree level by a Z-boson and a photon, as shown in Figure 10.2(a). At one-loop order, the box and vertex graphs furnish propagator-like contributions every time the Ward identity of Eq. (10.8) is triggered by a pinching momentum. Specifically, the term in Eq. (10.8) proportional to the inverse of the internal fermion propagator gives rise to a propagator-like term whose coupling to the external fermions f and $\bar{f}$ (with $f = f_1, f_2$) is proportional to an effective vertex $C_{W_\alpha f \bar{f}}$ given by (see also Figure 10.2(b))

$$C_{W_\alpha f \bar{f}} = -\mathrm{i}\left(\frac{g_w}{2}\right)\gamma_\alpha P_L. \tag{10.10}$$

Note that this effective vertex is unphysical in the sense that it does not correspond to any of the elementary vertices appearing in the electroweak Lagrangian. However, it can be written as a linear combination of the two *physical* tree-level vertices $\Gamma_{A_\alpha f \bar{f}}$ and $\Gamma_{Z_\alpha f \bar{f}}$ given by

$$\Gamma_{A_\alpha f \bar{f}} = -\mathrm{i}Q_f\gamma_\alpha$$
$$\Gamma_{Z_\alpha f \bar{f}} = -\mathrm{i}[(s_w^2 Q_f - T_z^f)P_L + s_w^2 Q_f P_R], \tag{10.11}$$

as follows:

$$C_{W_\alpha f \bar{f}} = \left(\frac{s_w}{2T_z^f}\right)\Gamma_{A_\alpha f \bar{f}} - \left(\frac{c_w}{2T_z^f}\right)\Gamma_{Z_\alpha f \bar{f}}. \tag{10.12}$$

In the preceding formulas, Q_f is the electric charge of the fermion f and T_z^f is its z-component of the weak isospin. The identity established in Eq. (10.12) allows one to combine the propagator-like parts with the conventional self-energy graphs by writing $1 = d_i(q^2)d_i^{-1}(q^2)$.

Next we will describe how the cancellation of the gauge-fixing parameter proceeds at the one-loop level for the simple case in which f_1 is a charged lepton, to be denoted by ℓ, and f_2 is a neutrino, denoted by v. Of course, on the basis of general field-theoretic principles, one knows in advance that the entire amplitude will be gauge-fixing parameter independent. What is important to recognize, however, is that this cancellation goes through without having to carry out any of the integrations over virtual loop momenta, exactly as happened in the case of QCD. From the practical point of view, the extensive gauge cancellations that are implemented through the pinch technique finally amount to the statement that one may start out in the Feynman gauge, that is, set directly $\xi_w = 1$ and $\xi_z = 1$, with no loss of generality.

The cancellation of ξ_z is easy to demonstrate. The box diagrams containing two Z-bosons (direct and crossed) form a ξ_z-independent subset. The way this works is completely analogous to the QED case, in which the two boxes contain photons: the ξ_z dependence of the direct box cancels exactly against the gauge-fixing parameter dependence of the crossed one. The only other set of graphs with a ξ_z dependence is the self-energy graphs; it is easy to show, by employing the simple algebraic identity

$$\frac{1}{k^2 - \xi_i M_i^2} = \frac{1}{k^2 - M_i^2} - \frac{(1 - \xi_i)M_i^2}{(k^2 - M_i^2)(k^2 - \xi_i M_i^2)}, \tag{10.13}$$

that their sum is independent of ξ_z, separately for ZZ and AZ.

Demonstrating the cancellation of ξ_w is significantly more involved. In what follows, we set $\lambda_w \equiv 1 - \xi_w$ and suppress a factor $g_w^2 \int_k$. We also define

$$I_3 \equiv \left\{ (k^2 - \xi_w M_w^2)(k^2 - M_w^2)[(k+q)^2 - M_w^2] \right\}^{-1}$$

$$I_4 \equiv \left\{ (k^2 - \xi_w M_w^2)[(k+q)^2 - \xi_w M_w^2](k^2 - M_w^2)[(k+q)^2 - M_w^2] \right\}^{-1}. \tag{10.14}$$

Note that terms proportional to q_α or q_β may be dropped directly because the external currents are conserved (massless fermions).

To get a feel of how the pinch technique organizes the various gauge-dependent terms, consider the box graphs shown in Figure 10.3. We have

$$(a) = (a)_{\xi_w=1} + \mathcal{V}_{W^\alpha \ell \bar{\ell}} \left(\lambda_w^2 I_4 k_\alpha k_\beta - 2\lambda_w I_3 g_{\alpha\beta} \right) \mathcal{V}_{W^\beta v \bar{v}}, \tag{10.15}$$

where the vertices $\mathcal{V}$ are defined according to

$$\mathcal{V}_{W_{\alpha ff}} = \bar{v}_f C_{W_{\alpha ff}} u_f$$

$$\mathcal{V}_{V_{\alpha ff}} = \bar{v}_f \Gamma_{V_{\alpha ff}} u_f; \qquad V = A, \, Z. \tag{10.16}$$

The first term on the right-hand side (rhs) of Eq. (10.15) is the pure box, that is, the part that does not contain any propagator-like structures, whereas the second term

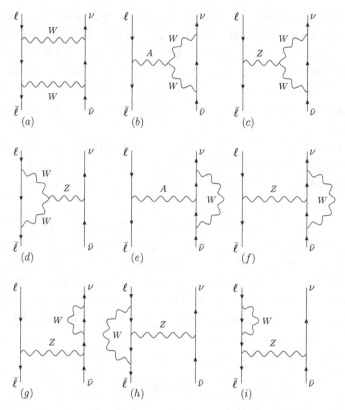

Figure 10.3. The box and vertex diagrams that depend on the gauge-fixing parameter ξ_w.

is the propagator-like contribution that must be combined with the conventional propagator graphs of Figure 10.4. To accomplish this, we employ Eq. (10.12) to write the unphysical vertices $\mathcal{V}_{W\ell\bar{\ell}}$ and $\mathcal{V}_{Wv\bar{v}}$ in terms of the physical vertices, $\mathcal{V}_{A\ell\bar{\ell}}$, $\mathcal{V}_{Z\ell\bar{\ell}}$, and $\mathcal{V}_{Zv\bar{v}}$. Specifically, using that in our case $T_z^\ell = -1/2$ and $T_z^v = 1/2$, we have

$$C_{W^\alpha\ell\bar{\ell}} = -s_w\Gamma_{A^\alpha\ell\bar{\ell}} + c_w\Gamma_{Z^\alpha\ell\bar{\ell}}$$

$$C_{W^\alpha v\bar{v}} = -c_w\Gamma_{Z^\alpha v\bar{v}}. \qquad (10.17)$$

Equation (10.17) determines unambiguously the parts that must be appended to $\Pi_{Z_\alpha Z_\beta}$ and $\Pi_{A_\alpha Z_\beta}$ self-energies. To make this separation manifest, one must do the extra step of writing $d_z(q^2)d_z^{-1}(q^2) = d_A(q^2)d_A^{-1}(q^2) = 1$ to force the external tree-level propagators to appear explicitly (see Figure 10.5). Thus, from the propagator-like part of the box, we finally obtain

$$(a)_{A_\alpha Z_\beta} = s_w c_w q^2(q^2 - M_z^2)\left(\lambda_w^2 I_4 k_\alpha k_\beta - 2\lambda_w I_3 g_{\alpha\beta}\right)$$

$$(a)_{Z_\alpha Z_\beta} = -c_w^2(q^2 - M_z^2)^2\left(\lambda_w^2 I_4 k_\alpha k_\beta - 2\lambda_w I_3 g_{\alpha\beta}\right). \qquad (10.18)$$

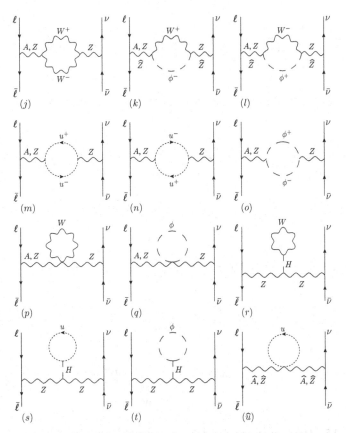

Figure 10.4. The ξ_w-dependent diagrams contributing to $\Pi_{Z_\alpha Z_\beta}$ and $\Pi_{Z_\alpha Z_\beta}$ and the corresponding diagrams in the BFM.

A similar procedure must be followed for the vertex graphs shown in Figure 10.3. Then all propagator-like terms identified from the boxes and the vertex graphs must be added to the conventional self-energy diagrams given in Figure 10.4. At this point, it would be a matter of straightforward algebra to verify that all ξ_w-dependent terms cancel; in doing that, Eq. (10.13) is useful. Of course, this cancellation proceeds completely independently for the ZZ and AZ contributions. Note that the inclusion of the tadpole graphs, namely, diagrams (r), (s), and (t) of Figure 10.4, is crucial for the final cancellation of the gauge-fixing parameter-dependent contributions that do not depend on q^2. Exactly as happened in the QCD case, the gauge-fixing parameter cancellations amount effectively to choosing the Feynman gauge, $\xi_w = \xi_z = 1$.

The next step is to consider the action of the remaining pinching momenta stemming from the three-gauge-boson vertices inside the non-Abelian diagrams (b), (c), and (d) of Figure 10.3, exposed after employing the standard PT decomposition of Eqs. (1.41). The propagator-like contributions that will emerge from the action of

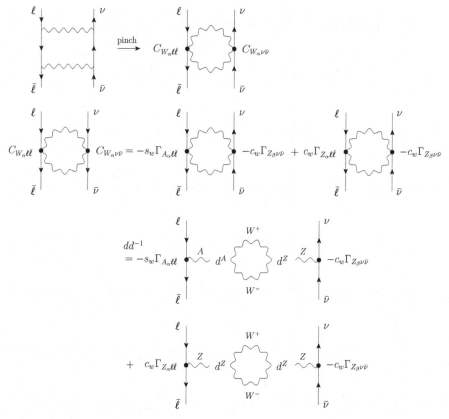

Figure 10.5. The procedure needed for splitting the propagator-like pieces coming from the WW box among the different AZ and ZZ self-energies.

Γ^P must then be reassigned to the conventional self-energy graphs, thus giving rise to the one-loop PT self-energies, which in this case are $\widehat{\Pi}_{Z_\alpha Z_\beta}$ and $\widehat{\Pi}_{A_\alpha Z_\beta}$. The part of the vertex graph containing the Γ^F, together with the Abelian graph which, in the Feynman gauge, remains unchanged, constitutes the one-loop PT vertices $A\nu\bar\nu$, $Z\nu\bar\nu$, and $Z\ell\bar\ell$, to be denoted by $\widehat{\Gamma}_{A\nu\bar\nu}$, $\widehat{\Gamma}_{Z\nu\bar\nu}$, and $\widehat{\Gamma}_{Z\ell\bar\ell}$, respectively.

The PT self-energies $\widehat{\Pi}_{Z_\alpha Z_\beta}$ and $\widehat{\Pi}_{Z_\alpha Z_\beta}$ are simply the sum of all propagator-like contributions, namely,

$$\widehat{\Pi}_{Z_\alpha Z_\beta}(q) = \Pi_{Z_\alpha Z_\beta}^{(\xi_w=1)}(q) + 4g_w^2 c_w^2 (q^2 - M_Z^2) g_{\alpha\beta} I_{WW}(q)$$

$$\widehat{\Pi}_{A_\alpha Z_\beta}(q) = \Pi_{A_\alpha Z_\beta}^{(\xi_w=1)}(q) - 2g_w^2 s_w c_w (2q^2 - M_Z^2) g_{\alpha\beta} I_{WW}(q), \qquad (10.19)$$

with

$$I_{WW}(q) = \int_k \frac{1}{(k^2 - M_w^2)[(k+q)^2 - M_w^2]}. \qquad (10.20)$$

It is relatively straightforward to prove that the ξ_W-independent PT self-energies given in Eqs. (10.19) coincide with their BFM counterparts computed at $\xi_W^Q = 1$. In particular, notice that (1) in the BFM, there is no $\widehat{A}W^{\pm}\phi^{\mp}$ interaction, and therefore graphs (k) and (l) of Figure 10.4 are absent in $\Pi_{\widehat{A}_\alpha \widehat{Z}_\beta}$, and (2) diagram ($\widehat{u}$) of the same figure, corresponding to the characteristic BFM four-field coupling $\widehat{V}\widehat{V}uu$, has been generated dynamically from the simple rearrangement of terms.

With a small extra effort, we can now obtain the closed expressions for $\widehat{\Pi}_{Z_\alpha Z_\beta}$ and $\widehat{\Pi}_{A_\alpha Z_\beta}$ in terms of the Passarino–Veltman functions [4]. We will only focus on the parts of the self-energies originating from Feynman graphs containing W propagators, together with the associated Goldstone boson and ghosts. The contributions coming from the rest of the diagrams (e.g., containing loops with fermions or Z- and H-bosons) are common to the conventional and PT self-energies, i.e., $\Pi_{AZ}^{(ff)} = \widehat{\Pi}_{AZ}^{(ff)}$ and $\Pi_{ZZ}^{(ff)} = \widehat{\Pi}_{ZZ}^{(ff)}$, and we do not report them here. Therefore, the only Passarino–Veltman function that will appear is $B_0(q^2, M_W^2, M_W^2)$.

To that end, one may use the closed expressions for $\Pi_{Z_\alpha Z_\beta}^{(\xi_W=1)}$ and $\Pi_{A_\alpha Z_\beta}^{(\xi_W=1)}$ given by Denner [5] and add to them the pinch terms given in Eqs. (10.19) and (10.20). Using the identity $iB_0(q^2, M_W^2, M_W^2) = 16\pi^2 I_{WW}(q)$, we finally obtain

$$\widehat{\Pi}_{AZ}^{(WW)}(q^2) = \frac{\alpha}{4\pi} \frac{1}{3s_w c_w}$$
$$\times \left\{ \left[\left(21c_w^2 + \frac{1}{2} \right) q^2 + (12c_w^2 - 2)M_W^2 \right] B_0(q^2, M_W^2, M_W^2) \right.$$
$$\left. - (12c_w^2 - 2)M_W^2 B_0(0, M_W^2, M_W^2) + \frac{1}{3}q^2 \right\}$$

$$\widehat{\Pi}_{ZZ}^{(WW)}(q^2) = -\frac{\alpha}{4\pi} \frac{1}{6s_w^2 c_w^2}$$
$$\times \left\{ \left[\left(42c_w^4 + 2c_w^2 - \frac{1}{2} \right) q^2 + (24c_w^4 - 8c_w^2 - 10)M_W^2 \right] \right.$$
$$\times B_0(q^2, M_W^2, M_W^2) - (24c_w^4 - 8c_w^2 + 2)M_W^2 B_0(0, M_W^2, M_W^2)$$
$$\left. + \frac{1}{3}(4c_w^2 - 1)q^2 \right\}. \tag{10.21}$$

Note that $\widehat{\Pi}_{AZ}^{(WW)}(0) = 0$, exactly as happens for the corresponding subset of fermionic corrections: evidently, as a result of the PT rearrangement, bosonic and fermionic radiative corrections are treated on the same footing. This last property is important for phenomenonogical applications, as, for example, the unambiguous definition of the physical charge radius of the neutrinos (see Chapter 11).

10.2.1 The unitary gauge

In the previous sections, we have applied the pinch technique in the framework of the linear renormalizable R_ξ gauges, and we have obtained ξ-independent one-loop self-energies for the gauge bosons. What would happen, however, if one were to work *directly* in the unitary gauge? The unitary gauge is reached after gauging away the would-be Goldstone bosons through an appropriate field redefinition (which, at the same time, corresponds to a gauge transformation) $\phi(x) \rightarrow \phi'(x) = \phi(x)\exp(-i\zeta(x)/v)$, where $\zeta(x)$ denotes generically the Goldstone fields. Note that the unitary gauge is defined completely independently of the R_ξ gauges; of course, operationally, it is identical to the $\xi_W, \xi_Z \rightarrow \infty$ limit of the latter. In particular, in the unitary gauge, the W and Z propagators are given by Eq. (10.5), where $i = W, Z$.

Given that the contributions of unphysical scalars and ghosts cancel in this gauge, the unitarity of the theory becomes *manifest* (hence its name). In the language employed before, *manifest unitarity* means that the optical theorem (a direct consequence of unitarity) holds in its strong version. The most immediate way to realize this is by noticing that the unitary gauge propagators Eq. (10.5) and the expression for the sum over the polarization vectors of a massive spin one vector boson (see Eq. (10.26)) are practically identical.

Since the early days of spontaneously broken non-Abelian gauge theories, the unitary gauge has been known to give rise to nonrenormalizable Green's functions in the sense that their divergent parts cannot be removed by the usual mass and field-renormalization counterterms. It is easy to deduce from the tree-level expressions of the gauge-boson propagators why this happens: the longitudinal contribution in Eq. (10.5) is divided by a squared mass instead of a squared momentum, i.e., $q^\mu q^\nu/M_i^2$ instead of $q^\mu q^\nu/q^2$, and therefore $U_{\mu\nu}^i(q) \sim 1$ as $q \rightarrow \infty$. As a consequence, when $U_{\mu\nu}^i(q)$ is inserted inside quantum loops (and q is the virtual momentum that is being integrated over), it gives rise to highly divergent integrals. If dimensional regularization is applied, this hard short-distance behavior manifests itself in the occurrence of divergences proportional to high powers of q^2. Thus, at one loop, the divergent part of the W or Z self-energies proportional to $g_{\mu\nu}$ has the general form

$$\Pi_{WW}^{\mathrm{div}}(q^2) = \frac{1}{\epsilon}(c_1 q^6 + c_2 q^4 + c_3 q^2 + c_4), \qquad (10.22)$$

where the coefficients c_i, of appropriate dimensionality, depend on the gauge coupling and combinations of M_w^2 and M_z^2. The important point is that, whereas the last two terms on the rhs of Eq. (10.22) can be absorbed into mass and wave-function renormalization, as usual, the first two cannot be absorbed into a redefiniton of the parameters in the original Lagrangian because they are proportional to q^6 and q^4.

As was shown in a series of papers [6, 7, 8], when one puts together the individual Green's functions to form S-matrix elements, an extensive cancellation of all nonrenormalizable divergent terms takes place, and the resulting S-matrix element can be rendered finite through the usual mass and gauge coupling renormalization. Actually, in retrospect, this cancellation is nothing but another manifestation of the pinch technique (of course, the papers mentioned predate the pinch technique). Even though this situation may be considered acceptable from the practical point of view, in the sense that S-matrix elements may still be computed consistently, the inability to define renormalizable Green's functions has always been a theoretical shortcoming of the unitary gauge.

The actual demonstration of how to construct renormalizable Green's functions at one loop starting from the unitary gauge was given in [9]. The methodology is identical to that used in the context of the R_ξ gauges: the propagator-like parts of vertices and boxes are identified and subsequently redistributed among the various gauge-boson self-energies. Evidently, the pinch contributions themselves contain divergent terms proportional to q^6 and q^4, which, when added to the analogous contributions contained in the conventional propagators, cancel exactly. After this cancellation, the remaining terms reorganize themselves in such a way as to give rise exactly to the *unique* PT gauge-boson self-energies, viz. Eqs. (10.21).

10.2.2 Absorptive construction in the electroweak sector

We will now study with an explicit example how the PT subamplitudes of the electroweak theory satisfy the strong version of the optical theorem [10, 11]. As in the case of QCD, the fundamental reason for this may be traced back to a characteristic s-t cancellation operating also in the presence of tree-level symmetry breaking.

Consider the process $f(p_1)\bar{f}(p_2) \to W^+(k_1)W^-(k_2)$, with $q = p_1 + p_2 = k_1 + k_2$ and $s = q^2 = (p_1 + p_2)^2 = (k_1 + k_2)^2 > 4M_W^2$. The corresponding tree-level amplitude, $\mathcal{T}^{\mu\nu}$, is given by two s-channel graphs, one mediated by a photon and the other by a Z-boson, to be denoted by $T_A^{\mu\nu}$ and $T_Z^{\mu\nu}$, respectively, and one t-channel graph, to be denoted by $T_t^{\mu\nu}$, i.e. (see also Figure 10.6),

$$\mathcal{T}^{\mu\nu} = T_{s,A}^{\mu\nu} + T_{s,Z}^{\mu\nu} + T_t^{\mu\nu}, \tag{10.23}$$

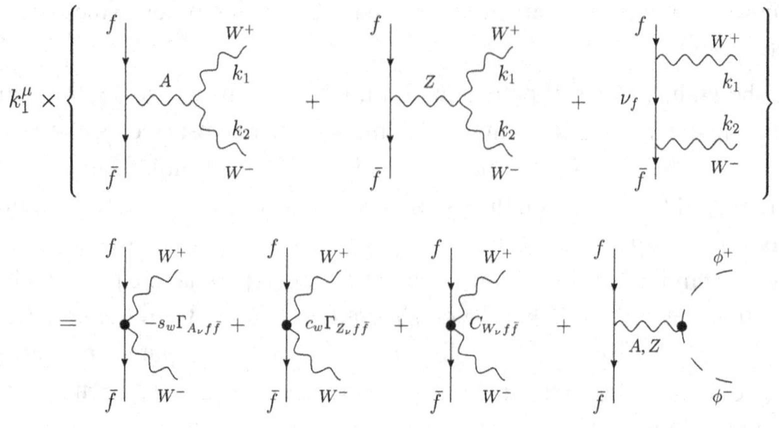

Figure 10.6. The fundamental s-t cancellation for the process $f(p_1)\bar{f}(p_2) \to W^+(k_1)W^-(k_2)$.

where

$$T^{\mu\nu}_{s,A} = -\mathcal{V}_{A^\alpha f\bar{f}}d_A(q^2)g_w s_w \Gamma^{\mu\nu}_\alpha(q,k_1,k_2),$$

$$T^{\mu\nu}_{s,Z} = \mathcal{V}_{Z^\alpha f\bar{f}}d_Z(q^2)g_w c_w \Gamma^{\mu\nu}_\alpha(q,k_1,k_2),$$

$$T^{\mu\nu}_t = -\frac{g^2_w}{2}\bar{v}_f(p_2)\gamma^\mu P_L\, S^{(0)}_{f'}(p_1-k_1)\gamma^\nu P_L u_f(p_1). \tag{10.24}$$

Note that we have already used current conservation to eliminate the (gauge-fixing parameter-dependent) longitudinal parts of the tree-level photon and Z-boson propagators. Then,

$$\mathcal{M} = [T_A + T_Z + T_t]^{\mu\nu}\, L_{\mu\mu'}(k_1)L_{\nu\nu'}(k_2)\left[T^*_A + T^*_Z + T^*_t\right]^{\mu'\nu'}, \tag{10.25}$$

where the polarization tensor $L^{\mu\nu}(k)$ corresponds to a massive gauge boson, that is,

$$L_{\mu\nu}(k) = \sum_{\lambda=1}^3 \varepsilon^\lambda_\mu(k)\varepsilon^\lambda_\nu(k) = -g_{\mu\nu} + \frac{k_\mu k_\nu}{M^2_w}. \tag{10.26}$$

On shell ($k^2 = M^2_w$), we have that $k^\mu L_{\mu\nu}(k) = 0$. Therefore, as in the QCD case, when the two non-Abelian vertices are decomposed as in Eq. (1.41), the Γ^P parts vanish, and only the Γ^F parts contribute in the s-channel graphs; we denote them by $T^{\mu\nu}_{A,F}$ and $T^{\mu\nu}_{Z,F}$, respectively.

Let us then study what happens when $T_{\mu\nu}$ is contracted by a longitudinal momentum, k^μ_1 or k^ν_2, coming from the polarization tensors. Employing the appropriate tree-level Ward identities, we obtain

$$k_{1\mu} T_{A,F}^{\mu\nu} = -s_w \mathcal{V}_{A^\nu f f} + \mathcal{S}_A^\nu,$$

$$k_{1\mu} T_{Z,F}^{\mu\nu} = c_w \mathcal{V}_{Z^\nu f f} + \mathcal{S}_Z^\nu,$$

$$k_{1\mu} T_t^{\mu\nu} = \mathcal{V}_{W^\nu f f}, \tag{10.27}$$

with

$$\mathcal{S}_A^\nu = -\mathcal{V}_{A^\alpha f f} d_A(q^2) g_w s_w (k_1 - k_2)_\alpha k_2^\nu$$

$$\mathcal{S}_Z^\nu = \mathcal{V}_{Z^\alpha f f} d_Z(q^2) g_w c_w \left[(k_1 - k_2)_\alpha k_2^\nu - M_Z^2 g_{\alpha\nu} \right]. \tag{10.28}$$

Adding by parts both sides of Eq. (10.27), we see that a major cancellation takes place: the pieces containing the vertices $\mathcal{V}_{A^\nu f f}$ and $\mathcal{V}_{Z^\nu f f}$ cancel against $\mathcal{V}_{W^\nu f f}$ by virtue of Eq. (10.17), and one is left on the rhs with a purely s-channel contribution, namely,

$$k_{1\mu} T^{\mu\nu} = \mathcal{S}_A^\nu + \mathcal{S}_Z^\nu. \tag{10.29}$$

An exactly analogous cancellation takes place when one contracts with k_2^ν.

After the implementation of the preceding cancellations, we can isolate, e.g., the part of the squared amplitude that is purely s-channel–mediated by a Z-boson, to be denoted by $\widehat{\mathcal{M}}_{ZZ}$. It is composed of the sum of the following terms:

$$\widehat{\mathcal{M}}_{ZZ} = T_Z^F \cdot T_Z^{F*} - 2\frac{\mathcal{S}_Z \cdot \mathcal{S}_Z^*}{M_W^2} + \frac{(k_2 \cdot \mathcal{S}_Z) \cdot (k_2 \cdot \mathcal{S}_Z^*)}{M_W^4}. \tag{10.30}$$

After elementary algebra, we find

$$\widehat{\mathcal{M}}_{ZZ} = \mathcal{V}_{Z_\alpha f f} d_Z(q^2) K_{ZZ}^{\alpha\beta} d_Z(q^2) \mathcal{V}_{Z_\beta f, f}, \tag{10.31}$$

with

$$K_{ZZ}^{\alpha\beta} = -\frac{g_w^2}{c_w^2} \left[\left(8q^2 c_w^4 - 2M_W^2 \right) g^{\alpha\beta} + \left(3c_w^4 - c_w^2 + \frac{1}{4} \right) (k_1 - k_2)^\alpha (k_1 - k_2)^\beta \right]. \tag{10.32}$$

We must then integrate this last expression over the available phase space and isolate its coefficient proportional to $g^{\alpha\beta}$, to be denoted by K_{ZZ}. Then, if the optical theorem holds at the level of the $\widehat{\Pi}_{ZZ}^{(WW)}(q)$, we should have

$$2\Im m \widehat{\Pi}_{ZZ}^{(WW)}(q) = K_{ZZ}, \tag{10.33}$$

where the left-hand side (lhs) of Eq. (10.33) must be obtained from Eq. (10.21) by taking its imaginary part.

Using for the lhs the elementary result

$$\Im m\, B_0(q^2, M_W^2, M_W^2) = 8\pi^2 \int_{\text{PS}_{WW}},$$ (10.34)

with $\int_{\text{PS}_{WW}}$ the two Ws phase space integral (see Section 1.7.2), and for the rhs that

$$\int_{\text{PS}_{WW}} (k_1 - k_2)^\alpha (k_1 - k_2)^\beta = -\frac{1}{3}(q^2 - 4M_W^2)g^{\alpha\beta} \int_{\text{PS}_{WW}},$$ (10.35)

it is easy to verify that Eq. (10.33) is indeed satisfied.

At this point, one could go a step further and employ a twice-subtracted dispersion relation to reconstruct the real part of $\widehat{\Pi}_{ZZ}^{(WW)}(q)$. The end result of this procedure will coincide with the corresponding expression obtained from Eq. (10.21) after appropriate renormalization (for a detailed derivation, see [11]).

Finally, we return to the nonrenormalizability of the unitary gauges, now seen from the absorptive point of view. As mentioned in the previous subsection, in the unitary gauge, the strong version of the optical theorem is satisfied; in relation to this section, what this means is that the optical theorem is satisfied diagram by diagram, without having to resort explicitly to the s-t cancellation. For example, the imaginary part of the conventional self-energy $\Pi_{ZZ}^{(WW)}(s)$ in the unitary gauge is given by

$$\Im m\, \Pi_{ZZ}^{(WW)}(s) \sim (s - M_Z)^2 \int_{\text{PS}_{WW}} T_Z^{\mu\nu} L_{\mu\mu'}(k_1) L_{\nu\nu'}(k_2) T_Z^{*\mu'\nu'}.$$ (10.36)

What is the price one pays for *not* implementing explicitly the s-t cancellation? Simply, the conventional subamplitudes, such as the one given earlier, contain terms that grow as s^2 or as s^3 (see, e.g., [11, 12]); indeed, the s-t cancellation eliminates precisely terms of this type. Consequently, if one were to substitute the $\Im m\, \Pi_{ZZ}^{(WW)}(s)$ obtained from the rhs of Eq. (10.36) into a twice-subtracted dispersion relation – the maximum number of subtractions allowed by renormalizability – ultraviolet divergent real parts proportional to q^4 or q^6 would be encountered. Of course, these are precisely the nonrenormalizable terms encountered in Eq. (10.22), now obtained not from a direct one-loop calculation but rather from the combined use of unitarity and analyticity.

10.2.3 Background field method away from $\xi_Q = 1$: physical versus unphysical thresholds

As mentioned already, the fact that the BFM Green's functions satisfy the same QED-like Ward identities for every value of the quantum gauge-fixing parameter ξ_Q does not mean that the PT Green's functions, reproduced from the background field method at $\xi_Q = 1$, are simply some among an infinity of physically equivalent choices, parametrized by ξ_Q. This interpretation is not correct: the BFM Green's functions obtained away from $\xi_Q = 1$ are not physically equivalent to the privileged case of $\xi_Q = 1$. The following basic observation clarifies this point beyond any doubt: for $\xi_Q \neq 1$, the imaginary parts of the BFM electroweak self-energies include terms with unphysical thresholds [11, 10]. For example, for the one-loop contributions of the W and its associated would-be Goldstone boson and ghost to $\tilde{\Pi}_{ZZ}^{(WW)}(\xi_Q, s)$, one obtains

$$\Im m\,\tilde{\Pi}_{ZZ}^{(WW)}(s, \xi_Q) = \Im m\,\widehat{\Pi}_{ZZ}^{(WW)}(s) + \frac{\alpha}{24 s_w^2 c_w^2}\left(\frac{s - M_Z^2}{s M_Z^4}\right)$$
$$\times \left[W_1(s) + W_2(s, \xi_Q) + W_3(s, \xi_Q)\right], \qquad (10.37)$$

with

$$W_1(s) = f_1(s)\theta(s - 4M_w^2)$$
$$W_2(s, \xi_Q) = f_2(s, \xi_Q)\lambda^{1/2}(s, \xi_Q M_w^2, \xi_Q M_w^2)\theta(s - 4\xi_Q M_w^2)$$
$$W_3(s, \xi_Q) = f_3(s, \xi_Q)\lambda^{1/2}(s, M_w^2, \xi_Q M_w^2)\theta(s - M_w^2(1 + \sqrt{\xi_Q})^2), \qquad (10.38)$$

where $\lambda(x, y, z) = (x - y - z)^2 - 4yz$ and

$$f_1(s) = \left(8M_w^2 + s\right)\left(M_Z^2 + s\right) + 4M_w^2\left(4M_w^2 + 3M_Z^2 + 2s\right)$$
$$f_2(s, \xi_Q) = f_1(s) - 4\left(\xi_Q - 1\right)M_w^2\left(4M_w^2 + M_Z^2 + s\right)$$
$$f_3(s, \xi_Q) = -2\left[8M_w^2 + s - 2\left(\xi_Q - 1\right)M_w^2 + \left(\xi_Q - 1\right)^2\frac{M_w^4}{s}\right]\left(M_Z^2 + s\right).$$

$$(10.39)$$

These gauge-dependent unphysical thresholds (see the arguments of the θ functions) are artifacts of the BFM gauge-fixing procedure; in the calculation of any physical process, they cancel exactly against unphysical contributions from the imaginary parts of the one-loop vertices and boxes. After these cancellations have been implemented, one is left just with the contribution proportional to the tree-level cross section for the on-shell physical process $f\bar{f} \to W^+W^-$, with thresholds only at $q^2 = 4M_w^2$. In fact, by obtaining in the previous subsection the full W-related contribution to the PT self-energy, namely, $\widehat{\Pi}_{WW}^{(ZZ)}(s)$, directly from the on-shell physical process $f\bar{f} \to W^+W^-$, we have shown explicitly that in the background field method at $\xi_Q = 1$, the thresholds that occur at $q^2 = 4M_w^2$ are due solely to the physical W^+W^- pair. We therefore conclude that the particular value $\xi_Q = 1$ in the background field method is distinguished on physical grounds from all other values of ξ_Q.

10.3 Nonconserved currents and Ward identities

We now discuss some important issues related to the application of the pinch technique when the fermions are massive, as discussed in Section 2.2.5. Consider the elastic process $e^-(r_1)\nu_e(p_1) \to e^-(p_2)\nu_e(r_2)$, and concentrate on the charged channel, which, at tree level, is shown in Figure 10.7. The momentum transfer q is defined as $q = p_1 - p_2 = r_2 - r_1$, and we will consider the electrons to be massive, with a mass m_e, whereas the neutrinos will be treated for simplicity as massless. The tree-level propagators of the W and the corresponding Goldstone boson are those given in Eq. (10.1) and Eq. (10.3) (for $i = W$); the index W will be suppressed in what follows. The elementary vertices describing the coupling of the charged bosons with the external fermions are $\Gamma_\alpha \equiv \Gamma_{W_\alpha^+ \bar\nu_e e} = \Gamma_{W_\alpha^- \bar e \nu_e}$, $\Gamma_+ \equiv \Gamma_{\phi^+ \bar\nu_e e}$, and $\Gamma_- \equiv \Gamma_{\phi^- \bar e \nu_e}$ and are given by

$$\Gamma_\alpha = \frac{ig_w}{\sqrt{2}} \gamma_\alpha P_L; \qquad \Gamma_{+(-)} = -\frac{ig_w}{\sqrt{2}} \frac{m_e}{M_w} P_{R(L)}. \tag{10.40}$$

When sandwiched between the external spinors, they are denoted by $\Gamma_1^\alpha = \bar u_{\nu_e}(r_2) \Gamma^\alpha u_e(r_1)$, $\Gamma_2^\alpha = \bar u_e(p_2) \Gamma^\alpha u_{\nu_e}(p_1)$, $\Gamma_1 = \bar u_{\nu_e}(r_2) \Gamma_+ u_e(r_1)$, and $\Gamma_2 = \bar u_e(p_2) \Gamma_- u_{\nu_e}(p_1)$. The elementary identities

$$q_\alpha \Gamma_{1,2}^\alpha = M_w \Gamma_{1,2}$$
$$i\Gamma_{1,2} = M_w q^\beta \Delta_{\beta\alpha}(q) \Gamma_{1,2}^\alpha + q^2 D(q) \Gamma_{1,2}, \tag{10.41}$$

valid for every ξ, are also useful.

We will start by considering the S-matrix at tree level (Figure 10.7), to be denoted by T_0:

$$T_0 = \Gamma_1^\alpha \Delta_{\alpha\beta}(q) \Gamma_2^\beta + \Gamma_1 D(q) \Gamma_2. \tag{10.42}$$

Of course, T_0 must be ξ independent, and it is easy to demonstrate that this is indeed so. Using Eqs. (10.4) and (10.41), it is elementary to verify that T_0 can be written as

$$T_0 = \Gamma_1^\alpha U_{\alpha\beta}(q) \Gamma_2^\beta. \tag{10.43}$$

Thus, even though one works in the R_ξ gauge, making no assumption on the value of ξ (in particular, not taking the limit $\xi \to \infty$), one is led effectively to the unitary gauge, with no (unphysical) would-be Goldstone bosons present.

There is an alternative way of writing T_0 that makes manifest the role of the massless Goldstone bosons. It is well known that the Ward identities (or Slavnov–Taylor identities) in theories with spontaneous symmetry breaking maintain the same form as in the unbroken theory at the expense of introducing massless longitudinal poles. The role of these massless poles is obscured because, through the process of gauge

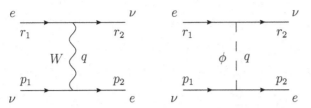

Figure 10.7. The process $e\nu_e \to e\nu_e$ at tree level in the standard model.

fixing, they can be changed to poles of arbitrary mass (as explained earlier). These massless poles do not appear in the S-matrix to the extent that they are absorbed by gauge bosons. However, simple algebra can recast the tree-level amplitude into a form in which the presence of the massless poles becomes manifest. Specifically, using the algebraic identity

$$\frac{1}{M^2} = \frac{1}{q^2} + \frac{q^2 - M^2}{q^2 M^2},$$ (10.44)

we can write $U_{\alpha\beta}(q)$ as

$$U_{\alpha\beta}(q) = P_{\alpha\beta}(q)d_w(q^2) + \frac{q_\alpha q_\beta}{M_w^2}\frac{i}{q^2},$$ (10.45)

where we have used the transverse projector $P_{\alpha\beta}(q)$ defined in Eq. (1.26). Then Eq. (10.43) can be rewritten as

$$T_0 = \Gamma_1^\alpha P_{\alpha\beta}(q)d_w(q^2)\Gamma_2^\beta + \Gamma_1 \frac{i}{q^2}\Gamma_2.$$ (10.46)

It turns out that the PT rearrangement of the physical amplitude allows the generalization of Eq. (10.43) and Eq. (10.46) to higher orders. To see how this happens, assume that the PT procedure has been carried out as usual (with the additional operational complications mentioned earlier), giving rise to the gauge-fixing parameter-independent self-energies $\Pi_{W_\alpha W_\beta}$, $\Pi_{W_\alpha \phi}$, $\Pi_{\phi W_\beta}$, and $\Pi_{\phi\phi}$, to be denoted by $\widehat{\Pi}_{\alpha\beta}$, $\widehat{\Theta}_\alpha$, $\widehat{\Theta}_\beta$, and $\widehat{\Omega}$, respectively (Figure 10.8). During this construction, it becomes clear that the cancellations of the gauge-fixing parameter-dependent loop contributions proceed without interfering with the the gauge-fixing parameter dependence of the tree-level propagators connecting the one-loop graphs to the external fermions. Of course, any residual gauge-fixing parameter dependence coming from these tree-level propagators (see Eqs. (10.1) and (10.3)) must also cancel to obtain fully gauge-fixing parameter-independent subamplitudes $\widehat{T}_1$ and $\widehat{T}_2$ ($\widehat{T}_3$, being boxlike, does not have external propagators and is already fully gauge-fixing parameter independent). It turns out that, quite remarkably, the requirement of this final gauge-fixing parameter cancellation imposes a set of nontrivial Ward identities on the one-loop PT self-energies and vertices [1, 3].

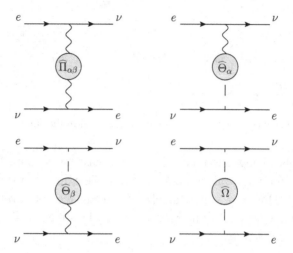

Figure 10.8. The ξ-independent PT self-energies (grey blobs); the tree-level propagators are still ξ-dependent. Requiring that any gauge-fixing parameter dependence coming from these tree-level propagators must cancel imposes a set of non-trivial Ward identities on the one-loop PT self-energies (and vertices).

We see how the Ward identities for the self-energies are derived from the requirement of the full gauge-fixing parameter independence of $\widehat{T}_1$. Neglecting tadpole contributions, we have that $\widehat{T}_1$ is given by

$$\widehat{T}_1 = \Gamma_1^\mu \Delta_{\mu\alpha}(q)\widehat{\Pi}^{\alpha\beta}(q)\Delta_{\beta\nu}(q)\Gamma_2^\nu + \Gamma_1 D(q)\,\widehat{\Omega}(q)\,D(q)\Gamma_2$$
$$+ \Gamma_1^\mu \Delta_{\mu\alpha}(q)\widehat{\Theta}^\alpha(q)D(q)\Gamma_2 + \Gamma_1 D(q)\widehat{\Theta}^\beta(q)\Delta_{\beta\nu}(q)\Gamma_2^\nu, \quad (10.47)$$

or, after using Eq. (10.4),

$$\widehat{T}_1 = \Gamma_1^\mu \left[U_{\mu\alpha}(q) - \frac{q_\mu q_\alpha}{M_w^2} D(q) \right] \widehat{\Pi}^{\alpha\beta}(q) \left[U_{\beta\nu}(q) - \frac{q_\beta q_\nu}{M_w^2} D(q) \right] \Gamma_2^\nu$$

$$+ \Gamma_1^\mu \left[U_{\mu\alpha}(q) - \frac{q_\mu q_\alpha}{M_w^2} D(q) \right] \widehat{\Theta}^\alpha(q)D(q)\Gamma_2$$

$$+ \Gamma_1 D(q)\, \widehat{\Theta}^\beta(q) \left[U_{\beta\nu}(q) - \frac{q_\beta q_\nu}{M_w^2} D(q) \right] \Gamma_2^\nu + \Gamma_1 D(q)\widehat{\Omega}(q)D(q)\Gamma_2.$$

$$(10.48)$$

This way of writing $\widehat{T}_1$ has the advantage of isolating all residual ξ dependence inside the propagators $D(q)$. Demanding that $\widehat{T}_1$ be ξ independent, we obtain as a condition for the cancellation of the quadratic terms in $D(q)$

$$q^\beta q^\alpha \widehat{\Pi}_{\alpha\beta}(q) - 2M_w q^\alpha \widehat{\Theta}_\alpha(q) + M_w^2 \widehat{\Omega}(q) = 0, \quad (10.49)$$

whereas for the cancellation of the linear terms, we must have

$$q^\alpha \widehat{\Pi}_{\alpha\beta}(q) - M_W \widehat{\Theta}_\beta(q) = 0. \tag{10.50}$$

From Eqs. (10.49) and (10.50), it follows that

$$q^\beta q^\alpha \widehat{\Pi}_{\alpha\beta}(q) = M_w^2 \widehat{\Omega}(q) \tag{10.51}$$

$$q^\alpha \widehat{\Theta}_\alpha(q) = M_w \widehat{\Omega}(q). \tag{10.52}$$

Equations (10.49) and (10.52) are the announced Ward identities. To be sure, they are identical to those obtained formally within the background field method but are derived through a procedure that has no apparent connection with the latter; indeed, all that one evokes is the full gauge-fixing parameter independence of the S-matrix. Applying an identical procedure for $\widehat{T}_2$, one obtains the corresponding Ward identity relating the higher-order PT vertices $\widehat{\Gamma}_\alpha$ and $\widehat{\Gamma}_\pm$.

Finally, the gauge-fixing parameter–independent $\widehat{T}_1$ is given by

$$\widehat{T}_1 = \Gamma_1^\mu U_{\mu\alpha}(q) \, \widehat{\Pi}^{\alpha\beta}(q) U_{\beta\nu}(q) \Gamma_2^\nu. \tag{10.53}$$

Notice that Eq. (10.53) is the higher-order generalization of Eq. (10.43).

We can now use the Ward identities derived previously to reformulate the S-matrix in a very particular way; specifically, we will show that the higher-order physical amplitude given earlier may be cast in the tree-level form of Eq. (10.46). Such a reformulation gives rise to a new transverse gauge-fixing parameter-independent W self-energy $\widehat{\Pi}_{\alpha\beta}^t$ with a gauge-fixing parameter-independent longitudinal part, exactly as in Eq. (10.46). Of course, the cost of such a reformulation is the appearance of massless Goldstone poles in our expressions. However, inasmuch as both the old and new quantities originate from the same unique S-matrix, all poles introduced by this reformulation will cancel against each other because the S-matrix contains no massless poles to begin with.

To see how this works out, write $\widehat{\Theta}_\alpha$ in the form

$$\widehat{\Theta}_\alpha(q) = q_\alpha \widehat{\Theta}(q); \tag{10.54}$$

from Eq. (10.52), it follows that

$$\widehat{\Theta}(q) = \frac{M_w}{q^2} \widehat{\Omega}(q). \tag{10.55}$$

Then we can define $\widehat{\Pi}_{\alpha\beta}^t(q)$ in terms of $\widehat{\Pi}^{\alpha\beta}(q)$ and $\widehat{\Theta}(q)$ as follows:

$$\widehat{\Pi}_{\alpha\beta}^t(q) = \widehat{\Pi}_{\alpha\beta}(q) - \frac{q_\alpha q_\beta}{q^2} M_w \widehat{\Theta}(q). \tag{10.56}$$

Evidently $\widehat{\Pi}^t_{\alpha\beta}(q)$ is transverse, e.g., $q^\alpha \widehat{\Pi}^t_{\alpha\beta}(q) = q^\beta \widehat{\Pi}^t_{\alpha\beta}(q) = 0$. Moreover, using Eqs. (10.50) and (10.55),

$$\widehat{\Pi}^t_{\alpha\beta}(q) = P_{\alpha\mu}(q)\widehat{\Pi}^{\mu\nu}(q)P_{\beta\nu}(q). \tag{10.57}$$

We may now reexpress $\widehat{T}_1$ of Eq. (10.53) in terms of $\widehat{\Pi}^t_{\alpha\beta}$ and $\widehat{\Omega}$; using Eq. (10.45) and (10.41), we have

$$\widehat{T}_1 = \Gamma^\alpha_1 d_w(q^2)\widehat{\Pi}^t_{\alpha\beta}(q)d_w(q^2)\Gamma^\beta_2 + \Gamma_1\frac{\mathrm{i}}{q^2}\widehat{\Omega}(q)\frac{\mathrm{i}}{q^2}\Gamma_2. \tag{10.58}$$

Equation (10.58) is the generalization of Eq. (10.46): $\widehat{T}_1$ is the sum of two self-energies, one corresponding to a *transverse massive vector field* and one to a *massless Goldstone boson*. It is interesting to notice that the preceding rearrangements have removed the mixing terms $\widehat{\Theta}_\alpha$ and $\widehat{\Theta}_\beta$ between W and ϕ, thus leading to the generalization of the well-known tree-level property of the R_ξ gauges to higher orders. It is important to emphasize again that the massless poles in the preceding expressions would not have appeared had we not insisted on the transversality of the W self-energy (or the vertex); notice in particular that they are not related to any particular gauge choice, such as the Landau gauge ($\xi = 0$). A completely analogous procedure may be followed for the one-loop (and beyond) vertex [1], yielding the corresponding Abelian-like Ward identity; as in the case of the self-energy studied earlier, the Ward identity of the vertex is realized by means of massless Goldstone bosons.

10.4 The all-order construction

The all-order extension of the pinch technique in theories with spontaneous symmetry breaking can be achieved by resorting to the same algorithm described in Chapter 3 for QCD; however, now the construction is significantly more involved. First, depending on the nature of the line carrying the physical momentum q, there are four vertices to be constructed: two for the gauge-boson sector (charged ($W^\pm$) and neutral (A, Z)) and, as described in Section 10.1, two for the scalar sector (charged ($\phi^\pm$) and neutral (χ, H)). In addition, the BRST symmetry, and therefore the Slavnov–Taylor identities, are now realized through Goldstone bosons; thus the different identities one needs to derive will have a richer structure than that shown in the QCD case (Eq. (3.5)). Finally, the comparison of the PT Green's functions with those of the background Feynman gauge is more laborious because of the proliferation of couplings, e.g., tri- and quadrilinear mixed gauge-boson–scalar vertices.

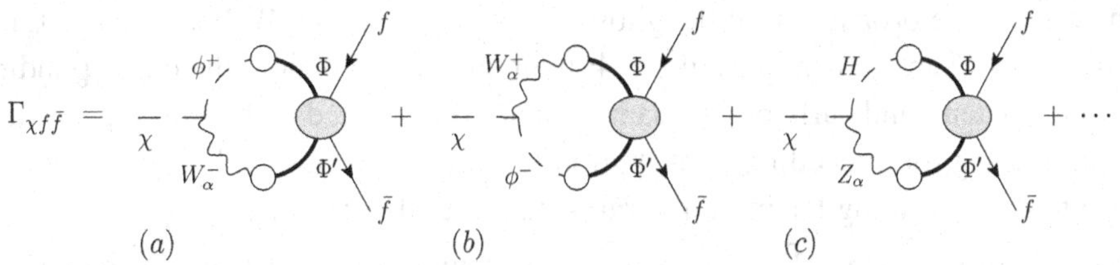

Figure 10.9. The subset of all the diagrams that contribute to the vertex $\Gamma_{\chi f\bar{f}}$ (in the R_ξ gauge) and receive the action of pinching momenta. Here Φ and Φ' denote all fields allowed by the different couplings. The ellipses represent all remaining graphs, where pinching cannot take place. Finally, Bose-symmetric terms are not shown.

We illustrate some of the preceding points through a specific example, namely, the construction of the vertex $\widehat{\Gamma}_{\chi f\bar{f}}$. This case is particularly instructive because it exposes a new type of PT-driven cancellation not encountered so far in the book.

As usual, we choose the Feynman gauge as our starting point, thus receiving pinching momenta only from the tree-level vertices. Figure 10.9 shows all the graphs that must be contracted by the longitudinal momentum contained in the term $\Gamma^P_\alpha(q, k_1, k_2)$. We will concentrate only on diagram (a), which is proportional to a tree-level $\Gamma^{(0)}_{\chi\phi^+W_\alpha^-}$ vertex. Diagram (b), proportional to $\Gamma^{(0)}_{\chi W_\alpha^+\phi^-}$, will give rise to very similar structures, whereas diagram (c), proportional to $\Gamma^{(0)}_{\chi HZ_\alpha}$, does not mix with the previous two, and the corresponding analysis can be carried out separately.

The pinching part of diagram (a) reads

$$(a)^P = \frac{g_w}{2} \int k_2^\alpha \left[\mathcal{T}_{\phi^+W_\alpha^-}(k_1, k_2) \right]_{t,I}, \tag{10.59}$$

where $\mathcal{T}_{\phi^+W_\alpha^-}$ represents the amplitude $\phi^+W_\alpha^- \to f\bar{f}$, where the fermions are taken to be on shell. Then, similar to the QCD case, the pinching action amounts to the replacement

$$k_2^\alpha \left[\mathcal{T}_{\phi^+W_\alpha^-} \right]_{t,I} \to \left[k_2^\alpha \mathcal{T}_{\phi^+W_\alpha^-} \right]^{PT}_{t,I} = \left[\mathcal{S}_{\phi^+W^-} \right]_{t,I}. \tag{10.60}$$

In this case, the (on-shell) Slavnov–Taylor identity gives [13]

$$\begin{aligned}
\mathcal{S}_{\phi^+W^-}(k_1, k_2) = {} & M_w D_c(k_1) D_c(k_2) \mathcal{G}_{\bar{c}^+c^-}(k_1, k_2) \\
& - M_w D_w(k_1) D_w(k_2) \mathcal{G}_{\phi^+\phi^-}(k_1, k_2) \\
& + \frac{g_w}{2} D_c(k_2) \left[i\mathcal{Q}_{\{\chi c^+\}c^+}(k_1, k_2) + \mathcal{Q}_{\{Hc^+\}c^+}(k_1, k_2) \right] \\
& - g_w s_w D_c(k_2) \mathcal{Q}_{\{\phi^+c^A\}c^+}(k_1, k_2) \\
& + g_w \frac{c_w^2 - s_w^2}{2c_w} D_c(k_2) \mathcal{Q}_{\{\phi^+c^Z\}c^+}(k_1, k_2), \tag{10.61}
\end{aligned}$$

where D_c is the propagator of the ghosts associated with the W-bosons. The functions $\mathcal{G}$ and $\mathcal{Q}$ are shown in Figure 3.4; their subindices denote the corresponding particle content, and curly brackets enclose fields evaluated at the same space-time point. Adding the preceding contribution (and the ones not explicitly considered here) to the remaining PT-inert diagrams furnishes the PT vertex $\widehat{\Gamma}_{\chi f \bar{f}}$.

One can next compare $\widehat{\Gamma}_{\chi f \bar{f}}$ with the background Feynman gauge vertex $\Gamma_{\widehat{\chi} f \bar{f}}$. First, it is fairly straightforward to show that the last three terms in Eq. (10.61) will generate the needed ghost-scalar quadrilinear couplings ($\widehat{\chi} H \bar{c}^- c^+$, $\widehat{\chi}\chi\bar{c}^- c^+$, $\widehat{\chi}\phi^+ \bar{c}^- c^A$, and $\widehat{\chi}\phi^+ \bar{c}^- c^Z$).

The remaining PT terms must be appropriately combined with some of the other R_ξ diagrams contributing to $\Gamma_{\chi f \bar{f}}$. Consider first the trilinear scalar-ghost sector. In the R_ξ gauge, it reads

$$-\frac{g_w}{2} M_W \int_{k_2} D_c(k_1) D_c(k_2) \left[\mathcal{G}_{\bar{c}^+ c^-}\right]_{t,I}, \qquad (10.62)$$

and cancels precisely against the corresponding PT term (first term in Eq. (10.61)). This is the new type of PT cancellation mentioned earlier: within the PT, the absence of tree level coupling between a background field $\widehat{\chi}$ and two ghosts (see, e.g., Denner et al. [14]) is obtained dynamically. Similarly, had we chosen to construct the PT Higgs-fermion vertex $\widehat{\Gamma}_{H f \bar{f}}$ instead, these two terms would have added up, furnishing the correct background Higgs-ghost coupling.

Finally, consider the scalar-scalar trilinear sector. The second term in Eq. (10.61) gives a contribution to an effective PT vertex of the type $\chi\phi^+\phi^-$. A similar contribution is generated from Figure 10.9(b), whose pinching action is proportional to $[k_1^\alpha \mathcal{T}_{W_\alpha^+ \phi^-}(k_1, k_2)]_{t,I} = [\mathcal{S}_{W^+\phi^-}]_{t,I}$. It turns out that these two contributions exactly cancel out in accordance with the absence of the (tree-level) couplings $\chi\phi^+\phi^-$ and $\widehat{\chi}\phi^+\phi^-$. Once again, when constructing the Higgs vertex $\widehat{\Gamma}_{H f \bar{f}}$, the two contributions would add up, thus providing the correct BFM coupling $\widehat{H}\phi^+\phi^-$ when summed with the corresponding R_ξ diagram. All remaining diagrams are identical to those of the background Feynman gauge because of the equality of the corresponding tree-level couplings. This concludes our (partial) proof of the equality $\widehat{\Gamma}_{\chi f \bar{f}} = \Gamma_{\widehat{\chi} f \bar{f}}$.

References

[1] J. Papavassiliou, Gauge invariant proper self-energies and vertices in gauge theories with broken symmetry, *Phys. Rev.* **D41** (1990) 3179.
[2] G. Degrassi and A. Sirlin, Gauge invariant self-energies and vertex parts of the standard model in the pinch technique framework, *Phys. Rev.* **D46** (1992) 3104.

[3] J. Papavassiliou, Gauge independent transverse and longitudinal self-energies and vertices via the pinch technique, *Phys. Rev.* **D50** (1994) 5958.

[4] G. Passarino and M. J. G. Veltman, One loop corrections for $e^+ e^-$ annihilation into $\mu^+ \mu^-$ in the Weinberg model, *Nucl. Phys.* **B160** (1979) 151.

[5] A. Denner, Techniques for calculation of electroweak radiative corrections at the one-loop level and results for W physics at LEP-200, *Fortschr. Phys.* **41** (1993) 307.

[6] S. Weinberg, Physical processes in a convergent theory of the weak and electromagnetic interactions, *Phys. Rev. Lett.* **27** (1971) 1688.

[7] S. Y. Lee, Finite higher-order weak and electromagnetic corrections to the strangeness-conserving hadronic beta decay in Weinberg's theory of weak interactions, *Phys. Rev.* **D6** (1972) 1803.

[8] T. Appelquist and H. R. Quinn, Divergence cancellations in a simplified weak interaction model, *Phys. Lett.* **B39** (1972) 229.

[9] J. Papavassiliou and A. Sirlin, Renormalizable W self-energy in the unitary gauge via the pinch technique, *Phys. Rev.* **D50** (1994) 5951.

[10] J. Papavassiliou and A. Pilaftsis, Gauge-invariant resummation formalism for two point correlation functions, *Phys. Rev.* **D54** (1996) 5315.

[11] J. Papavassiliou, E. de Rafael, and N. J. Watson, Electroweak effective charges and their relation to physical cross sections, *Nucl. Phys.* **B503** (1997) 79.

[12] W. Alles, C. Boyer, and A. J. Buras, W boson production in $e^+ e^-$ collisions in the Weinberg-Salam model, *Nucl. Phys.* **B119** (1977) 125.

[13] D. Binosi, Electroweak pinch technique to all orders, *J. Phys.* **G30** (2004) 1021.

[14] A. Denner, G. Weiglein, and S. Dittmaier, Application of the background field method to the electroweak standard model, *Nucl. Phys.* **B440** (1995) 95.

11

Other applications of the pinch technique

11.1 Introduction

The pinch technique (PT) makes it possible to understand many questions in a variety of gauge theories gauge invariantly, as we have already seen for QCD in Chapters 7, 8, and 9 and for the electroweak sector of the standard model (SM) theory in Chapter 10. This chapter goes into more detail on some physical questions in electroweak theory and thermal NAGTs[1] that were difficult to interpret in the conventional framework of Feynman graphs and sometimes resulted in unfounded attempts to find physical properties in gauge-dependent calculations. Different authors used different gauges and got different results, sometimes not even agreeing on the sign. The pinch technique has resolved these issues. We also mention some interesting results for NAGTs embedded in supersymmetric theories, where the pinch technique confirms a number of supersymmetry nonrenormalization relations among the contributions of scalars and fermions to the PT three-gluon vertex, already discussed in Chapter 2. At the level of off-shell Green's functions, these relations hold only for the pinch technique and not for the conventional gauge-dependent Green's functions.

In this chapter, we cover the following subjects: (1) non-Abelian effective charges, (2) physical renormalization schemes versus $\overline{\text{MS}}$, (3) non-Abelian off-shell form factors, (4) the neutrino charge radius, (5) making gauge-particle resonance widths gauge invariant, (6) finite-temperature NAGTs, and (7) hints of supersymmetry in the PT Green's functions.

11.2 Non-Abelian effective charges

The extension of the concept of the effective charge from QED to non-Abelian gauge theories is, as the reader has already appreciated in Chapter 6, of

---

[1] The three-dimensional NAGTs of Chapter 9 carry the nonperturbative infrared singularities of finite-temperature gauge theories in four dimensions.

fundamental interest. This concept is even more important in theories involving unstable particles – for example, in the SM electroweak sector or disparate energy scales (e.g., grand unified theories). In the former, the Dyson summation of (appropriately defined) self-energies is needed to regulate the kinematic singularities of the corresponding tree-level propagators in the vicinity of resonances. In the latter, instead, the extraction of accurate low-energy predictions requires an exact treatment of threshold effects due to heavy particles: the construction of an effective charge, valid for all momenta and not just the asymptotic regime governed by the β function, constitutes the natural way to account for such threshold effects. As we know from Chapter 6, because the pinch technique cures the problem of the gauge-fixing parameter dependence of the conventionally defined gauge-boson self-energies, it is an ideal tool for the definition of physical effective charges in NAGTs.

11.2.1 Electroweak effective charges

In the electroweak sector of the SM, the various PT self-energies organize themselves into appropriate renormalization group (RG)-invariant combinations, essentially for the same fundamental reasons as in QCD [1]. The effective weak mixing angle $\bar{s}_w^2(q^2)$ corresponds to the RG-invariant combination

$$\bar{s}_w^2(q^2) = (s_w^0)^2 \left[1 - \left(\frac{c_w^0}{s_w^0} \right) \frac{\widehat{\Pi}_{AZ}^0(q^2)}{q^2 + \widehat{\Pi}_{AA}^0(q^2)} \right] = s_w^2 \left[1 - \left(\frac{c_w}{s_w} \right) \frac{\widehat{\Pi}_{AZ}(q^2)}{q^2 + \widehat{\Pi}_{AA}(q^2)} \right].$$

$$(11.1)$$

Using the fact that $\widehat{\Pi}_{AZ}(0) = 0$, we may write $\widehat{\Pi}_{AZ}(q^2) = q^2 \widehat{\Pi}_{AZ}(q^2)$; then, at the one-loop level, $\bar{s}_w^2(q^2)$ reduces to

$$\bar{s}_w^2(q^2) = s_w^2 \left[1 - \left(\frac{c_w}{s_w} \right) \widehat{\Pi}_{AZ}(q^2) \right].$$

$$(11.2)$$

Evidently, $\bar{s}_w^2(q^2)$ constitutes a universal modification to the effective vertex of the charged fermion.

Similarly, one may demonstrate that the combinations

$$g_w^2 \widehat{\Delta}_{WW}(q^2); \qquad \frac{g_w^2}{c_w^2} \widehat{\Delta}_{ZZ}(q^2); \qquad \frac{g_w^2}{M_W^2} \widehat{\Delta}_H(q^2),$$

$$(11.3)$$

are RG-invariant. The analog of Eq. (6.53) may be defined for the first two combinations. Specifically, retaining only the real parts of the corresponding self-energies, and casting $\Re e \widehat{\Pi}(q^2)_{ii}$ in the form $\Re e \widehat{\Pi}_{ii}(q^2) = \Re e \widehat{\Pi}_{ii}(M_i^2) + (q^2 - M_i^2) \Re e \widehat{\Pi}_{ii}(M_i^2)$, and then pulling out a common factor $(q^2 - M_i^2)$, we obtain

$$\alpha_{w,\text{eff}}(q^2) = \frac{\alpha_w}{1 + \Re e \widehat{\Pi}_{WW}(q^2)}; \qquad \alpha_{z,\text{eff}}(q^2) = \frac{\alpha_z}{1 + \Re e \widehat{\Pi}_{ZZ}(q^2)}, \qquad (11.4)$$

where, as with $\widehat{\Pi}_{AZ}$, we factor out a mass-shell factor

$$\widehat{\Pi}_{ii}(q^2) = \frac{\widehat{\Pi}_{ii}(q^2) - \widehat{\Pi}_{ii}(M_i^2)}{q^2 - M_i^2}, \qquad i = W, Z, \qquad (11.5)$$

and $\alpha_w = g_w^2/4\pi$ and $\alpha_z = \alpha_w/c_w^2$.

Interestingly enough, the third RG-invariant combination in Eq. (11.3) leads to the concept of the *Higgs boson effective charge* [1]: the SM Higgs boson H couples universally to matter with an effective charge inversely proportional to its VEV.

11.2.2 Relation to physical cross sections

We consider the QED effective charge introduced in Chapter 6. This quantity displays a nontrivial dependence on the fermion masses m_f, which allows its reconstruction from physical amplitudes by resorting to the optical theorem and analyticity (i.e., dispersion relations). Given a particular contribution to the photon spectral function $\Im m\,\Pi(s)$, the corresponding contribution to $\Pi(q^2)$ can be reconstructed via a *once-subtracted dispersion relation* (see, e.g., de Rafael [2]). For example, for the one-loop contribution of the fermion f, choosing the on-shell renormalization scheme,

$$\Pi_{f\bar f}(q^2) = \frac{1}{\pi} q^2 \int_{4m_f^2}^{\infty} ds\, \frac{\Im m\,\Pi_{f\bar f}(s)}{s(s - q^2)}. \qquad (11.6)$$

For $f \neq e$, $\Im m\,\Pi_{f\bar f}(s)$ is measured directly in the tree-level cross section for $e^+e^- \to f^+f^-$. For $f = e$, it is necessary to isolate the self-energy-like component of the tree-level Bhabha cross section. This is indeed possible because the self-energy-, vertex, and boxlike components of the Bhabha differential cross section are *linearly independent functions* of $\cos\theta$; they may therefore be projected out by convoluting the differential cross section with appropriately chosen polynomials in $\cos\theta$. Thus, in QED, knowledge of the spectral function $\Im m\,\Pi_{f\bar f}(s)$, determined from the tree-level $e^+e^- \to f^+f^-$ cross sections, together with a measurement of the fine structure constant α, enables the construction of the one-loop effective charge $\alpha_{\text{eff}}(q^2)$ for all q^2.

Keeping the QED example in mind, let us now turn to the case of the PT electroweak effective charges and study the procedure that would allow, at least in principle, the extraction of $\alpha_{z,\text{eff}}(q^2)$ from experiment [3]. In general, the renormalization of

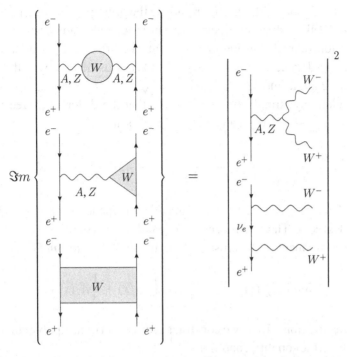

Figure 11.1. The relation between the imaginary parts of the subset of the W-related one-loop corrections to $e^+e^- \to e^+e^-$ and the tree-level process $e^+e^- \to W^+W^-$.

$\widehat{\Pi}_{ZZ}$ requires two subtractions: for mass and field renormalization. If we denote the subtraction point by s_0, then the twice-subtracted dispersion relation corresponding to the renormalized W^+W^- contributions reads

$$\widehat{\Pi}_{ZZ}^{(WW)}(q^2) = \frac{1}{\pi}(q^2 - s_0)^2 \int_{4M_W^2}^{\infty} ds \frac{\Im m \widehat{\Pi}_{ZZ}^{(WW)}(s)}{(s - q^2)(s - s_0)^2}. \tag{11.7}$$

The property instrumental for the observability of $\alpha_{z,\mathrm{eff}}(q^2)$ is that, in contrast to the conventional gauge-dependent self-energies, the absorptive parts of the PT self-energies appearing on the right-hand side (rhs) of Eq. (11.7) are directly related to components of the physical cross section $e^+e^- \to W^+W^-$ that are experimentally observable (see Figure 11.1). Indeed, as we have already seen in Chapter 10, the characteristic s-t cancellation, triggered by the longitudinal momenta of the on-shell polarization tensors, rearranges the tree-level cross section $e^+e^- \to W^+W^-$ into subamplitudes, which, through the use of the optical theorem, can be connected unambiguously with the absorptive parts of the one-loop PT Green's functions.

To simplify the algebra without compromising the principle, let us consider the limit of $e^+e^- \to W^+W^-$ when the electroweak mixing angle vanishes: $s_w^2 = 0$. In this limit, all photon-related contributions are switched off, and the two massive gauge bosons become degenerate ($M_Z = M_W \equiv M$). Let us denote by θ the center-of-mass scattering angle and set $x = \cos\theta$, $\beta = \sqrt{1 - 4M^2/s}$, and $z = (1 + \beta^2)/2\beta$. Then it is relatively straightforward to show that the differential tree-level cross section for $e^+e^- \to W^+W^-$ can be cast in the form [3]

$$(z - x)^2 \left(\frac{d\sigma}{dx}\right)_{s_w=0} = \frac{g^4}{64\pi} \frac{s\beta}{(s - M^2)^2} \theta(s - 4M^2) \sum_{i=1}^{5} A_i(s) F_i(s, x), \quad (11.8)$$

where the $F_i(s, x)$, $i = 1, 2 \ldots 5$, are linearly independent polynomials in x of maximum degree 4. The coefficients $A_1(s)$ and $A_2(s)$ contribute only to the self-energy-like component of the cross section, being related to $\Im m \, \widehat{\Pi}_{ZZ}^{(WW)}(s)$ by

$$\Im m \, \widehat{\Pi}_{ZZ}^{(WW)}(s)\big|_{s_w=0} = \frac{g^2}{4\pi} \beta s \left(A_1(s) + \frac{1}{3} A_2(s)\right). \quad (11.9)$$

To project the functions $A_i(s)$, we construct a further set of five polynomials $\widetilde{F}_i(s, x)$ satisfying the orthogonality conditions

$$\int_{-1}^{1} dx \, F_i(s, x) \widetilde{F}_j(s, x) = \delta_{ij}. \quad (11.10)$$

The coefficient functions $A_i(s)$ may then be projected from the observable formed by taking the product of the differential cross section with the kinematic factor $(z - x)^2$:

$$\int_{-1}^{1} dx \, \widetilde{F}_i(s, x) (z - x)^2 \left(\frac{d\sigma}{dx}\right)_{s_w=0} = \frac{g^4}{64\pi} \frac{s\beta}{(s - M^2)^2} A_i(s). \quad (11.11)$$

Thus it is possible to extract $\Im m \, \widehat{\Pi}_{ZZ}^{(WW)}(s)\big|_{s_w=0}$ directly from $d\sigma(e^+e^- \to W^+W^-)/dx\big|_{s_w=0}$.

Of course, to use the dispersion relation of Eq. (11.7) to compute $\widehat{\Pi}_{ZZ}^{(WW)}(q^2)$, one needs to integrate the spectral density $\Im m \, \widehat{\Pi}_{ZZ}^{(WW)}(s)$ over a large number of values of s. This in turn means that one needs experimental data for the process $e^+e^- \to W^+W^-$ for a variety of center-of-mass energies s, and for each value of s, one must repeat the procedure described earlier. Regardless of whatever practical difficulties this might entail, it does not constitute a problem of principle. Finally, the general case with $s_w^2 \neq 0$ requires, in addition, the observation of spin density matrices [4]; though technically more involved, the procedure is in principle the same.

11.3 Physical renormalization schemes versus $\overline{\text{MS}}$

It is no secret that the popular renormalization schemes, such as $\overline{\text{MS}}$ and $\overline{\text{DR}}$, convenient as they may be for formal manipulations, are plagued with persistent threshold and matching errors. The origin of these errors can be understood by noting that the aforementioned (unphysical) schemes implicitly integrate out all masses heavier than the physical energy scale until they are crossed, and then they turn them back on abruptly by means of a step function. Integrating out heavy fields, however, is only valid for energies well below their masses. This procedure is conceptually problematic because it does not correctly incorporate the finite probability that the uncertainty principle gives for a particle to be pair produced below threshold [5]. In addition, complicated matching conditions must be applied when crossing thresholds to maintain consistency for such desert scenarios. In principle, these schemes are only valid for theories in which all particles have zero or infinite mass or if one knows the full field content of the underlying physical theory.

Instead, in the physical renormalization scheme defined with the pinch technique, gauge couplings are defined directly in terms of physical observables, namely, the effective charges. The latter run smoothly over spacelike momenta and have non-analytic behavior only at the expected physical thresholds for timelike momenta; as a result, the thresholds associated with heavy particles are treated with their correct analytic dependence. In particular, particles will contribute to physical predictions even at energies below their threshold [5].

Historically, the gauge-invariant parametrization of physics offered by the pinch technique has been first systematized by Hagiwara et al. [6] and has led to an alternative framework for confronting the precision electroweak data with theoretical predictions. This approach resorts to the pinch technique to separate the one-loop corrections into gauge-fixing, parameter-independent universal (process independent) and process-specific pieces; the former are parametrized using the PT effective charges $\alpha_{\text{eff}}(q^2)$, $\bar{s}_w^2(q^2)$, $\alpha_{w,\text{eff}}(q^2)$, and $\alpha_{z,\text{eff}}(q^2)$, defined earlier. There is a total of nine electroweak parameters that must be determined in this approach: the eight universal parameters M_W, M_Z, $\alpha_{\text{eff}}(0)$, $\bar{s}_w^2(0)$, $\alpha_{w,\text{eff}}(0)$, $\alpha_{z,\text{eff}}(0)$, $\bar{s}_w^2(M_Z^2)$, and $\alpha_{z,\text{eff}}(M_Z^2)$ and one process-dependent parameter $\delta_b(M_Z^2)$, related to the form factor of the $Z\bar{b}_L b_L$ vertex.

Reference [6] explains in detail the advantage of their approach over the $\overline{\text{MS}}$ scheme. In particular, they emphasize that the nondecoupling nature of the $\overline{\text{MS}}$ forces one to adopt an effective field theory approach in which the heavy particles are integrated out of the action. The couplings of the effective theories are then related to each other by matching conditions ensuring that all effective theories give identical results at zero momentum transfer because the effects of heavy particles in the effective light field theory must be proportional to q^2/M^2, where M is the heavy mass scale. This procedure, however, is not only impractical in the presence of many quark and lepton mass scales but introduces errors because of the mistreatment of the threshold effects. In addition, direct use of the $\overline{\text{MS}}$ couplings leads to expressions in which the masses used for the light quarks are affected by sizable nonperturbative QCD effects.

The relevance of the effective charges in the quantitative study of threshold corrections due to heavy particles in grand unified theories (GUTs) was already recognized in [6], but it was not until a decade later that this was actually accomplished by Binger and Brodsky [5]. As was shown by these authors, the effective charges defined with the pinch technique furnish a conceptually superior and calculationally more accurate framework for studying the important issue of gauge-coupling unification. The main advantage of the effective charge formalism is that it provides a template for calculating all mass threshold effects for any given GUT; such threshold corrections may be instrumental in making the measured values of the gauge couplings consistent with unification.

In [5], the effective charges $\alpha_{\text{eff}}(q^2)$, $\alpha_{s,\text{eff}}(q^2)$, and the effective mixing angle $\bar{s}_w^2(q^2)$ were used to define new effective charges $\tilde{\alpha}_1(q^2)$, $\tilde{\alpha}_2(q^2)$, and $\tilde{\alpha}_3(q^2)$, which correspond to the standard combinations of gauge couplings used to study gauge-coupling unification. Specifically,

$$\tilde{\alpha}_1(q^2) = \left(\frac{5}{3}\right)\frac{\alpha_{\text{eff}}(q^2)}{1 - \bar{s}_w^2(q^2)}; \qquad \tilde{\alpha}_2(q^2) = \frac{\alpha_{\text{eff}}(q^2)}{\bar{s}_w^2(q^2)}; \qquad \tilde{\alpha}_3(q^2) = \alpha_{s,\text{eff}}(q^2).$$

(11.12)

The preceding couplings were used to obtain novel heavy and light threshold corrections, and the resulting impact on the unification predictions for a general GUT model was studied. Notice that even in the absence of new physics, i.e., using only the known SM spectrum, there are appreciable numerical discrepancies between the values of the conventional and PT couplings at M_Z (see Table I of [5]). Given that these values are used as initial conditions for the evolution of the couplings to the GUT scale, these differences alone may affect the unification properties of the couplings.

11.4 Gauge-independent off-shell form factors

It is well known that renormalizability and gauge invariance restrict severely the type of interaction vertices that can appear at the level of the fundamental Lagrangian. Thus, the tensorial possibilities allowed by Lorentz invariance are drastically reduced to relatively simple tree-level vertices. Beyond tree level, the tensorial structures that have been so excluded appear due to quantum corrections; that is, they are generated from loops. This fact does not conflict with renormal-izability and gauge invariance provided that the tensorial structures generated, not present at the level of the original Lagrangian, are UV finite; that is, no counterterms need be introduced to the fundamental Lagrangian proportional to the forbidden structures.

To fix the ideas, let us consider a concrete textbook example. In standard QED, the tree-level photon-electron vertex is simply proportional to γ_μ, whereas kine-matically, one may have, in addition (for massive on-shell electrons, using the Gordon decomposition), a term proportional to $\sigma_{\mu\nu}q^\nu$ that would correspond to a nonrenormalizable interaction. Of course, the one-loop photon-electron vertex generates such a term: one has

$$\Gamma_\mu(q) = \gamma_\mu F_1(q^2) + \sigma_{\mu\nu}q^\nu F_2(q^2), \tag{11.13}$$

where the scalar cofactors multiplying the two tensorial structures are the cor-responding form factors; they are in general nontrivial functions of the photon momentum transfer. $F_1(q^2)$ is the electric form factor, whereas $F_2(q^2)$ is the mag-netic form factor. $F_1(q^2)$ is UV divergent and becomes finite after carrying out the standard vertex renormalization. On the other hand, $F_2(q^2)$ comes out UV finite, as it should, given that there is no term proportional to $\sigma_{\mu\nu}q^\nu$ (in configuration space) in the original Lagrangian, where a potential UV divergence could be absorbed. Of course, in the limit of $q^2 \to 0$, the magnetic form factor $F_2(q^2)$ reduces to the famous Schwinger anomaly [7].

At the level of an Abelian theory, such as QED, the preceding discussion exhausts more or less the theoretical complications associated with the calculation of off-shell form factors. However, in NAGTs, such as the electroweak sector of the SM, there is an additional important complication: the off-shell form factors obtained from the conventional one-loop vertex (and beyond) depend explicitly on the gauge-fixing parameter. This dependence disappears when going to the on-shell limit of the incoming gauge boson ($q^2 \to 0$ for a photon, $q^2 \to M_Z^2$ for a Z-boson, etc.) but is present for any other value of q^2. This fact becomes phenomenologically relevant because one often wants to study the various form factors of particles that are produced in high-energy collisions, where the gauge boson mediating the interaction is far off shell. In the case of e^+e^- annihilation into heavy fermions, the value of q^2 must be above the heavy fermion threshold. For example, top quarks may be pair produced through the reaction $e^+e^- \to t\bar{t}$, with center-of-mass energy $s = q^2 \geq 4m_t^2$. In such a case, the intermediate photon and Z are far off-shell, and therefore the form factors F_i^V, appearing in the standard

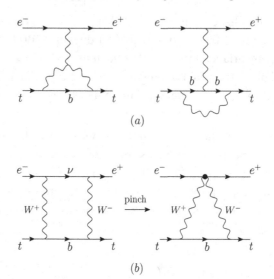

Figure 11.2. (*a*) The conventional one-loop vertex and (*b*) the vertexlike piece extracted from the box for $\xi \neq 1$.

decompositions

$$\Gamma_\mu^V(q^2) = \gamma_\mu F_1^V(q^2, \xi) + \sigma_{\mu\nu} q^\nu F_2^V(q^2, \xi) + \gamma_\mu \gamma_5 F_3^V(q^2, \xi)$$
$$+ \gamma_5 \sigma_{\mu\nu} q^\nu F_4^V(q^2, \xi), \tag{11.14}$$

depend explicitly on ξ, which stands collectively for ξ_W, ξ_Z, ξ_A, and $V = A, Z$.

The situation described is rather general and affects most form factors; very often, the residual gauge dependences have serious physical consequences. For example, the form factors display unphysical thresholds, have bad high-energy behavior, and sometimes are UV and IR divergent. The way out is to use the PT construction and extract the physical, gauge-independent form factors from the corresponding off-shell one-loop PT vertex (and beyond). Applying the pinch technique to the case of the form factors amounts to saying that one has to identify vertexlike contributions (with the appropriate tensorial structure corresponding to the form factor considered) contained in box diagrams, as shown in Figure 11.2. The latter, when added to the usual vertex graphs, render all form factors ξ independent and well behaved in all respects.

A particularly interesting case in which the pinch technique has been successfully applied is the study of the three-boson vertices AW^+W^- and ZW^+W^-, with the neutral gauge bosons off shell and the W pair on shell, or off shell and subsequently decaying to on-shell particles. Historically, the main motivation for exploring their properties was that they were going to be tested at LEP2 by direct W-pair production, proceeding through the process $e^+e^- \rightarrow W^+W^-$; their experimental scrutiny could provide invaluable information on the non-Abelian nature of the electroweak sector of the SM. Particularly appealing in this quest has been the possibility of measuring anomalous gauge-boson couplings, that is, the appearance of contributions to AW^+W^- and ZW^+W^- not encoded in the fundamental Lagrangian of the SM. Linear combinations of these form factors are related to the magnetic dipole and the electric quadrupole moments of the W boson. Such contributions may originate from two sources: (1) from radiative corrections within the SM, (2) from physics beyond the SM, or both. Therefore, the first theoretical task is to carry out the necessary calculations for completing part (1).

Calculating the one-loop expressions for these anomalous form factors is a nontrivial task, both from technical and conceptual points of view. We focus for concreteness on the photon case. If one calculates just the Feynman diagrams contributing to the AW^+W^- vertex and then extracts from them the contributions to the relevant form factors, one arrives at expressions that are plagued with several pathologies, gauge-fixing parameter dependence being one of them. Indeed, even if the two Ws are considered to be on-shell ($p_1^2 = p_2^2 = M_W^2$) because the incoming photon is not, there is no a priori reason why a gauge-fixing parameter-independent answer need emerge. Indeed, in the context of the renormalizable R_ξ gauges, the final answer depends on the choice of the gauge-fixing parameter ξ, which enters into the one-loop calculations through the gauge-boson propagators (W, Z, A, and unphysical would-be Goldstone bosons). In addition, as shown by an explicit calculation performed in the Feynman gauge ($\xi = 1$), the answer is infrared divergent and violates perturbative unitarity, that is, it grows monotonically for $q^2 \to \infty$ [8].

All the preceding pathologies may be cured if one uses the PT definition of the relevant (off-shell) gauge-boson vertices [9]. The application of the pinch technique identifies vertexlike contributions from the box graphs, as shown in Figure 11.3, which are subsequently distributed, in a unique way, among the various form factors. Thus one arrives finally at new expressions that are gauge-fixing parameter-independent, ultraviolet and infrared finite, and monotonically decreasing for large momentum transfers q^2.

11.4.1 Neutrino charge radius

The neutrino electromagnetic form factor and the neutrino charge radius have constituted an important theoretical puzzle for more than three decades. Since well before the SM, it has been pointed out that radiative corrections will induce an effective one-loop $A^*(q^2)\nu\nu$ vertex, to be denoted by $\Gamma^\mu_{A\nu\bar\nu}$, with $A^*(q^2)$ an off-shell photon. Such a vertex would in turn give rise to a small but nonvanishing charge radius. Traditionally (and, of course, nonrelativistically and rather heuristically), this charge radius has been interpreted as a measure of the size of the neutrino

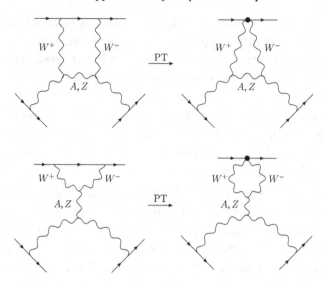

Figure 11.3. Two of the graphs contributing pinching parts to the gauge independent $V W^+ W^-$ vertex.

ν_i when probed electromagnetically, owing to its classical definition (in the static limit) as the second moment of the spatial neutrino charge density $\rho_\nu(\mathbf{r})$, i.e.,

$$\langle r_\nu^2 \rangle \sim \int d\mathbf{r} \, r^2 \rho_\nu(\mathbf{r}). \tag{11.15}$$

From the quantum field theory point of view, the neutrino charge radius is defined as follows. If we write $\Gamma^\mu_{A\nu\bar{\nu}}$ in the form

$$\Gamma^\mu_{A\nu\bar{\nu}}(q^2) = \gamma_\mu(1 - \gamma_5) F_D(q^2), \tag{11.16}$$

where $F_D(q^2)$ is the (dimensionless) Dirac electromagnetic form factor, then the neutrino charge radius is given by

$$\langle r_\nu^2 \rangle = 6 \left. \frac{\partial F_D(q^2)}{\partial q^2} \right|_{q^2=0}. \tag{11.17}$$

Gauge invariance (if not compromised) requires that in the limit $q^2 \to 0$, $F_D(q^2)$ must be proportional to q^2; that is, it can be cast in the form $F_D(q^2) = q^2 F(q^2)$, with the dimensionful form factor $F(q^2)$ being regular as $q^2 \to 0$. As a result, the q^2 contained in $F_D(q^2)$ cancels against the $(1/q^2)$ coming from the propagator of the off-shell photon, and one obtains effectively a contact interaction between the neutrino and the sources of the (background) photon, as would be expected from classical considerations.

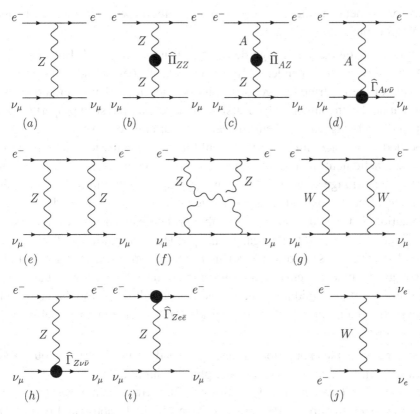

Figure 11.4. The electroweak diagrams contributing to the entire electron-neutrino scattering process at one loop. Diagram (j) (and the corresponding dressing) is absent when the neutrino species is muonic.

Even though, in the SM, the one-loop computation of the entire S-matrix element describing the electron-neutrino scattering, shown in Figure 11.4, is conceptually straightforward, the identification of a subamplitude that would serve as the effective $\Gamma^\mu_{A\nu\bar{\nu}}$ has been faced with serious complications, associated with the simultaneous reconciliation of crucial requirements such as gauge invariance, finiteness, and target independence. Various attempts to define the value of the neutrino charge radius within the SM from the one-loop $\Gamma^\mu_{A\nu\bar{\nu}}$ vertex calculated in the renormalizable (R_ξ) gauges reveal that the corresponding electromagnetic form factor depends explicitly on the gauge-fixing parameter ξ. In particular, even though in the static limit of zero momentum transfer, $q^2 \to 0$, the Dirac form factor vanishes and therefore is independent of ξ, its first derivative with respect to q^2 (which corresponds to the definition of the neutrino charge radius) continues to depend on it. Similar (and sometimes worse) problems occur in the context of other gauges (e.g., the unitary gauge). These complications have obscured the entire concept of a charge radius for the neutrino and have cast serious doubt on whether it can be regarded as a genuine physical observable.

Of course, if a quantity is gauge dependent, it is not physical. But that the off-shell vertex is gauge dependent only means that it does not serve as a physical definition of the neutrino charge radius; it does not mean that an effective charge radius cannot be encountered that satisfies all necessary physical properties, gauge independence being one of them. Indeed, several authors have attempted to find a modified vertexlike amplitude that would lead to a consistent definition of the electromagnetic neutrino charge radius. The common underlying idea in all these works is to rearrange somehow the Feynman graphs contributing to the scattering amplitude of neutrinos with charged particles in an attempt to find a vertexlike combination that would satisfy all desirable properties. What makes this exercise so difficult is that in addition to gauge independence, a multitude of other crucial physical requirements need to be satisfied as well. For example, one should not enforce gauge independence at the expense of introducing target dependence. Therefore, a definite guiding principle is needed, allowing for the construction of physical subamplitudes with definite kinematic structure (i.e., self-energies, vertices, boxes).

The guiding principle in question has been provided by the pinch technique. As was shown in [10], the rearrangement of the physical amplitude $f^\pm \nu \to f^\pm \nu$, where $f^\pm$ are the target fermions, into PT self-energies, vertices, and boxes conclusively settles the issue: the proper PT vertex with an off-shell photon and two on-shell neutrinos, denoted by $\widehat{\Gamma}^\mu_{A\nu_i \bar\nu_i}$, furnishes unambiguously and uniquely the physical neutrino charge radius.

Most recently, the issue of the neutrino charge radius was revisited in a series of papers [11, 12, 13]. There three important conceptual points have been conclusively settled:

1. As explained in [10], the box diagrams furnish gauge-dependent (propagator-like) contributions that are crucial for the gauge cancellations, but once these contributions have been identified and extracted, the remaining pure box cannot form part of the neutrino charge radius because it would introduce process dependence (in view of its nontrivial dependence on the target fermion masses, for one thing). The most convincing way to understand why the pure box could not possibly enter into the neutrino charge radius definition is to consider the case of right-handed polarized target fermions that do not couple to the Ws: in that case, the box diagram is not even there. The gauge cancellations proceed differently because the coupling of the Z-boson to the target fermions is also modified [11, 12, 13].

2. The mixing self-energy $\widehat{\Pi}_{AZ}(q^2)$ should not be included in the definition of the neutrino charge radius either. The reason for this is more subtle: $\widehat{\Pi}_{AZ}(q^2)$ is not an RG-invariant quantity; adding it to the finite contribution coming from the proper vertex would convert the resulting neutrino charge radius into a μ-dependent, and therefore unphysical, quantity. Instead, $\widehat{\Pi}^{AZ}(q^2)$ must be combined with the appropriate Z-mediated tree-level contributions (which evidently do not enter into the definition of the charge radius) to form with them the RG-invariant combination $\bar{s}_w^2(q^2)$ of Eq. (11.2), whereas the ultraviolet-finite neutrino charge radius will be determined from the proper vertex only. Writing $\widehat{\Gamma}^\mu_{A\nu_i\bar{\nu}_i} = q^2\widehat{F}_i(q^2)\gamma_\mu(1 - \gamma_5)$, the physical neutrino charge radius is then defined as $\langle r^2_{\nu_i}\rangle = 6\widehat{F}_i(0)$, and the explicit calculation yields

$$\langle r^2_{\nu_i}\rangle = \frac{G_F}{4\sqrt{2}\pi^2}\left[3 - 2\log\left(\frac{m_i^2}{M_W^2}\right)\right], \qquad (11.18)$$

where $i = e, \mu, \tau$, m_i denotes the mass of the charged isodoublet partner of the neutrino under consideration and G_F is the Fermi constant.

3. The neutrino charge radius defined through the pinch technique may be extracted from experiment, at least in principle. One may express a set of experimental electron-neutrino cross sections in terms of the finite neutrino charge radius and the two additional gauge- and RG-invariant quantities, corresponding to the electroweak effective charge $\alpha_{z,\mathrm{eff}}(q^2)$ and mixing angle $\bar{s}_w^2(q^2)$, defined earlier.

11.5 Resummation formalism for resonant transition amplitudes

The physics of unstable particles and the computation of resonant transition amplitudes have attracted significant attention in recent years because they are both phenomenologically relevant and theoretically challenging. The practical interest in the problem is related to the resonant production of various particles in all sorts of accelerators, most notably LEP1 and LEP2, the TEVATRON, and the LHC. From the theoretical point of view, the issue comes up every time fundamental resonances, i.e., unstable particles that appear as basic degrees of freedom in the original Lagrangian of the theory (as opposed to composite bound states), can be produced resonantly. The presence of such fundamental resonances makes it impossible to compute physical amplitudes for arbitrary values of the kinematic parameters, unless a resummation has taken place first. Simply stated, perturbation theory breaks down in the vicinity of resonances, and information about the dynamics to all orders needs to be encoded already at the level of Born amplitudes. The difficulty arises that in the context of NAGTs, the standard Breit–Wigner resummation used for regulating physical amplitudes near resonances is at odds with gauge invariance, unitarity, and the equivalence theorem [14]. Consequently, the resulting Born-improved amplitudes in general fail to capture faithfully the underlying dynamics. It is therefore important to devise a self-consistent calculational scheme that manifestly preserves all relevant field-theoretic properties [1, 15, 16, 17, 18].

The mathematical expressions for computing transition amplitudes are ill defined in the vicinity of resonances because the tree-level propagator of the particle mediating the interaction (i.e., $\Delta = (s - M^2)^{-1}$) becomes singular as the center-of-mass energy approaches the mass of the resonance (i.e., as $\sqrt{s} \sim M$). The standard way to regulate this physical kinematic singularity is to use a Breit–Wigner type of propagator; this amounts essentially to the replacement (near the resonance) $(s - M^2)^{-1} \rightarrow (s - M^2 + iM\Gamma)^{-1}$, where Γ is the width of the unstable (resonating) particle. The presence of $iM\Gamma$ in the denominator prevents the amplitude from being divergent, even at physical resonance (i.e., when $s = M^2$).

The actual field-theoretic mechanism that justifies the apperance of the width is the Dyson resummation of the self-energy $\Pi(s)$ of the unstable particle, which amounts to the rigorous substitution $(s - M^2)^{-1} \rightarrow (s - M^2 + \Pi(s))^{-1}$. The running width of the particle is then defined as $M\Gamma(s) = \Im m \Pi(s)$, whereas the usual (on shell) width (see earlier) is simply its value at $s = M^2$.

It is relatively easy to realize that the Breit–Wigner procedure, as described earlier, is tantamount to a reorganization of the perturbative series. Resumming the self-energy $\Pi(s)$ amounts to removing a particular piece from each order of the perturbative expansion because from all the Feynman graphs contributing to a given order n, we only pick the part that contains the corresponding string of self-energy bubbles $\Pi(s)$ and then take $n \rightarrow \infty$. Notice, however, that the off-shell Green's functions contributing to a physical quantity, at any finite order of the conventional perturbative expansion, participate in a subtle cancellation that eliminates all unphysical terms. Therefore the act of resummation, which treats unequally the various Green's functions, is in general liable to distort these cancellations. To put it differently, if $\Pi(s)$ contains unphysical contributions (which would eventually cancel against other terms within a given order), resumming it naively would mean that these unphysical contributions would also undergo infinite summation (they now appear in the denominator of the propagator $\Delta(s)$). To remove them, one would have to add the remaining perturbative pieces to an infinite order, clearly an impossible task because the latter (boxes and vertices) do not constitute a resummable set. Thus, if the resummed $\Pi(s)$ contains such unphysical terms, one arrives at predictions plagued with unphysical artifacts. The crucial novelty introduced by the pinch technique is that the resummation of the (physical)

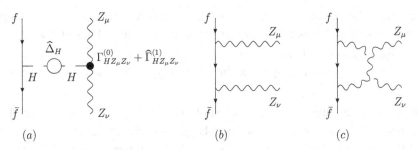

Figure 11.5. The amplitude for the process $f\bar{f} \to ZZ$. The s-channel graph (a) may become resonant and must be regulated by appropriate resummation of the Higgs propagator and dressing of the HZZ vertex.

self-energy graphs must take place only after the amplitude of interest has been cast via the PT algorithm into manifestly physical subamplitudes, with distinct kinematic properties, order by order in perturbation theory. Put in the language employed earlier, the PT ensures that all unphysical contributions contained inside $\Pi(s)$ have been identified and properly discarded before $\Pi(s)$ undergoes resummation.

11.5.1 An example

To get a flavor of the subtle interplay between the various physical constraints [1, 15, 16, 17, 18], we consider a concrete example. We study the process $f(p_1)\bar{f}(p_2) \to Z(k_1)Z(k_2)$, shown in Figure 11.5, and $s = (p_1 + p_2)^2 = (k_1 + k_2)^2$ is the center of mass energy squared. The tree-level amplitude of this process is the sum of an s- and t-channel contribution, denoted by $\mathcal{T}_s$ and $\mathcal{T}_t$, respectively, given by

$$\mathcal{T}_s^{\mu\nu} = \Gamma_{HZZ}^{\mu\nu}\Delta_H(s)\,\bar{v}(p_2)\Gamma_{Hf\bar{f}}u(p_1)$$

$$\mathcal{T}_t^{\mu\nu} = \bar{v}(p_2)\left[\Gamma_{Zf\bar{f}}^{\nu}S^{(0)}(p_1+k_1)\Gamma_{Zf\bar{f}}^{\mu} + \Gamma_{Zf\bar{f}}^{\mu}S^{(0)}(p_1+k_2)\Gamma_{Zf\bar{f}}^{\nu}\right]u(p_1),$$

$$(11.19)$$

where

$$\Gamma_{HZZ}^{\mu\nu} = \mathrm{i}g_w\frac{M_Z^2}{M_W}g^{\mu\nu}; \qquad \Gamma_{Hf\bar{f}} = -\mathrm{i}g_w\frac{m_f}{2M_W}$$

$$\Gamma_{Zf\bar{f}}^{\mu} = -\mathrm{i}\frac{g_w}{c_w}\gamma_\mu(T_z^f P_L - Q_f s_w^2),$$

$$(11.20)$$

are the tree-level HZZ, $Hf\bar{f}$, and $Zf\bar{f}$ couplings, respectively.

The s-channel contribution is mediated by the Higgs boson of mass M_H and becomes resonant if the kinematics are such that $\sqrt{s}$ lies in the vicinity of M_H;

in that case, the resonant amplitude must be properly regulated. The simplest way to accomplish this is to (1) Dyson-resum the one-loop PT self-energy of the (resonating) Higgs boson and (2) appropriately dress the tree-level vertex $\Gamma^{\mu\nu}_{HZZ}$, that is, by replacing in the amplitude the vertex $\Gamma^{\mu\nu}_{HZZ}$ with the one-loop PT vertex $\widehat{\Gamma}^{\mu\nu}_{HZZ}$.

We first see what happens if one attempts to regulate the resonant amplitude by means of the conventional one-loop Higgs self-energy in the R_ξ gauges. A straightforward calculation yields (tadpole and seagull terms omitted) [1, 18]

$$\Pi^{(WW)}_{HH}(s, \xi_W) = \frac{\alpha_w}{4\pi} \left(\frac{s^2}{4M^2_W} - s + 3M^2_W \right) B_0(s, M^2_W, M^2_W)$$

$$+ \frac{\alpha_w}{4\pi} \left(\frac{M^4_H - s^2}{4M^2_W} \right) B_0(s, \xi_W M^2_W, \xi_W M^2_W). \qquad (11.21)$$

We see that for $\xi_W \neq 1$, the term growing as s^2 survives and is proportional to the difference $B_0(s, M^2_W, M^2_W) - B_0(s, \xi_W M^2_W, \xi_W M^2_W)$. For any finite value of ξ_W, this term vanishes for sufficiently large s, that is, $s \gg M^2_W$ and $s \gg \xi_W M^2_W$. Therefore the quantity in Eq. (11.21) displays good high-energy behavior in compliance with high-energy unitarity. Notice, however, that the onset of this good behavior depends crucially on the choice of ξ_W. Because ξ_W is a free parameter and may be chosen to be arbitrarily large, but finite, the restoration of unitarity may be arbitrarily delayed as well. This fact poses no problem as long as one is restricted to the computation of physical amplitudes at a finite order in perturbation theory. However, if the preceding self-energy were to be resummed to regulate resonant transition amplitudes, it would lead to an artificial delay of unitarity restoration, which becomes numerically significant for large values of ξ_W. In addition, a serious pathology occurs for any value of $\xi_W \neq 1$, namely, the appearance of unphysical thresholds [15, 16, 17]. Such thresholds may be particularly misleading if ξ_W is chosen in the vicinity of unity, giving rise to distortions in the line shape of the unstable particle.

How does the situation change if, instead, we compute the corresponding part of the Higgs-boson self-energy in the BFM for an arbitrary value of ξ_Q? Denoting it by $\widetilde{\Pi}^{(WW)}_{HH}(s, \xi_Q)$, and using the appropriate set of Feynman rules [19], we obtain

$$\widetilde{\Pi}^{(WW)}_{HH}(s, \xi_Q) = \Pi_{HH}(s, \xi_W \to \xi_Q) - \frac{\alpha_w}{4\pi} \xi_Q (s - M^2_H) B_0(s, \xi_Q M^2_W, \xi_Q M^2_W).$$

$$(11.22)$$

Evidently, away from $\xi_Q = 1$, $\widetilde{\Pi}^{(WW)}_{HH}(s, \xi_Q)$ displays the same unphysical characteristics mentioned earlier for $\Pi^{(WW)}_{HH}(s, \xi_W)$. Therefore, when it comes to the study of resonant amplitudes, calculating in the BFM for general ξ_Q is as patho-logical as calculating in the conventional R_ξ gauges.

To solve these problems, one has simply to follow the PT procedure, within either gauge-fixing scheme R_ξ or background field method; identify the corresponding Higgs-boson-related pinch parts from the vertex and box diagrams; and add them to Eq. (11.21) or Eq. (11.22). Then a unique answer emerges, the PT one-loop Higgs boson self-energy, given by $\widehat{\Pi}_{HH}(q^2)$

$$\widehat{\Pi}_{HH}^{(WW)}(s) = \frac{\alpha_w}{16\pi} \frac{M_H^4}{M_W^2} \left[1 + 4\frac{M_W^2}{M_H^2} - 4\frac{M_W^2}{M_H^4}(2s - 3M_W^2)\right] B_0(s, M_W^2, M_W^2).$$

(11.23)

Setting $\xi_Q = 1$ in the expression of Eq. (11.22), we recover the full PT answer of Eq. (11.23), as expected. Clearly $\widehat{\Pi}_{HH}^{(WW)}(s)$ has none of the pathologies observed earlier.

We now turn to the way the PT-regulated amplitude satisfies the equivalence theo-rem [14]. This theorem states that at very high energies ($s \gg M_Z^2$), the amplitude for emission or absorption of a longitudinally polarized gauge boson becomes equal to the amplitude in which the gauge boson is replaced by the corresponding would-be Goldstone boson. The preceding statement is a consequence of the underlying local gauge invariance of the SM and holds to all orders in perturbation theory for multiple absorptions and emissions of massive vector bosons. Compliance with this theorem is a necessary requirement for any resummation algorithm because any Born-improved amplitude that fails to satisfy it is bound to be missing important physical information. The reason why most resummation methods are at odds with the equivalence theorem is that in the usual diagrammatic analysis, the underly-ing symmetry of the amplitudes is not manifest. Just as happens in the case of the optical theorem, the conventional subamplitudes, defined in terms of Feynman diagrams, do not satisfy the equivalence theorem individually. The resummation of such a subamplitude will in turn distort several subtle cancellations, thus giving rise to artifacts and unphysical effects. Instead, the PT subamplitudes satisfy the equivalence theorem individually; as usual, the only nontrivial step for establishing this is the proper exploitation of elementary Ward identities.

Turning to our explicit process $f(p_1)\bar{f}(p_2) \to Z(k_1)Z(k_2)$, the equivalence theo-rem states that the full amplitude $\mathcal{T} = \mathcal{T}_s + \mathcal{T}_t$ satisfies

$$\mathcal{T}(Z_L Z_L) = -\mathcal{T}(\chi\chi) - i\mathcal{T}(\chi z) - i\mathcal{T}(z\chi) + \mathcal{T}(\chi\chi),$$

(11.24)

where Z_L is the longitudinal component of the Z-boson, χ is its associated would-be Goldstone boson, and $z_\mu(k) = \varepsilon_\mu^L(k) - k_\mu/M_Z$ is the energetically suppressed

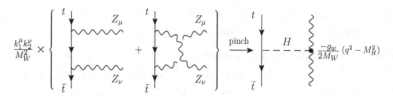

Figure 11.6. The Higgs-boson-related contribution extracted from the boxes through pinching; to get it, we must contract with both momenta.

part of the longitudinal polarization vector ε_μ^L. It is crucial to observe, however, that already at tree level, the conventional s- and t-channel subamplitudes $\mathcal{T}_s$ and $\mathcal{T}_t$ fail to satisfy the equivalence theorem individually [1, 18].

To verify this, one has to calculate $\mathcal{T}_s(Z_L Z_L)$, using explicit expressions for the longitudinal polarization vectors, and check if the answer obtained is equal to the Higgs-boson-mediated s-channel part of the left-hand side of Eq. (11.24). In particular, in the center-of-mass system, we have

$$z_\mu(k_1) = \varepsilon_\mu^L(k_1) - \frac{k_{1\mu}}{M_Z} = -2M_Z \frac{k_{2\mu}}{s} + \mathcal{O}\left(\frac{M_Z^4}{s^2}\right) \qquad (11.25)$$

and an exactly analogous expression for $z_\mu(k_2)$. The residual vector $z_\mu(k)$ has the properties $k^\mu z_\mu = -M_Z$ and $z^2 = 0$. After a straightforward calculation, we obtain a new term $\mathcal{T}_s^{\mathrm{P}} \sim (ig_w/2M_W)\bar{v}(p_2)\Gamma_{Hf\bar{f}}u(p_1)$ not found in Eq. (11.24)

$$\mathcal{T}_s(Z_L Z_L) = -\mathcal{T}_s(\chi\chi) - i\mathcal{T}_s(z\chi) - i\mathcal{T}_s(\chi z) + \mathcal{T}_s(zz) - \mathcal{T}_s^{\mathrm{P}}, \qquad (11.26)$$

where

$$\mathcal{T}_s(\chi\chi) = \Gamma_{H\chi\chi}\Delta_H(s)\bar{v}(p_2)\Gamma_{Hf\bar{f}}^{(0)}u(p_1)$$

$$\mathcal{T}_s(z\chi) + \mathcal{T}_s(\chi z) = [z_\mu(k_1)\Gamma_{HZ\chi}^\mu + z_\nu(k_2)\Gamma_{H\chi Z}^\nu]\Delta_H(s)\bar{v}(p_2)\Gamma_{Hf\bar{f}}u(p_1)$$

$$\mathcal{T}_s(zz) = z_\mu(k_1)z_\nu(k_2)\mathcal{T}_s^{\mu\nu}(ZZ), \qquad (11.27)$$

with $\Gamma_{H\chi\chi} = -ig_w M_H^2/(2M_W)$ and $\Gamma_{HZ\chi}^\mu = -g_w(k_1 + 2k_2)_\mu/(2c_w)$. Evidently the presence of the term $\mathcal{T}_s^{\mathrm{P}}$ prevents $\mathcal{T}_s^H(Z_L Z_L)$ from satisfying the equivalence theorem. This is, of course, not surprising given that an important Higgs-boson-mediated s-channel part has been omitted. The momenta k_1^μ and k_2^ν, stemming from the leading parts of the longitudinal polarization vectors $\varepsilon_L^\mu(k_1)$ and $\varepsilon_L^\nu(k_2)$, extract such a term from $\mathcal{T}_t(Z_L Z_L)$ (see Figure 11.6); this term is precisely $\mathcal{T}_s^{\mathrm{P}}$ and must be added to $\mathcal{T}_s(Z_L Z_L)$ to form a well-behaved amplitude at high energies. In other words, the amplitude

$$\widehat{\mathcal{T}}_s(Z_L Z_L) = \mathcal{T}_s(Z_L Z_L) + \mathcal{T}_s^{\mathrm{P}} \qquad (11.28)$$

satisfies the equivalence theorem by itself (see Eq. (11.24)).

In fact, this crucial property persists after resummation – thanks to the Ward identities satisfied by the PT vertices. As shown in Figure 11.5(a), the resummed amplitude, to be denoted by $\overline{T}_s(Z_L Z_L)$, is constructed from $T_s(Z_L Z_L)$ in Eq. (11.19) by replacing $\Delta_H(s)$ with the resummed Higgs-boson propagator $\widehat{\Delta}_H(s)$ and $\Gamma^{\mu\nu}_{HZZ}$ with the expression $\Gamma^{\mu\nu}_{HZZ} + \widehat{\Gamma}^{\mu\nu}_{HZZ}$, where $\widehat{\Gamma}^{\mu\nu}_{HZZ}$ is the one-loop HZZ vertex calculated within the pinch technique. It is then straightforward to show that the Higgs-mediated amplitude $\widetilde{T}_s(Z_L Z_L) = \overline{T}_s(Z_L Z_L) + T^P_s$ respects the equivalence theorem individually; to that end, we only need to employ PT Ward identities such as

$$k_{2\nu}\widehat{\Gamma}^{\mu\nu}_{HZZ}(q, k_1, k_2) + i M_Z \widehat{\Gamma}^{\mu}_{HZ\chi}(q, k_1, k_2) = -\frac{g_w}{2c_w}\widehat{\Pi}^{\mu}_{Z\chi}(k_1)$$

$$k_{1\mu}\widehat{\Gamma}^{\mu}_{HZ\chi}(q, k_1, k_2) + i M_Z \widehat{\Gamma}_{H\chi\chi}(q, k_1, k_2) = -\frac{g_w}{2c_w}\left[\widehat{\Pi}_{HH}(q^2) + \widehat{\Pi}_{\chi\chi}(k_2^2)\right].$$

$$(11.29)$$

In addition to the preceding issues, scattering amplitudes ought to be RG invariant; that is, they should not depend on the renormalization point μ chosen to carry out the subtractions nor on the renormalization scheme ($\overline{\text{MS}}$, on-shell scheme, momentum subtraction, etc.). This property must remain true in the vicinity of resonances, i.e., after resummation. To see how this happens for the process at hand, note that after the PT rearrangement, the resulting amplitude is decomposed into three individually RG-invariant parts:

1. A universal (process-independent) part, corresponding to the Higgs-boson effective charge, namely, the RG-invariant combination $(g_w^2/M_w^2)\widehat{\Delta}_H$, defined in Eq. (11.3); the line shape of the Higgs boson, being a universal quantity, must be obtained precisely from this part.
2. A process-dependent part, composed of the vertex corrections and the wave function renormalization of the external particles, which is RG invariant because of Abelian Ward identities.
3. A process-dependent part, coming from ultraviolet finite boxes; this is trivially RG invariant because it is ultraviolet finite and does not get renormalized.

Finally, on physical grounds, one expects that, far from the resonance, the Born-improved amplitude must behave exactly as its tree-level counterpart. In fact, a self-consistent resummation formalism should have this property built in; that is, far from resonance, one should recover the correct high-energy behavior without having to reexpand the Born-improved amplitude perturbatively. Recovering the correct asymptotic behavior is particularly tricky, however, when the final particles are gauge bosons. The exact mechanism that enforces the correct high-energy behavior of the Born-improved amplitude, when the PT width and vertex are used, has been studied in detail in [20] for the specific process considered here.

11.6 The pinch technique at finite temperature

Finite-temperature gauge theories are a large and complicated subject (see, e.g., Gross et al. [21]) for which the pinch technique is useful. In Chapter 9 we mentioned the relationship of $d = 3$ gauge theories to $d = 4$ gauge theories at very high temperature, where a hierarchy of scales, based on the smallness of the coupling, made it possible to ignore chromoelectric fields and other phenomena. There are, in principle, three scales (besides momenta) in a thermal $SU(N)$ NAGT: the temperature T itself, $(Ng^2)^{1/2}T$, and Ng^2T. (We ignore factors such as $1/4\pi$ that may or may not occur in particular applications, although these can be very important.) By g we mean the four-dimensional coupling as a function of T (and other variables, if needed). The so-called magnetic mass of a thermal NAGT scales with Ng^2T,[2] and the scale $(Ng^2)^{1/2}T$ appears as the scale of mass of the longitudinal electric degrees of freedom (the Debye or plasmon mass). In QCD, where the coupling decreases as T increases, or in EW theory, with its small coupling, these three scales should obey $T > (Ng^2)^{1/2}T > Ng^2T$, although this is not necessarily the case in any particular real-world application. Clearly the smallest scale, Ng^2T, sets the scale for infrared-dominated phenomena. Nonperturbative infrared phenomena at high T occur even for gauge theories normally thought of as weakly coupled such as the electroweak part of the standard model. Indeed, this theory is weakly coupled at low temperatures because of Higgs mass generation but strongly coupled at large T, where the Higgs VEV vanishes and the electroweak gauge bosons are perturbatively massless.

In the period 1980–1995, there were many attempts at calculating such quantities as the thermal β function and thermal plasmon[3] damping rate with standard Feynman-graph techniques, nearly all of which were plagued with gauge dependence. In an attempt to resolve this and other problems, some people argued for using the Batalin–Vilkovisky approach – in the Landau gauge – and others argued for using the background field method – in an arbitrary covariant gauge (see Elmfors and Kobes [22] for such calculations and references to other authors). But because PT principles were not invoked, the methods used were dependent on gauge, just as

---

[2] Just as in $d = 4$, the mass runs and decreases at large momentum, as signaled by a magnetic condensate $\langle \mathrm{Tr}\, G_{ij}^2 \rangle$. Recall that in Chapter 9 we proved that such a condensate exists in three dimensions.

[3] The plasmon is essentially the longitudinal electric degree of freedom with its Debye mass.

were the usual Feynman-gauge approaches. One notable exception is the work of Braaten and Pisarski [23, 24, 25] on hard thermal loops, which we discuss briefly later on.

In Section 11.7, we briefly review the Matsubara decomposition of a thermal field theory into an infinite set of coupled $d = 3$ field theories labeled by an integer K for bosons or $K + 1/2$ for fermions. In each $d = 3$ theory, the corresponding fields have mass $2\pi T |K|$ (bosons) or $2\pi T |K + 1/2|$ (fermions). It follows that for all except the $K = 0$ bosonic sector, the basic scale of these field theories is T itself; all infrared nonperturbative phenomena come from the $K = 0$ sector of an NAGT. Moreover, as we will see, the coupling of any of these theories is $g^2 T$. In particular, the $K = 0$ sector is just the $d = 3$ NAGT of Chapter 9, with the replacement of the coupling g_3^2 by $g^2 T$, the lowest available scale. The characteristic dimensionless parameter of any of these field theories is $N g^2 T / k$ at momentum scale k. Because the minimum momentum scale is a particle mass, it follows that the $K = 0$ sector is strongly coupled (dimensionless parameter of $\mathcal{O}(1)$), but this parameter is $\mathcal{O}(N g^2)$ for all other sectors, possibly allowing for a perturbative expansion.[4]

11.7 Basic principles of thermal field theory

Consider the partition function of a bosonic quantum theory:

$$\mathcal{Z} = \text{Tr} \, e^{-\beta H} = \int [d\phi(\vec{x})] \langle \phi(\vec{x}) | e^{-\beta H} | \phi(\vec{x}) \rangle, \tag{11.30}$$

where $\beta = 1/T$, and write this as a path integral for a generic field theory over field coordinates at zero time. The matrix element has a standard path integral representation:

$$\langle \phi(\vec{x}) | e^{-\beta H} | \phi(\vec{x}) \rangle = \int [d\Phi] \exp\left[- \int \mathcal{L}_E \right], \tag{11.31}$$

where $\mathcal{L}_E$ is the Euclidean Lagrangian corresponding to the Hamiltonian H, and the integral sign means

$$\int \rightarrow \int d^3 x \int_0^\beta d\tau \tag{11.32}$$

for Euclidean time τ. Because the trace sums diagonal matrix elements, the boundary conditions on the Φ path integral are

$$\Phi(\vec{x}, \tau = 0) = \phi(\vec{x}) = \Phi(\vec{x}, \tau = \beta). \tag{11.33}$$

---

[4] Except for strong effects coming from the coupling of a massive theory to the $K = 0$ sector, including effects from the Debye mass scale.

(For fermions, there is an extra minus sign.) So the quantum fields Φ are periodic (or antiperiodic, for fermions) in the time coordinate τ with period β. We can therefore write Φ as a Fourier sum,

$$\Phi(\vec{x}, \tau) = \sum_{-\infty}^{\infty} \phi_K(\vec{x}) e^{-i\omega_K \tau}, \tag{11.34}$$

with frequencies (called Matsubara frequencies) $\omega_K = 2\pi KT$. All Green's functions of Φ are similarly periodic. Inserting this periodic decomposition into $\mathcal{L}_E$ exposes it as the sum over Euclidean field theories with K-dependent masses, as we said earlier. Moreover, the τ integral introduces a Kronecker delta function that conserves frequencies and an overall factor of β. When combined with the $1/g^2$ factor in $\mathcal{L}_E$, the $d = 3$ coupling becomes $g^2 T$.

The well-known Feynman rules for a thermal field theory differ in the treatment of the energy component of momentum, with the replacement

$$\frac{i}{2\pi} \int dk_0 \to T \sum_K \tag{11.35}$$

$$2\pi\delta \left(\sum k_0(j) \right) \to \frac{1}{T} \delta_{0, \sum \omega_j}.$$

The free-field thermal propagator for a massless scalar field is (aside from an irrelevant constant factor)

$$\Delta(\vec{x}, \tau) = \frac{T}{(2\pi)^3} \int d^3 p \sum_K e^{i\vec{p}\cdot\vec{x} - i\omega_K \tau} \frac{1}{-(i\omega_K)^2 + \omega_p^2}, \tag{11.36}$$

with $\omega_p = +\sqrt{|\vec{p}|^2 + m^2}$; this is a sum of Euclidean propagators with masses $\omega_K = 2\pi KT$. The sum over Matsubara frequencies yields

$$\Delta(\vec{x}, \tau) = \frac{1}{(2\pi)^3} \int \frac{d^3 p}{2\omega_p} e^{i\vec{p}\cdot\vec{x}} \{e^{-\omega_p \tau}[1 + n(\vec{p})] + e^{\omega_p \tau} n(\vec{p})\}, \tag{11.37}$$

where

$$n(\vec{p}) = \frac{1}{e^{\beta\omega_p} - 1} \tag{11.38}$$

is the Bose–Einstein occupation number. Similar formulas hold for fermions; we need not record them here.

11.7.1 The pinch technique in the zero-Matsubara-frequency sector

The first PT calculations for thermal field theory were done, as reported in Chapter 9, for $d = 3$ gauge theory or, in other words, the zero-Matsubara-frequency sector

[26, 27]. Note that when the Matsubara frequency is fixed, there is no question of the dependence of the result on τ, which can only be reliably estimated by a sum over all frequencies. In any case, the zero-frequency sector gives no τ dependence to the periodic thermal fields.

The computations were actually done in the light-cone gauge, which causes no problems because all dependence on the gauge-fixing vector n_μ cancels out before any integrations or sums are done. We translate the one-loop PT propagator of Eq. (9.1) to the thermal regime, with the result:

$$\widehat{d}(q, T) = \frac{1}{q^2 - \pi b_3 g^2 T q} \qquad b_3 = \frac{15N}{32\pi}. \qquad (11.39)$$

In later years, a number of people attempted to extract a running charge from their calculations, as we describe in the next section. The usual procedure is to choose a definition (which is not necessarily unambiguous) for a thermal running charge $g_T(q, T)$ and to define a beta function by

$$\beta_T = T \frac{dg_T(q, T)}{dT}. \qquad (11.40)$$

From Eq. (11.39), we extract a zero-Matsubara-frequency running charge in one-loop perturbation theory as

$$g_T^2(q, T) = q^2 g^2 \widehat{d}(q, T) = \frac{g^2}{1 - 15g^2 T/(32q)}. \qquad (11.41)$$

This yields

$$T \frac{dg_T(q, T)}{dT} = +\frac{15NTg_T^3}{64q}. \qquad (11.42)$$

The derivative is positive because the running charge depends on T/q and the q derivative has the negative sign associated with infrared slavery. This means that the coupling constant runs away as T increases. Of course, this is equivalent to the infrared limit $q \to 0$, where we expect infrared-slavery diseases to arise.

This thermal β function, based as it is on one-loop perturbation theory in the zero-Matsubara-frequency sector, does not account for many important phenomena that are beyond the scope of this book. In particular, accounting for gluon electric masses in resummed internal propagators could, in principle, give rise to a term of $\mathcal{O}(g^4 T^2/q^2)$, which is of higher order in the infrared limit $T \gg q$. Other corrections come from including a magnetic mass. These have been discussed by Elmfors and Kobes [22] using a general covariant background-field gauge; one important result of this work is that for any gauge parameter ξ, the $\mathcal{O}(g^4 T^2/q^2)$ term vanishes. Although these authors did not realize that to find a gauge-invariant result, all we need to do is choose the Feynman background-field gauge $\xi = 1$, we do realize it.

When this is done and corrections from the magnetic mass are omitted, exactly the result of our Eq. (11.42) is found.

11.7.2 Developments in the full thermal field theory

Reference [28] gave the first PT calculation of the full thermal NAGT propagator at one-loop order. Again, the computations were done in the light-cone gauge. The result for the PT proper self-energy is (omitting the seagull graph that vanishes by dimensional integration)

$$\widehat{\Pi}_{\mu\nu} = \frac{1}{2} g^2 NT \sum_K \int \frac{d^3k}{(2\pi)^3} \frac{1}{k^2(q+k)^2}$$

$$\times \left[8(q^2 \delta_{\mu\nu} - q_\mu q_\nu) + 2(2k+q)_\mu (2k+q)_\nu \right]. \tag{11.43}$$

The time component of the Euclidean four-vector k is $k_4 = 2\pi TK$. This self-energy is conserved and has two independent scalar pieces multiplying two tensorial structures; these are equivalent to calculating $\widehat{\Pi}_{44}$ and $\widehat{\Pi}_{ij}$. Most authors give results only for $q_4 = 0$, and we do the same; in the following results, q is a three-vector. After doing the integrals, one finds the renormalized propagator

$$(g^2 \widehat{\Delta})_{44}^{-1} = bq^2 \ln \left(\frac{q^2}{\Lambda^2} \right) + \frac{NT^2}{\pi^2} P(\epsilon) \tag{11.44}$$

$$(g^2 \widehat{\Delta})_{ij}^{-1} = \left(\delta_{ij} - \frac{q_i q_j}{q^2} \right) \left[bq^2 \ln \left(\frac{q^2}{\Lambda^2} \right) + \frac{NT^2}{\pi^2} Q(\epsilon) \right],$$

where

$$b = \frac{11N}{48\pi^2}; \qquad \epsilon = \frac{q}{2T}, \tag{11.45}$$

and

$$P(\epsilon) = \frac{\pi^2}{3} + \frac{1}{2\epsilon} \int_0^\infty \frac{dy}{e^y - 1} \left[(y^2 - 4\epsilon^2) \ln \left| \frac{y+\epsilon}{y-\epsilon} \right| - 2y\epsilon \right] \tag{11.46}$$

$$Q(\epsilon) = \frac{1}{2} \int_0^\infty \frac{dy}{e^y - 1} \left[y - \frac{y^2 + 7\epsilon^2}{2\epsilon} \ln \left| \frac{y+\epsilon}{y-\epsilon} \right| \right].$$

In the infrared limit $q = 0$ (or $T = \infty$), we have $P(0) = \pi^2/3$ and $Q(0) = 0$. These correspond to the often-quoted perturbative values $m_e^2 = Ng^2 T^2/3$ for the electric mass and zero for the magnetic mass m_m, respectively. However, the kinetic term $q^2 \ln(q^2/\Lambda^2)$ is not well behaved at small q at the one-loop level, and the interpretation of the electric mass requires resummations that replace $\ln q^2$ by something like $\ln(q^2 + 4m^2)$, which accounts for masses on the internal lines. We will not discuss that interesting problem any further. In any event, as is well known, an electric mass arises at the one-loop level, but the magnetic mass vanishes to all orders in perturbation theory, despite that $m_m^2 \sim (Ng^2)^2 T^2$ looks like a fourth-order effect. Of course, the magnetic mass is nothing but the $d = 3$ dynamical gluon mass for which we argued in Chapter 9.

The next work that explicitly invoked the pinch technique to achieve gauge invariance in a thermal NAGT was an attempt to calculate the plasmon damping rate gauge invariantly [29]. Earlier gauge-dependent calculations gave a negative damping rate, which is a physical impossibility in any covariant gauge. Other calculations with other methods and in other gauges gave a positive damping coefficient. Finally, Nadkarni [29], using the results of Cornwall et al. [28], as stated in Eq. (11.43), found a one-loop PT damping rate that was unambiguously gauge invariant and also unambiguously negative. It was clear that this negative sign was precisely that arising from asymptotic freedom and that – as Nadkarni suggested – other, possibly nonperturbative effects needed to be included to get a positive rate. In an independent development, Braaten and Pisarski [23, 24, 25] developed an algorithm for resumming so-called hard thermal loops and using it to find a positive and gauge-invariant plasmon damping rate. An amplitude with external lines whose momenta p are soft (of order of the electric mass $\mathcal{O}(gT)$), coupled to a loop with hard loop momenta (of order $\mathcal{O}(T)$), has terms characterized by a dimensionless parameter $\mathcal{O}(g^2 T^2/p^2)$ that is of order unity and contributes at the same order as the soft tree-level amplitude. Here a factor $g^2 T$ comes from the coupling, and another power of T in the numerator comes from hard loop momenta. It is perhaps not surprising that the sum of all such hard loops is gauge invariant, as Braaten and Pisarksi proved, because they are all of the same order in the coupling. With this process of resummation, these authors found a positive and gauge-invariant plasmon damping rate.

Braaten and Pisarski made no reference to the pinch technique, although their arguments would strongly suggest PT principles to those familiar with them. A few years later, Sasaki [30, 31, 32] made this connection. First, he calculated [30, 31] a thermal β function using the pinch technique, checking that he found the same result in four distinct gauge families, including the background-field gauges. Then he [32] showed that using the one-loop PT propagator with resummed internal propagator lines also coming from the PT actually yielded precisely Braaten and Pisarski's result for the plasmon damping rate.

We quote here Sasaki's result for the thermal β function, as defined through the running charge g_T of Eq. (11.41). He finds that

$$T\frac{dg_{TS}}{dT} = +\frac{14NTg^3}{64q}, \tag{11.47}$$

which is hardly different from what was found from the $d = 3$ PT propagator, as shown in Eq. (11.42), which has 15 rather than 14 in the numerator. This perhaps surprisingly small difference may arise because Sasaki used the full thermal PT propagator rather than just its zero-frequency part, as we did in Eq. (11.42). (But remember that this β function is infrared dominated, and so the zero-frequency sector should give the largest contribution.) The development of the plasmon damping rate would take us too far afield here, and we refer the reader to Sasaki's papers.

In thermal NAGTs, just as in NAGTs at zero temperature, the pinch technique does not solve difficult physics problems, but it does make it possible to separate true physics issues from gauge artifacts. No one should be surprised that mere use of the pinch technique itself at some low order of perturbation theory [29] does not give physical results in an asymptotically free theory; it is this unphysicality that ultimately drives the formation of a dynamical gluon mass and requires a self-consistent formulation of these gauge theories, such as we have argued for throughout this book. The demonstration that the resummation of hard thermal loops is equivalent to a PT resummation should not be a surprise either. The fact that Braaten and Pisarski did not recognize the connection of their earlier work to the pinch technique should not lull the reader into thinking that there is no connection to the pinch technique, as Sasaki showed.

11.8 Hints of supersymmetry in the pinch technique Green's functions

Let us now focus on a very interesting property of the one-loop PT three-gluon vertex discovered recently by Binger and Brodsky [33]. These authors first added quark and scalar loops to $\widehat{\Gamma}_{\alpha\mu\nu}^{amn}(q_1, q_2, q_3)$; this is straightforward from the point of view of gauge independence and gauge invariance because these loops are automatically gauge-fixing parameter independent and satisfy the Ward identity (Eq. (1.92)). All resulting one-loop integrals, including those of Eqs. (1.85) and (1.86), were evaluated for the first time, thus determining the precise tensorial decomposition of $\widehat{\Gamma}_{\alpha\mu\nu}^{amn}(q_1, q_2, q_3)$. Then, after choosing a convenient tensor basis, $\widehat{\Gamma}_{\alpha\mu\nu}^{amn}(q_1, q_2, q_3)$ was expressed as a linear combination of 14 independent tensors, each multiplied by its own scalar form factor. Every form factor receives, in general, contributions from gluons (G), quarks (Q), and scalars (S). It turns out that these three types of contributions satisfy very characteristic relations that are closely linked to supersymmetry and conformal symmetry and, in particular, the $\mathcal{N} = 4$ nonrenormalization theorems. For all form factors F (in d-dimensions), it was shown that

$$F_G + 4F_Q + (10 - d)F_S = 0, \qquad (11.48)$$

which encodes the vanishing contribution of the $\mathcal{N} = 4$ supermultiplet in $d = 4$. Similar relations have been found in the context of supersymmetric scattering amplitudes [34, 35].

It should be emphasized that relations such as Eq. (11.48) do not exist for the gauge-dependent three-gluon vertex (see, e.g., Davydychev et al. [36]) because the gluon contributions depend on the gauge-fixing parameter, whereas the quarks and scalars do not. Indeed, it is uniquely the PT (or, equivalently, in the background Feynman gauge, $\xi_Q = 1$) Green's function that satisfies this homogeneous sum rule. Most important, calculating in the background field method with $\xi_Q \neq 1$ leads to a nonzero rhs of Eq. (11.48).

As was explained in detail by Binger and Brodsky, this type of relation hints at supersymmetry. To appreciate this point, it is useful to consider various supersymmetries in $d = 4$, as was done in [33]. Specifically, one may distinguish the following three cases, depending on the number $\mathcal{N}$ of supersymmetries:

1. $\mathcal{N} = 1$: From the preceding definitions, it is clear that a vector superplet V_1 (gluons plus gluinos) contributes $ig^2 N_c(F_G + F_Q) \equiv ig^2 N_c F_{V_1}$ to a generic form factor F, whereas N_Φ chiral superplets contribute $ig^2 N_\Phi(\frac{1}{2}F_Q + F_S) \equiv ig^2 N_\Phi F_\Phi$. By the sum rule Eq. (11.48) in $d = 4$, we have $F_{V_1} + 6F_\Phi = 0$. Thus, any form factor can be written as

$$F = ig^2(N_c F_{V_1} + N_\Phi F_\Phi) = \frac{ig^2}{3}\beta_0^{(N=1)}F_{V_1}, \qquad (11.49)$$

   where $\beta_0^{(N=1)} = 3N_c - 1/2N_\Phi$ is the first coefficient of the β function. Hence the contributions of vector and chiral superplets have precisely the same functional form for each form factor. Furthermore, every form factor is proportional to β_0, even though all but one of them are ultraviolet finite.

2. $\mathcal{N} = 2$: In this case, the vector superplet gives $ig^2 N_c(F_G + 2F_Q + 2F_S) \equiv ig^2 N_c F_{V_2}$, whereas N_h hyperplets (a Weyl fermion of each helicity plus a doublet of complex scalars) yield $ig^2 N_h(F_Q + 2F_S) \equiv ig^2 N_h F_h$. The sum rule of Eq. (11.48) can be written as $F_{V_2} + 2F_h = 0$, and thus

$$F = ig^2(N_c F_{V_2} + N_h F_h) = \frac{ig^2}{2}\beta_0^{(N=2)}F_{V_2}, \qquad (11.50)$$

   where $\beta_0^{(N=2)} = 2N_c - N_h$.

3. $\mathcal{N} = 4$: Now the vector superplet (the only multiplet allowed) contributes $2ig^2 N_c(F_G + 4F_Q + 6F_S) \equiv N_c F_{V_4}$, which is identically zero by the sum rule, which, of course, is a consequence of $\beta_0^{(N=4)} = 0$.

Thus, the similarities between form factors in $d = 4$ are related to supersymmetric nonrenormalization theorems. In particular, the exact conformal invariance of $\mathcal{N} = 4$ implies that the gauge-invariant, three-gluon Green's function is not renormalized at any order in perturbation theory. In fact, at one-loop order, there are not even finite corrections, as reflected in Eq. (11.48).

References

[1] J. Papavassiliou and A. Pilaftsis, Effective charge of the Higgs boson, *Phys. Rev. Lett.* **80** (1998) 2785.

[2] E. de Rafael, Lectures on quantum electrodynamics: 1, GIFT lectures given at Barcelona, Spain (1976).

[3] J. Papavassiliou, E. de Rafael, and N. J. Watson, Electroweak effective charges and their relation to physical cross sections, *Nucl. Phys.* **B503** (1997) 79.

[4] M. Bilenky, J. L. Kneur, F. M. Renard, and D. Schildknecht, Trilinear couplings among the electroweak vector bosons and their determination at LEP-200, *Nucl. Phys.* **B409** (1993) 22.

[5] M. Binger and S. J. Brodsky, Physical renormalization schemes and grand unification, *Phys. Rev.* **D69** (2004) 095007.

[6] K. Hagiwara, S. Matsumoto, D. Haidt, and C. S. Kim, A novel approach to confront electroweak data and theory, *Z. Phys.* **C64** (1994) 559.

[7] J. S. Schwinger, On quantum electrodynamics and the magnetic moment of the electron, *Phys. Rev.* **73** (1948) 416.

[8] E. N. Argyres, G. Katsilieris, A. B. Lahanas, C. G. Papadopoulos, and V. C. Spanos, One loop corrections to three vector boson vertices in the standard model, *Nucl. Phys.* **B391** (1993) 23.

[9] J. Papavassiliou and K. Philippides, Gauge invariant three boson vertices in the standard model and the static properties of the W, *Phys. Rev.* **D48** (1993) 4255.

[10] J. Papavassiliou, Gauge invariant proper self-energies and vertices in gauge theories with broken symmetry, *Phys. Rev.* **D41** (1990) 3179.

[11] J. Bernabeu, L. G. Cabral-Rosetti, J. Papavassiliou, and J. Vidal, On the charge radius of the neutrino, *Phys. Rev.* **D62** (2000) 113012.

[12] J. Bernabeu, J. Papavassiliou, and J. Vidal, On the observability of the neutrino charge radius, *Phys. Rev. Lett.* **89** (2002) 101802.

[13] J. Bernabeu, J. Papavassiliou, and J. Vidal, The neutrino charge radius is a physical observable, *Nucl. Phys.* **B680** (2004) 450.

[14] J. M. Cornwall, D. N. Levin, and G. Tiktopoulos, Derivation of gauge invariance from high-energy unitarity bounds on the S-matrix, *Phys. Rev.* **D10** (1974) 1145 [Erratum, ibid. **D11** (1975) 972].

[15] J. Papavassiliou and A. Pilaftsis, Gauge invariance and unstable particles, *Phys. Rev. Lett.* **75** (1995) 3060.

[16] J. Papavassiliou and A. Pilaftsis, A gauge independent approach to resonant transition amplitudes, *Phys. Rev.* **D53** (1996) 2128.

[17] J. Papavassiliou and A. Pilaftsis, Gauge-invariant resummation formalism for two point correlation functions, *Phys. Rev.* **D54** (1996) 5315.

[18] J. Papavassiliou, The pinch technique approach to the physics of unstable particles, paper presented at the 6th Hellenic School and Workshop on Elementary Particle Physics, Corfu, Greece (1998).

[19] A. Denner, G. Weiglein, and S. Dittmaier, Application of the background field method to the electroweak standard model, *Nucl. Phys.* **B440** (1995) 95.

[20] J. Papavassiliou, Asymptotic properties of Born-improved amplitudes with gauge bosons in the final state, *Phys. Rev.* **D60** (1999) 056001.

[21] D. J. Gross, R. D. Pisarski, and L. G. Yaffe, QCD and instantons at finite temperature, *Rev. Mod. Phys.* **53** (1981) 43.

[22] P. Elmfors and R. Kobes, The thermal beta function in Yang-Mills theory, *Phys. Rev.* **D51** (1995) 774.

[23] E. Braaten and R. D. Pisarski, Resummation and gauge invariance of the gluon damping rate in hot QCD, *Phys. Rev. Lett.* **64** (1990) 1338.

[24] E. Braaten and R. D. Pisarski, Soft amplitudes in hot gauge theories: A general analysis, *Nucl. Phys.* **B337** (1990) 569.

[25] E. Braaten and R. D. Pisarski, Deducing hard thermal loops from Ward identities, *Nucl. Phys.* **B339** (1990) 310.

[26] J. M. Cornwall, Nonperturbative mass gap in continuum QCD, paper presented at the French–American Seminar on Theoretical Aspects of Quantum Chromodynamics, Marseille, France (1981).

[27] J. M. Cornwall, Dynamical mass generation in continuum QCD, *Phys. Rev.* **D26** (1982) 1453.

[28] J. M. Cornwall, W. S. Hou, and J. E. King, Gauge-invariant calculations in finite-temperature QCD: Landau ghost and magnetic mass, *Phys. Lett.* **B153** (1985) 173.

[29] S. Nadkarni, Linear response and plasmon decay in hot gluonic matter, *Phys. Rev. Lett.* **61** (1988) 396.

[30] K. Sasaki, Gauge-independent thermal β function in Yang-Mills theory, *Phys. Lett.* **B369** (1996) 117.

[31] K. Sasaki, Gauge-independent thermal β function in Yang-Mills theory via the pinch technique, *Nucl. Phys.* **B472** (1996) 271.

[32] K. Sasaki, Gauge-independent resummed gluon self-energy in hot QCD, *Nucl. Phys.* **B490** (1997) 472.

[33] M. Binger and S. J. Brodsky, The form factors of the gauge-invariant three-gluon vertex, *Phys. Rev.* **D74** (2006) 054016.

[34] Z. Bern, L. J. Dixon, D. C. Dunbar, and D. A. Kosower, One-loop n-point gauge theory amplitudes, unitarity and collinear limits, *Nucl. Phys.* **B425** (1994) 217.

[35] Z. Bern, L. J. Dixon, D. C. Dunbar, and D. A. Kosower, Fusing gauge theory tree amplitudes into loop amplitudes, *Nucl. Phys.* **B435** (1995) 59.

[36] A. I. Davydychev, P. Osland, and O. V. Tarasov, Three-gluon vertex in arbitrary gauge and dimension, *Phys. Rev.* **D54** (1996) 4087 [Erratum, ibid. **D59** (1999) 109901].

Appendix

Feynman rules

A.1 R_ξ and BFM gauges

The Feynman rules for QCD in R_ξ gauges are given in Figure A.1. In the case of the background field gauge, because the gauge-fixing term is quadratic in the quantum fields, apart from vertices involving ghost fields, only vertices containing exactly two quantum fields might differ from the conventional ones. Thus, the vertices $\Gamma_{\widehat{A}\psi\bar{\psi}}$ and $\Gamma_{\widehat{A}AAA}$ have, to lowest order, the same expression as the corresponding R_ξ vertices $\Gamma_{A\psi\bar{\psi}}$ and Γ_{AAAA} (to higher order, their relation is described by the corresponding BQIs). Feynman rules for background fields are shown in Figure A.2.

A.2 Antifields

The couplings of the antifields Φ^* with fields is entirely encoded in the BRST Lagrangian of Eq. (4.4). When choosing the BFM gauge, the additional coupling $gf^{amn}A_\mu^{*m}\widehat{A}_\nu^n c^a$ will arise in the BRST Lagrangian $\mathcal{L}_{\text{BRST}}$ as a consequence of the background field method splitting $A \to \widehat{A} + A$. One then gets the Feynman rules given in Figure A.3.

A.3 BFM sources

Feynman rules involving the background field source Ω_μ^m can be easily derived by the one involving the (gluon) antifield A_μ^{*m} through the replacements $A_\mu^{*m} \to \Omega_\mu^m$ and $c^m \to \bar{c}^m$.

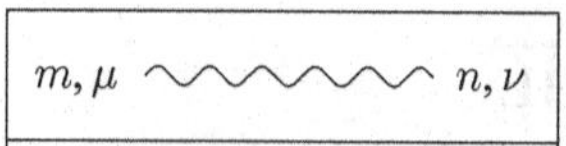 m, μ $\sim\!\!\sim\!\!\sim\!\!\sim\!\!\sim$ n, ν	$-\mathrm{i}\delta^{mn}\frac{1}{q^2}\left[g_{\mu\nu} - (1-\xi)\frac{q_\mu q_\nu}{q^2}\right]$
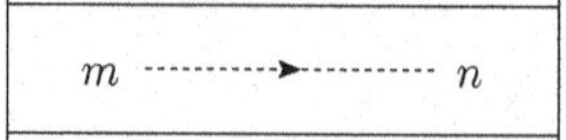 m - - - - $\blacktriangleright$ - - - - n	$\mathrm{i}\delta^{mn}\frac{1}{k^2}$
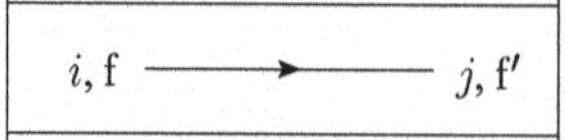 i, f —————▶——— j, f'	$\mathrm{i}\delta^{ij}\delta^{\mathrm{ff}'}\frac{1}{k^\mu\gamma_\mu - m_{\mathrm{f}}}$
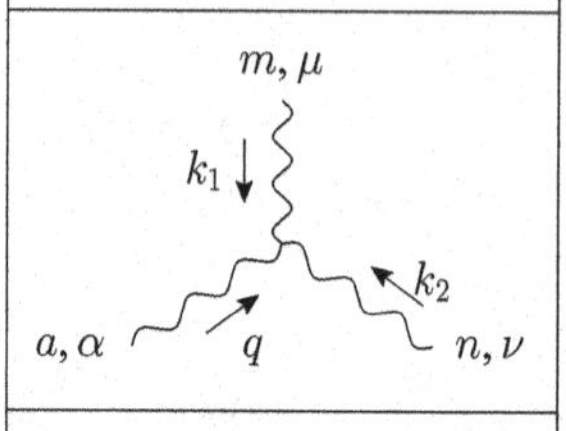	$gf^{amn}\left[g_{\mu\nu}(k_1 - k_2)_\alpha + g_{\alpha\nu}(k_2 - q)_\mu \right.$ $\left. + g_{\alpha\mu}(q - k_1)_\nu\right]$
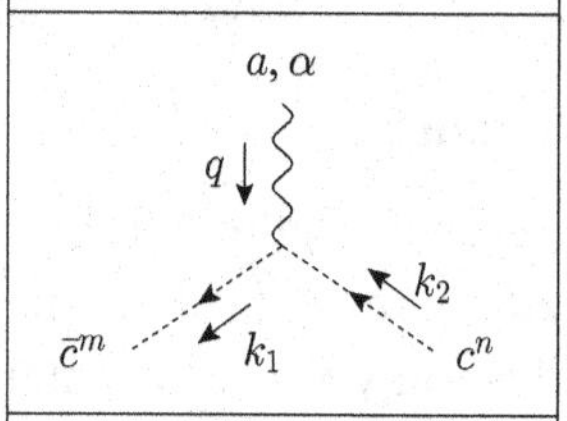	$gf^{amn}k_{1\alpha}$
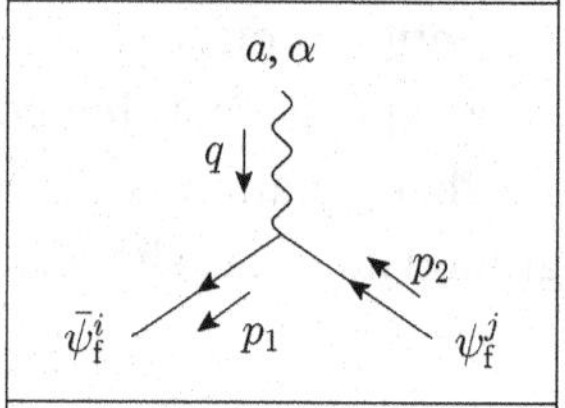	$\mathrm{i}g\gamma^\alpha t^a_{ij}$
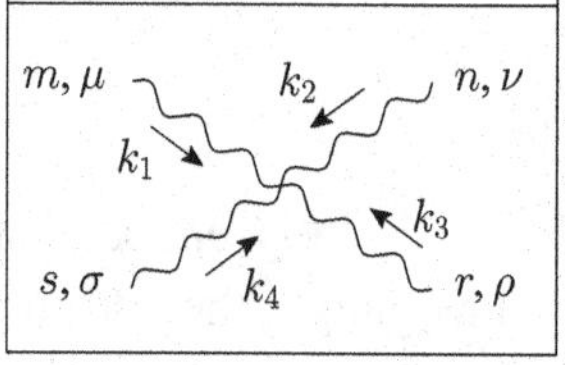	$-\mathrm{i}g^2\left[f^{mse}f^{ern}\left(g_{\mu\rho}g_{\nu\sigma} - g_{\mu\nu}g_{\rho\sigma}\right)\right.$ $+ f^{mne}f^{esr}\left(g_{\mu\sigma}g_{\nu\rho} - g_{\mu\rho}g_{\nu\sigma}\right)$ $\left. + f^{mre}f^{esn}\left(g_{\mu\sigma}g_{\nu\rho} - g_{\mu\nu}g_{\rho\sigma}\right)\right]$

Figure A.1. Feynman rules for QCD in the R_ξ gauges.

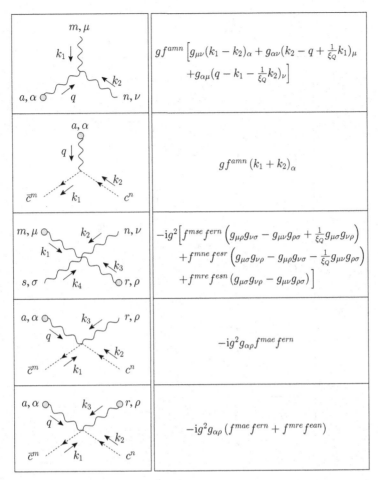

Figure A.2. Feynman rules for QCD in the background field gauge. We include only those rules that are different from the R_ξ rules to lowest order. As usual, a shaded circle on a gluon line indicates a background field.

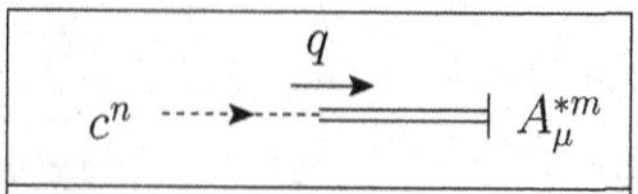

$$\delta^{mn} q_\mu$$

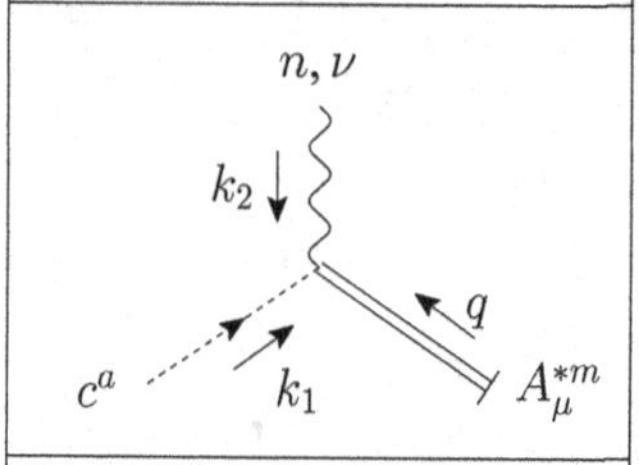

$$\mathrm{i} g f^{amn} g_{\mu\nu}$$

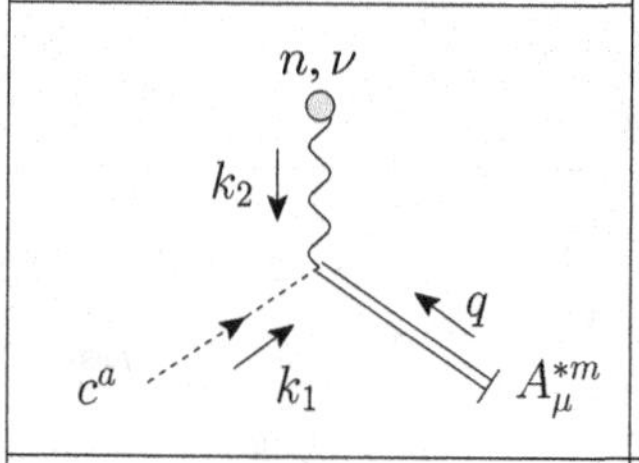

$$\mathrm{i} g f^{amn} g_{\mu\nu}$$

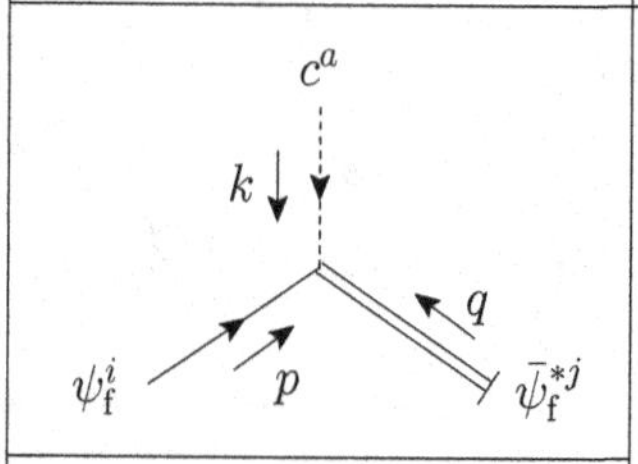

$$-g t_{ji}^a$$

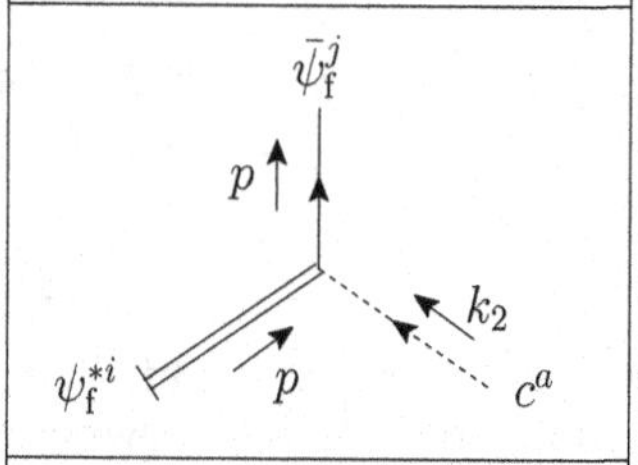

$$g t_{ji}^a$$

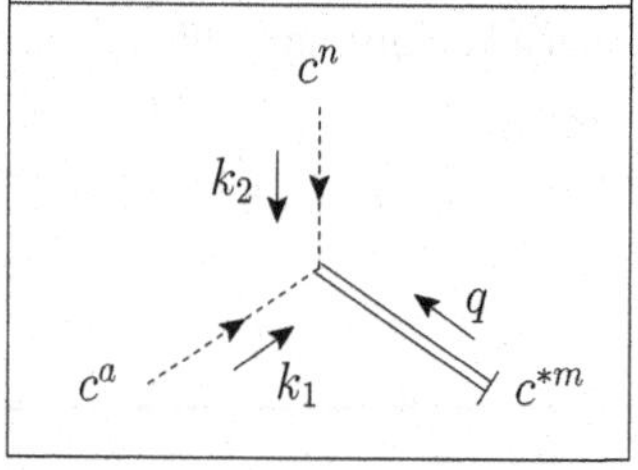

$$-\mathrm{i} g f^{amn}$$

Figure A.3. Feynman rules for QCD antifields.

Index

antifields, 86
antighost equation, 99
asymptotic freedom, 1, 34, 36, 42
  and infrared slavery, *see* infrared slavery

background-field method, 19, 45
  gauge fields, 66–69
  Slavnov–Taylor identities functional, 92
  sources, 92
  Ward identities functional, 93
Background-quantum identities, 95
Batalin–Vilkovisky
  formalism, 88
  pinch technique formulation of, *see* pinch technique

center vortices, 152–156
  homotopy, and, 151
  confinement, and, 159–163
  entropy, and, 147
  exceptional groups, and, 148
  screening, 163
Chern–Simons action, 48
  dynamical gluon mass, and, 209
    Yang–Mills–Chern–Simons solitons, 214–216
  phases of, with dynamical mass, 211
Chern–Simons number, 209
  half-integral
    knots, and, 220
condensates, 3
  gluon mass, and, 47, 146
  quantum solitons, and, 145
confinement, 45, 53, 147
  baryons in $SU(3)$, 148, 162
  chiral symmetry breaking, and, 186
  functional Schrödinger equation, and, 192
  Georgi–Glashow model, in, 167, 181

effective action, 63
  background fields, scalar, 63–66
  infrared, 51
  two-particle irreducible, 64
effective charge, 134
  electroweak, 251
  QCD, 135
  QED, 134
  relation to physical cross sections, 252
electroweak theory, 2
  all-orders pinch technique, in, 55

Faddeev–Popov equation, 93
Feynman gauge, 2
  background fields, 62
  gauge, ghost, and Goldstone masses, 61
finite-temperature gauge theory, 31
  magnetic mass, and, 148
form factors
  gauge independent, 256
  neutrino charge radius, 259
functional Schrödinger equation, 201
  $d = 4$ coupling, and, 208
  effective action, 202
  gauge technique, and, 203

gauge fixing, 5–6
  background gauge fields, 67
  FLS gauge, 55
  't Hooft–Feynman, 55
  unitary gauge, 61
gauge technique, 50
  Abelian, 105–109
  general properties, 104
  massless longitudinal poles, in, 108–109, 112
  non-Abelian, 109–112
gauged nonlinear sigma model, 51
  massless scalars, and, 54
Gauss link number, 160
  Hopf fibration, and, 222
  Wilson loop, and, 160
generalized pinch technique, 71
Georgi–Glashow model, 55, 56, 167, 168, 181
gluon mass, 1, 46
  condensates, and, 46
  gauge technique, and, 112

gluon mass (*cont.*)
  generation in QCD, 49–55
  hybrids, and, 164
  lattice evidence for, 115–117
  longitudinal massless fields, and, 48
  positivity, and, 43
  PT Schwinger–Dyson equations, and, 139
  quantum solitons, and, 145–146
  tachyon removal, and, 51
  three dimensions, in, 197–201
Green's functions, 88
  one-particle irreducible, 88
    generating functional, 88
Gribov ambiguity, 53

homotopy, 151
  first homotopy group, 151
  Gauss linking number, and, 159
Hopf fibration, 222
  Chern–Simons number, and, 222
hybrids, 164

infrared slavery, 1, 47–50, 105, 114, 190–199,
    210–213, 273

Kugo–Ojima function, 99

Landau gauge
  background formulation, 99
light-cone gauge, 31–34
  pinch technique, 54

pinch technique, 1, 45
  all-order, 75
  background-Feynman correspondence, 69
  in the background-field method, 71
  Batalin–Vilkovisky formulation, 100
  equal-time commutators, and, 59
  finite temperature, 270
    full thermal field theory, 274
    zero-Matsubara-frequency sector, 272
  four-gluon vertex, 30
  gauge boson mass, and, 47
  one-loop propagator, 6–17
  one-loop three-gluon vertex, 20–30
  resummation of resonant transition amplitudes, 263
  Schwinger–Dyson equations, 126

algorithm, 119
  construction, 120
  solution, 131
  truncation, 130
spontaneous symmetry breaking, with, 226
  absorptive construction, 237
  all-order construction, 246
  conserved currents case, 229
  differences with the symmetric case,
    227
  nonconserved current case, 242
  physical thresholds, 240
  unitary gauge case, 236
pinch technique, properties
  absorptive parts
    unitarity, and, 2, 34–36
  intrinsic construction, 19
  process independence, 17
  propagator transversality, 127
  and SUSY, 276
propagator positivity, 42–43

quantum solitons, 145–147

Slavnov–Taylor identities
  for 1PI functions, 93
  functional, 90
  for Schwinger–Dyson kernels, 77, 95
s-t cancellation
  all-order, 75
  tree-level, 39
string theory, 3
supersymmetry (SUSY), 250, 276
symmetry breaking, Higgs–Kibble, 48
  gauge boson mass, and, 55

three-dimensional gauge theories, 190
  chromomagnetic field, and, 195
  effective action, of, 193–195
  magnetic mass, and, 191
  one-loop gap equations, 197–201

Ward identities, 2
  fermion propagator, 10
  four-gluon vertex, 31
  quark-gluon vertex, 20–23
  three-gluon vertex, 8–10, 26, 34